FOURIER ACOUSTICS

Sound Radiation
and
Nearfield Acoustical
Holography

In memory of Mom and Ned whose love will live in us forever -
And to my Dearest who have been inspired by their lives' examples:
  My wife, Virginia
  Our children, Elizabeth and Ned Daniel
  My brother, Dan.

FOURIER ACOUSTICS

Sound Radiation
and
Nearfield Acoustical Holography

Earl G. Williams
Naval Research Laboratory
Washington, D.C.

ACADEMIC PRESS
San Diego London Boston New York
Sydney Tokyo Toronto

ISNG 0-12-753960-3

ACADEMIC PRESS
24-28 Oval Road
LONDON NW1 7DX
http://www.hbuk.co.uk/ap/

ACADEMIC PRESS
525 B Street. Suite 1900. San Diego,
California 92101-4495. USA
http://www.apnet.com

A catalogue record for this book is available from the British Library

Transferred to digital printing 2008.

Contents

Preface

This book is intended to serve both as a textbook and as a reference book. As a textbook it would be best suited for a graduate level course. In fact the book grew out of class notes written for a full year graduate course in radiation and scattering taught at The Catholic University of America. The reader need not have a background in acoustics, however. All of the necessary equations and concepts are included in an effort to make the text self-contained. It is assumed that the reader has good mathematical skills, especially a firm grounding in calculus, a knowledge of differential equations and a familiarity with the Fourier transform. Although an understanding of Fourier transforms is crucial, all the needed basic theorems are presented here, some with proof, some without. In fact, Chapter 1 covers a review of generalized functions, Fourier transforms, Fourier series and the discrete Fourier transform. Problems are included at the end of each chapter that test the material presented and provide additional concepts and theory.

How is this book different from the many books available in the field of acoustics? After thirty years of working in basic research in vibration, radiation and scattering of sound I have built my knowledge base not only from the standard acoustics textbooks, but also to a large extent from texts in other fields such as electromagnetism and optics. The integration of the materials from these different fields into my own research has led to a great deal of success. I hope that sharing this knowledge base will bring the reader a similar bonanza.

Chapter 2 begins with an overview of the basic and most important (at least for this book) equations of acoustics. The rest of the chapter discusses plane waves and how they can be used to derive some important theories of vibration and radiation. The initial material is presented quite simply, with illustrations to demonstrate the very physical concepts of plane and evanescent waves. Evanescent waves are emphasized, since they are rarely discussed in other books and because they are so important in the realm of underwater, structural acoustics. Expansions of plane and evanescent waves lead to the most important concept of the book: Fourier acoustics and the angular spectrum. The angular spectrum is used to derive some very powerful tools for the acoustician: the Rayleigh integrals, the Ewald sphere construction, plate radiation and supersonic intensity.

A similar approach is taken in Chapter 4 which deals with wave expansions in cylindrical coordinates and in Chapter 6 which presents spherical wave expansions. Thus some of the important theories discussed for plane waves in Chapter 2 are extended to cylindrical and spherical coordinates. Additionally, spherical coordinates present an excellent forum for some additional concepts such as multipoles, transient radiation and scattering from spheres to be introduced. These concepts round out Chapter 6.

Although these chapters lead to some higher mathematical functions, such as Bessel functions and spherical harmonics, it is assumed the reader is not familiar with them, and thus they are discussed and plotted in great detail. This is done with all of the higher mathematical functions found in this book. I have too often found scientists somewhat timid when it comes to working with higher level functions, and thus have made a conscious effort here to emphasize their details, especially with visual help. Throughout my own research I have found that an essential ingredient for success has been mastering the mathematics and the mathematical functions presented here. I have attempted to be rigorous whenever possible, and precise in the symbolic conventions.

The use of the Fourier transform and Fourier series in the analysis in the three geometries presented in Chapters 2, 4 and 6 motivated the subtitle of the book, Fourier Acoustics.

Chapter 8 provides a detailed look at the Helmholtz integral equation (HIE), an essential tool for anyone working in acoustics. The HIE is truly a modern and popular tool, to which the many commercial computer codes predicting vibration and radiation on the market will attest. Detailed derivations of the HIE are presented for both interior and exterior radiation problems as well as the scattering problem. This chapter also presents Green functions in depth, providing formulas not usually found in acoustics texts. For example, the evanescent Green function is introduced. Dirichlet and Neumann Green functions are presented for various geometries which simplify the Helmholtz integral equation.

Now to explain Chapters 3, 5, 7 and the last section of Chapter 8. This book is aimed at both the theoretician and the experimentalist, although it is basically a theoretical text. My early background in experimental acoustics, coupled with the desire to understand in detail the physics of vibration and radiation of sound, led to the invention and development of more illuminating and more sophisticated experimental techniques, especially nearfield acoustical holography (NAH). NAH provides a solution to an inverse problem, backtracking the sound field in time and space. It requires mastery of theoretical concepts and mathematical methods (all presented in this book) for its successful implementation. Furthermore, this mastery is also necessary for the interpretation and understanding of the experimental results. It is the synergism between theory and experiment that gave birth to the materials presented in this book. The extraordinary power of the NAH technique has been proven by many researchers throughout the world. However, this is the first book presenting NAH in detail, including all of the basic theory needed to implement it not only in planar coordinates but also in other geometries. As Fourier optics is to optical holography, Fourier acoustics is to nearfield acoustical holography.

NAH is discussed in Chapters 3, 5, 7 and in the last section of Chapter 8, presenting the technique in planar, cylindrical and spherical coordinate systems and finally for an arbitrary geometry, respectively. The implementation of NAH is discussed thoroughly so that the reader has all the necessary information to apply NAH in his own work, if he or she should so desire. To demonstrate the power of NAH, actual experimental results are shown from my own research for the cylindrical case.

Chapters 2, 4, 6 and 8 are completely self-consistent, that is, they do not rely on the NAH chapters, Chapters 3, 5 and 7, in any way. Thus, for the reader not interested in inverse problems, these latter chapters can be skipped without compromising the understanding of any of the material in the rest of the book. The inverse, however, is not true. Chapters 3, 5 and 7 rely heavily on the material in Chapters 2, 4 and 6.

I am indebted to many who have inspired me along the way, beginning with my PhD advisor at The Pennsylvania State University, Eugen Skudrzyk. Many enlightening discussions with Julian Maynard and Dean Aires followed in my postdoctoral work there. However, I am most indebted to an incredible research institution, the Naval Research Laboratory, which has allowed me to grow and mature through exciting and unperturbed basic research for the last sixteen years. This would not have been possible without the incredible support over these years of Joseph Bucaro, branch head, who also planted the idea and encouraged me to write a book on my work. And finally this work would have been possible without the support of a brilliant experimentalist, Brian Houston.

Great thanks go to my dedicated reviewers. To Joseph Kasper, who took such a serious interest and improved the work with his comments. To a close colleague, Anthony Romano, who also provided invaluable reviewing. To all my inquisitive students who sat through my lectures which formed the foundation of this book. And finally to my wonderful wife, Virginia, who has supported me throughout this task and who helped proofread the manuscript.

Dr. Earl G. Williams
Naval Research Laboratory
Washington D.C.
1998

Chapter 1

Fourier Transforms & Special Functions

1.1 Introduction

At the heart of Fourier acoustics is the Fourier transform which includes the concepts of the Fourier series and the Hankel transform. We present in this chapter much of the prerequisite mathematics needed to understand the concepts presented in this book. Special functions, like the Dirac delta function, are crucial and provide an elegant shorthand in the mathematics. The rectangle and comb functions are invaluable in understanding the formulation of nearfield acoustical holography, especially in regard to discretization of the formulation for coding on a computer. Essential in this discretization is the relationship between the DFT (discrete Fourier transform) and the integral (continuous) Fourier transform.

1.2 The Fourier Transform

The Fourier transform, $F(k_x)$ of a function $f(x)$ throughout this work will be defined as

$$F(k_x) = \int_{-\infty}^{\infty} f(x)e^{-ik_x x}dx. \tag{1.1}$$

The following shorthand notation will be useful. Let $\mathcal{F}_x$ represent the Fourier transform operator so that Eq. (1.1) becomes

$$\mathcal{F}_x[f(x)] \equiv F(k_x). \tag{1.2}$$

In this book we will use symbol $\equiv$ to mean 'definition of' to differentiate from an equality defined with $=$. The inverse transform corresponding to Eq. (1.1) is

$$f(x) = \frac{1}{2\pi} \int_{-\infty}^{\infty} F(k_x)e^{ik_x x}dk_x, \tag{1.3}$$

and the shorthand notation for this equation is

$$\mathcal{F}_x^{-1}[F(k_x)] \equiv f(x). \tag{1.4}$$

Equation (1.3) is verified by inserting it into Eq. (1.1) and using the delta function relation, Eq. (1.36), written as

$$\delta(k_x - k_x') = \frac{1}{2\pi} \int_{-\infty}^{\infty} e^{i(k_x' - k_x)x} dx. \tag{1.5}$$

Thus

$$F(k_x) = \frac{1}{2\pi} \int_{-\infty}^{\infty} dk_x' F(k_x') \int_{-\infty}^{\infty} e^{i(k_x' - k_x)x} dx = F(k_x).$$

The spatial transform pair given in Eq. (1.1) and Eq. (1.3) is the counterpart to the time-frequency pair:

$$F(\omega) = \int_{-\infty}^{\infty} f(t)e^{i\omega t} dt \tag{1.6}$$

and

$$f(t) = \frac{1}{2\pi} \int_{-\infty}^{\infty} F(\omega)e^{-i\omega t} d\omega. \tag{1.7}$$

Notice a subtle difference, however. The sign of the exponential term is reversed. This is necessary, as will be discussed in Chapter 2, to retain the meaning of a plane wave given by $\exp(i(k_x x + k_y y + k_z z - \omega t))$. Thus a function of space and time when expanded with the inverse transforms is

$$f(x,t) = \frac{1}{4\pi^2} \int_{-\infty}^{\infty} \int_{-\infty}^{\infty} F(k_x, \omega)e^{ik_x x} e^{-i\omega t} dk_x d\omega. \tag{1.8}$$

It is simple to determine the Fourier transform of $\frac{\partial f(x)}{\partial x}$ by taking the partial derivative of Eq. (1.3),

$$\frac{\partial f(x)}{\partial x} = \frac{1}{2\pi} \int_{-\infty}^{\infty} ik_x F(k_x)e^{ik_x x} dk_x \tag{1.9}$$

or

$$\frac{\partial f(x)}{\partial x} = \mathcal{F}_x^{-1}[ik_x F(k_x)],$$

from which we see, in view of Eq. (1.1), that

$$\mathcal{F}_x\left[\frac{\partial f(x)}{\partial x}\right] = ik_x F(k_x). \tag{1.10}$$

There are several important theorems regarding the Fourier transform which we need to review.

- The shift theorem states that

$$\int_{-\infty}^{\infty} f(x - x')e^{-ik_x x} dx = F(k_x)e^{-ik_x x'}. \tag{1.11}$$

This theorem is easily proven by a change of variables.

- The convolution theorem is

$$\int_{-\infty}^{\infty} \left[\int_{-\infty}^{\infty} f(x-x')g(x')dx' \right] e^{-ik_x x} dx = F(k_x)G(k_x). \qquad (1.12)$$

Using the shift theorem the latter is proven:

$$\int_{-\infty}^{\infty} \left[\int_{-\infty}^{\infty} f(x-x')g(x')\,dx' \right] e^{-ik_x x} dx$$

$$= \int_{-\infty}^{\infty} \left[\int_{-\infty}^{\infty} f(x-x')e^{-ik_x x}dx \right] g(x')\,dx'$$

$$= \int_{-\infty}^{\infty} \left[F(k_x)e^{-ik_x x'} \right] g(x')\,dx'$$

$$= F(k_x) \int_{-\infty}^{\infty} g(x')e^{-ik_x x'}\,dx'$$

$$= F(k_x)G(k_x).$$

In shorthand notion the convolution theorem is

$$\mathcal{F}_x \Big[f(x) * g(x) \Big] = F(k_x)G(k_x),$$

where the asterisk ($*$) denotes convolution:

$$f(x) * g(x) \equiv \int_{-\infty}^{\infty} f(x-x')g(x')dx'. \qquad (1.13)$$

Taking the inverse transform of both sides of Eq. (1.12) yields another form of the convolution theorem:

$$\mathcal{F}_x^{-1}[F(k_x)G(k_x)] = f(x) * g(x). \qquad (1.14)$$

- The convolution theorem for a product of two spatial functions is

$$\int_{-\infty}^{\infty} f(x)g(x)e^{-ik_x x}dx = \frac{1}{2\pi} \int_{-\infty}^{\infty} F(k'_x)G(k_x - k'_x)dk'_x, \qquad (1.15)$$

or

$$\mathcal{F}_x \Big[f(x)g(x) \Big] = \frac{1}{2\pi} F(k_x) * G(k_x).$$

Transition into two dimensions, dealing with functions of two variables, is straightforward. The two-dimensional function $f(x,y)$ has the two-dimensional Fourier transform $F(k_x, k_y)$, satisfying the following relations:

$$F(k_x, k_y) = \int_{-\infty}^{\infty} \int_{-\infty}^{\infty} f(x,y)e^{-i(k_x x + k_y y)}dxdy \qquad (1.16)$$

and

$$f(x,y) = \frac{1}{4\pi^2} \int_{-\infty}^{\infty} \int_{-\infty}^{\infty} F(k_x, k_y) e^{i(k_x x + k_y y)} dk_x dk_y. \tag{1.17}$$

If we define a two-dimensional convolution as

$$f(x,y) ** g(x,y) \equiv \int_{-\infty}^{\infty} \int_{-\infty}^{\infty} f(x - x', y - y') g(x', y') dx' dy', \tag{1.18}$$

then the two-dimensional convolution theorem is

$$\mathcal{F}_x \mathcal{F}_y \Big[f(x,y) ** g(x,y) \Big] = F(k_x, k_y) G(k_x, k_y), \tag{1.19}$$

or, equivalently,

$$f(x,y) ** g(x,y) = \mathcal{F}_x^{-1} \mathcal{F}_y^{-1} \Big[F(k_x, k_y) G(k_x, k_y) \Big]. \tag{1.20}$$

Similarly the transform of the product of two spatial functions becomes

$$\int_{-\infty}^{\infty} \int_{-\infty}^{\infty} f(x,y) g(x,y) e^{-i(k_x x + k_y y)} dx \, dy$$

$$= \frac{1}{4\pi^2} \int_{-\infty}^{\infty} \int_{-\infty}^{\infty} F(k_x', k_y') G(k_x - k_x', k_y - k_y') dk_x' dk_y', \tag{1.21}$$

or in shorthand

$$\mathcal{F}_x \mathcal{F}_y \Big[f(x,y) g(x,y) \Big] = \frac{1}{4\pi^2} F(k_x, k_y) ** G(k_x, k_y), \tag{1.22}$$

or, equivalently,

$$f(x,y) g(x,y) = \frac{1}{4\pi^2} \mathcal{F}_x^{-1} \mathcal{F}_y^{-1} \Big[F(k_x, k_y) ** G(k_x, k_y) \Big]. \tag{1.23}$$

1.3   Fourier Series

For problems in which the functions have circular symmetry, such as the vibrations of a circular plate or membrane, we will need the following relationships. The circular (polar) coordinates are given by ρ and ϕ, so that a function $f(\rho, \phi)$ can be represented in a Fourier series in the ϕ coordinate as

$$f(\rho, \phi) = \sum_{n=-\infty}^{\infty} f_n(\rho) e^{in\phi} \equiv \mathcal{F}_o^{-1} \Big[f_n(\rho) \Big], \tag{1.24}$$

where the coefficient functions, $f_n(\rho)$, are given by

$$f_n(\rho) = \frac{1}{2\pi} \int_0^{2\pi} f(\rho, \phi) e^{-in\phi} d\phi \equiv \mathcal{F}_o \Big[f(\rho, \phi) \Big]. \tag{1.25}$$

Note that the $1/2\pi$ could just as easily have been transferred to Eq. (1.24) instead of Eq. (1.25), but we use the former convention throughout this book.

The convolution relationship for Fourier series is easily derived given the completeness relationship[1] for the circumferential harmonics,

$$\frac{1}{2\pi} \sum_{n=-\infty}^{\infty} e^{in\phi} e^{-in\phi'} = \delta(\phi - \phi'). \tag{1.26}$$

Thus given the transforms, F_n and G_n of two functions $f(\phi)$ and $g(\phi)$,

$$F_n = \frac{1}{2\pi} \int_0^{2\pi} f(\phi) e^{-in\phi} d\phi,$$

$$G_n = \frac{1}{2\pi} \int_0^{2\pi} g(\phi) e^{-in\phi} d\phi,$$

we have

$$
\begin{aligned}
\sum_{n=-\infty}^{\infty} F_n G_n e^{in\phi} &= \frac{1}{4\pi^2} \int \int f(\phi') g(\phi'') \sum_n e^{in(\phi-\phi'-\phi'')} d\phi' d\phi'' \\
&= \frac{1}{2\pi} \int \int f(\phi') g(\phi'') \delta(\phi - \phi' - \phi'') d\phi' d\phi'' \\
&= \frac{1}{2\pi} \int_0^{2\pi} f(\phi') g(\phi - \phi') d\phi' \\
&= \frac{1}{2\pi} \int_0^{2\pi} f(\phi - \phi') g(\phi') d\phi'.
\end{aligned}
\tag{1.27}
$$

We define as usual

$$f(\phi) * g(\phi) = \int_0^{2\pi} f(\phi') g(\phi - \phi') d\phi' = \int_0^{2\pi} f(\phi - \phi') g(\phi') d\phi',$$

so that Eq. (1.27) is

$$\mathcal{F}_\phi^{-1}[F_n G_n] \equiv \sum_{n=-\infty}^{\infty} F_n G_n e^{in\phi} = \frac{1}{2\pi} f(\phi) * g(\phi). \tag{1.28}$$

1.4 Fourier–Bessel (Hankel) Transforms

Hankel transforms arise in problems in polar coordinates, and the forward and inverse Hankel transforms are analogous to the forward and inverse Fourier transforms for rectangular coordinates. The nth order Hankel transform is defined as

$$F_n(k_\rho) = \int_0^{\infty} f_n(\rho) J_n(k_\rho \rho) \rho \, d\rho, \tag{1.29}$$

---

[1] For a discussion of completeness see J. D. Jackson (1975). *Classical Electrodynamics*, 2nd ed. Wiley & Sons, pp. 65–68.

where k_ρ is the transform variable and the relationship to rectangular coordinates is $\rho = \sqrt{x^2 + y^2}$. J_n is a Bessel function described in Section 4.2.1. We use the following shorthand notation for the Hankel transform:

$$\mathcal{B}_n[f_n(\rho)] \equiv F_n(k_\rho). \tag{1.30}$$

To derive the inverse Hankel transform we draw on an important integral for the Dirac delta function valid for all n:[2]

$$\frac{\delta(\rho - \rho')}{\rho} = \int_0^\infty J_n(k_\rho \rho') J_n(k_\rho \rho) k_\rho \, dk_\rho. \tag{1.31}$$

Multiply both sides of Eq. (1.29) by $J_n(k_\rho \rho') k_\rho$ and integrate over k_ρ:

$$\int_0^\infty F_n(k_\rho) J_n(k_\rho \rho') k_\rho \, dk_\rho = \int_0^\infty \rho \, d\rho f_n(\rho) \int_0^\infty J_n(k_\rho \rho') J_n(k_\rho \rho) k_\rho \, dk_\rho.$$

Making use of Eq. (1.31) yields $f_n(\rho')$ on the right hand side. and we arrive at (writing ρ for ρ')

$$f_n(\rho) = \int_0^\infty F_n(k_\rho) J_n(k_\rho \rho) k_\rho \, dk_\rho, \tag{1.32}$$

the nth order inverse Hankel transform. We will use the shorthand notation

$$\mathcal{B}_n^{-1}[F_n(k_\rho)] \equiv \int_0^\infty F_n(k_\rho) J_n(k_\rho \rho) k_\rho \, dk_\rho \tag{1.33}$$

and $\mathcal{B}_n[f_n(\rho)]$ for the forward Hankel transform. Note that Eqs (1.29) and (1.32) define an infinite set of Hankel transforms pairs, one for each order n. Most common in practice is the 0th order Hankel transform pair:

$$F(k_\rho) = \int_0^\infty f(\rho) J_0(k_\rho \rho) \rho d\rho, \tag{1.34}$$

or $F[k_\rho] = \mathcal{B}[f(\rho)]$, and

$$f(\rho) = \int_0^\infty F(k_\rho) J_0(k_\rho \rho) k_\rho \, dk_\rho, \tag{1.35}$$

or $f(\rho) = \mathcal{B}^{-1}[F(k_\rho)]$.

1.5   The Dirac Delta Function

The following are important properties of the delta function, $\delta(x - x_0)$, drawn from the theory of generalized functions:[3]

---

[2] J. D. Jackson (1975), *Classical Electrodynamics*, 2nd ed. Wiley & Sons, p. 110.

[3] M. J. Lighthill (1958). *Introduction to Fourier Analysis and Generalised Functions*, Cambridge University Press.

- An important integral relation for the delta function is

$$\delta(x - x_0) = \frac{1}{2\pi} \int_{-\infty}^{\infty} e^{ik_x(x-x_0)} dk_x.$$ (1.36)

Other integral relations for the Delta function will be given throughout this book.

- The sifting property is

$$\int_{-\infty}^{\infty} \delta(x - x_0) f(x)\, dx = f(x_0),$$ (1.37)

so that the area under the delta function is unity:

$$\int_{-\infty}^{\infty} \delta(x - x_0)\, dx = 1.$$ (1.38)

Equation (1.36) shows that $\delta(x - x_0)$ is the inverse Fourier transform of $e^{-ik_x x_0}$:

$$\delta(x - x_0) = \mathcal{F}_x^{-1}\left(e^{-ik_x x_0}\right).$$

It follows from the forward Fourier transform that

$$\int_{-\infty}^{\infty} \delta(x - x_0) e^{-ik_x x} dx = e^{-ik_x x_0},$$ (1.39)

or

$$e^{-ik_x x_0} = \mathcal{F}_x\left(\delta(x - x_0)\right).$$

For finite limits we have

$$\int_{-\infty}^{\xi} \delta(x - x_0)\, dx = \begin{cases} 0 & \xi < x_0 \\ \frac{1}{2} & \xi = x_0 \\ 1 & \xi > x_0 \end{cases}.$$ (1.40)

1.6 The Rectangle Function

The rectangle function is defined by

$$\Pi(x/L) = \begin{cases} 1 & |x| < L/2 \\ \frac{1}{2} & |x| = L/2 \\ 0 & |x| > L/2. \end{cases}$$ (1.41)

The Fourier transform of the rectangle function is

$$\int_{-\infty}^{\infty} \Pi(x/L) e^{-ik_x x} dx = \frac{L \sin(k_x L/2)}{(k_x L/2)} = L \operatorname{sinc}(k_x L/2),$$ (1.42)

where

$$\operatorname{sinc}(x) \equiv \sin(x)/x.$$ (1.43)

1.7   The Comb Function

The comb function is an infinite series of delta functions, and is defined as

$$\text{III}(x/a) \equiv |a| \sum_{n=-\infty}^{\infty} \delta(x - na). \tag{1.44}$$

The Fourier transform of the comb function is another comb function,

$$\int_{-\infty}^{\infty} \text{III}(x/a)e^{-ik_x x}dx = a\text{III}(\frac{k_x}{2\pi/a}), \tag{1.45}$$

where, consistent with Eq. (1.44),

$$\text{III}(\frac{k_x}{2\pi/a}) \equiv \frac{2\pi}{|a|} \sum_{n=-\infty}^{\infty} \delta\left(k_x - n(2\pi/a)\right). \tag{1.46}$$

Since $\text{III}(x/a)$ is a periodic function with period a, it can be expanded in a Fourier series which will lead us to an important formula. Following Eq. (1.24) and defining $\phi \equiv 2\pi x/a$ and $f(\rho, \phi) = f(\phi) = \text{III}(x/a)$, then

$$\text{III}(x/a) = \sum_{m=-\infty}^{\infty} f_m e^{im(2\pi x/a)},$$

which reflects the period of $x = a$. The constants f_m are obtained from Eq. (1.25) with $d\phi = \frac{2\pi}{a}dx$:

$$f_m = \frac{1}{a} \int_0^a \text{III}(x/a)e^{-im(2\pi x/a)}dx.$$

The delta functions at each end of the integration range contribute a factor of $1/2$ (see Eq. (1.40)). Thus $f_m = 1$. Inserting this value in Eq. (1.7) leads to the important formula (called the Poisson sum formula)

$$\frac{1}{a}\text{III}(x/a) \equiv \sum_{n=-\infty}^{\infty} \delta(x - na) = \frac{1}{a} \sum_{m=-\infty}^{\infty} e^{2\pi imx/a}. \tag{1.47}$$

The comb function is a dimensionless quantity.

1.8   Continuous Fourier Transform and the DFT

The discrete Fourier transform (DFT) is defined by the forward and inverse relations

$$F_m = \sum_{q=0}^{N-1} f_q e^{-2\pi iqm/N} \tag{1.48}$$

and

$$f_q = \frac{1}{N} \sum_{m=0}^{N-1} F_m e^{2\pi i q m/N}, \tag{1.49}$$

respectively. The equivalence between Eqs (1.48) and (1.49) can be proven with the following relation, the discrete analog of the Dirac delta function: Eq. (1.36),

$$\frac{1}{N} \sum_{m=0}^{N-1} e^{2\pi i m(q-q')/N} = \delta_{qq'} \qquad \text{where} \quad \left\{ \begin{array}{ll} \delta_{qq'} = 1 & \text{if } q = q' \\ \delta_{qq'} = 0 & \text{otherwise.} \end{array} \right. \tag{1.50}$$

1.8.1 Discretization of the Fourier Transform

We assume for the particular problem of interest that the infinite integral of the continuous Fourier transform can be approximated accurately by the finite integral:

$$F(k) = \int_{-\infty}^{\infty} f(x)e^{-ikx}dx \approx \int_{-L/2}^{L/2-\Delta x} f(x)e^{-ikx}dx. \tag{1.51}$$

(Finite aperture effects will be discussed in Section 3.9.)

This finite integral can be transformed to look like a DFT by using a simple rectangular integration rule to replace the integral with a summation. Discretize the function $f(x)$ with equally spaced samples separated by Δx and let

$$x = q\Delta x, \quad q = -N/2, -N/2 + 1, \cdots, N/2 - 1,$$

where N is the total number of samples, and the last sample is just shy of the right end of the interval as indicated in Eq. (1.51). We must have that

$$\Delta x = L/N. \tag{1.52}$$

At the same time assume that we are interested in the positive and negative wavenumbers of the continuous Fourier transform, which we also quantize as

$$k = m\Delta k, \quad m = -N/2, -N/2 + 1, \cdots, N/2 - 1, \tag{1.53}$$

using the same number of points. In order to obtain a DFT we must restrict Δk to

$$\Delta k = 2\pi/L. \tag{1.54}$$

This is equivalent to one wavelength over the total extent of the aperture L.

With these relations the rectangular quadrature rule is used to approximate Eq. (1.51) to yield

$$\begin{aligned} F(m\Delta k) &\approx \sum_{q=-N/2}^{N/2-1} f(q\Delta x)e^{-imq\Delta x\Delta k}\Delta x \\ &= \frac{L}{N} \sum_{q=-N/2}^{N/2-1} f(q\Delta x)e^{-i2\pi mq/N}. \end{aligned} \tag{1.55}$$

To arrive at Eq. (1.48) we make the following change of indices: $m' = m + N/2$ and $q' = q + N/2$, so that $m' = 0, 1, \cdots, N - 1$ and $q' = 0, 1, \cdots, N - 1$, and substitute in Eq. (1.55):

$$F((m' - N/2)\Delta k) \approx \frac{L}{N} \sum_{q'=0}^{N-1} f((q' - N/2)\Delta x) e^{-i2\pi(m' - N/2)(q' - N/2)/N} \tag{1.56}$$

$$= (-1)^{m'} \frac{L}{N} \underbrace{\sum_{q'=0}^{N-1} \underbrace{(-1)^{q'} f((q' - N/2)\Delta x)}_{f_{q'}} e^{-i2\pi m' q'/N}}_{F_{m'}} e^{-i\pi N/2}.$$

As indicated by the underbraces, this last expression is equivalent to Eq. (1.48) if we define

$$f_{q'} = (-1)^{q'} f((q' - N/2)\Delta x), \tag{1.57}$$

and

$$F((m' - N/2)\Delta k) = (-1)^{m'} \frac{L}{N} F_{m'}. \tag{1.58}$$

If we choose N to be a power of 2 (as in a fast Fourier transform (FFT) algorithm), then the term $e^{-i\pi N/2} = 1$. (For all practical purposes N is greater than 2.) Equations (1.57) and (1.58) provide the recipe for using the forward DFT to approximate a forward continuous Fourier transform, Eq. (1.51).

As an example consider the DFT of a rectangle function $\Pi(x/L)$ with $N = 8$ shown on the left of Fig. 1.1. On the right is shown the modification necessary to construct the

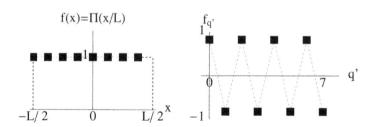

Figure 1.1: Left: Sampled version of $\Pi(x/L)$, with $N = 8$.
Right: Modification for the DFT according to Eq. (1.57).
Note that $(-1)^{q'}$ creates the sawtooth effect.

series $f_{q'}$. Note the oscillation due to the multiplication by $(-1)^{q'}$. The result from the DFT is shown in Fig. 1.2. The left plot results from the FFT algorithm (an optimized coding of the DFT for computations).[4] implementing Eq. (1.48) with $m' = m = 0, \cdots, 7$. The continuous transform produces the result on the right of the figure, which when sampled at intervals of $\Delta k = 2\pi/L$ exactly equals the FFT result. One obtains exact agreement, according to the Shannon sampling theorem,[5] when the Fourier transform

---

[4]E. Oran Brigham (1974). *The Fast Fourier Transform*, Prentice-Hall, N. J.
[5]A. Papoulis (1962). *The Fourier Integral and its Applications*, McGraw-Hill, N.Y., p. 50.

of the function $f(x)$ is zero above a specific frequency k_s, that is, when

$$F(k) = 0 \quad \text{for} \quad k \geq |k_s|. \tag{1.59}$$

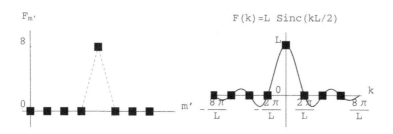

Figure 1.2: Left: Result from the FFT. Only point $m' = 4$ is nonzero. Right: Continuous Fourier transform of the rectangle function with sample interval shown. Since $\Delta k = 2\pi/L$ then the sinc function is sampled exactly at the zero crossings and the FFT provides an errorless result.

1.8.2 Discretization of the Inverse Fourier Transform

Using the identical quantization scheme and derivation, we will obtain the approximation of the continuous inverse Fourier transform using the DFT. Again we assume that the infinite limits can be replaced with finite ones without appreciable error:

$$f(x) = \frac{1}{2\pi} \int_{-\infty}^{\infty} F(k)e^{ikx}\,dk \approx \frac{1}{2\pi} \int_{-k_m}^{k_m - \Delta k} e^{ikx}\,dk, \tag{1.60}$$

where k_m represents the maximum wavenumber provided by Eqs (1.52–1.54),

$$k_m = (N/2)\Delta k = \pi N/L = \pi/\Delta x, \tag{1.61}$$

that is, one wavelength over two spatial samples. Inserting the quantizations given above into Eq. (1.60) leads to

$$
\begin{aligned}
f(q\Delta x) &\approx \frac{1}{2\pi} \sum_{m=-N/2}^{N/2-1} F(m\Delta k)e^{imq\Delta x\Delta k}\Delta k \\
&= \frac{1}{L} \sum_{m=-N/2}^{N/2-1} F(m\Delta k)e^{i2\pi mq/N}.
\end{aligned} \tag{1.62}
$$

Making the same change of indices given in the last section in order to obtain the form of the inverse DFT, Eq. (1.49), we obtain

$$f((q' - N/2)\Delta x) \approx \frac{1}{L} \sum_{m'=0}^{N-1} F((m' - N/2)\Delta k) e^{i2\pi(m'-N/2)(q'-N/2)/N}$$

$$= (-1)^{q'} \frac{N}{L} \frac{1}{N} \sum_{m'=0}^{N-1} \underbrace{(-1)^{m'} F((m' - N/2)\Delta k)}_{F_{m'}} e^{i2\pi m' q'/N} e^{i\pi N/2},$$

$$\underbrace{\qquad\qquad\qquad\qquad\qquad\qquad\qquad\qquad\qquad\qquad\qquad\qquad\qquad}_{f_{q'}}$$

which is related to Eq. (1.49) by equating

$$F_{m'} = (-1)^{m'} F((m' - N/2)\Delta k) \tag{1.63}$$

and (again $e^{i\pi N/2} = 1$)

$$f((q' - N/2)\Delta x) = (-1)^{q'} \frac{N}{L} f_{q'}. \tag{1.64}$$

Equations (1.63) and (1.64) provide the recipe for the approximation of the inverse Fourier transform using the DFT.

1.8.3   Circumferential Transforms: Fourier Series

The inverse and forward relations which make up the Fourier series are, respectively,

$$f(\phi) = \sum_{m=-\infty}^{\infty} F(m) e^{imo} \tag{1.65}$$

and

$$F(m) = \frac{1}{2\pi} \int_{-\pi}^{\pi} f(\phi) e^{-imo} d\phi. \tag{1.66}$$

where $f(\phi)$ is a cyclic function with period 2π.

Results identical to the Fourier transform discretization are obtained in this case noting that

$$L = 2\pi$$

and $x \to \phi$ (where $-\pi \le \phi \le \pi$), $k \to m$, $\Delta k = 1$, $\Delta x = 2\pi/N$, except that the $\frac{1}{2\pi}$ normalization is included with the forward transform instead of the inverse, which leads to an extra factor of 2π dividing Eq. (1.57) and the same factor multiplying Eq. (1.64).

Problems

1.1 Carry out the integration

$$\int_{-\infty}^{\infty} f(t')\delta(t - t' + t_0)dt'.$$

1.2 If $\mathcal{F}_x[f(x)] = J_1(\beta k_x)$, then what is $\mathcal{F}_x[f(x - a)]$?

1.3 Let the complex function $f(x)$ be written as a sum of an even $e(x)$ and an odd function $o(x)$. Prove that if $F(k_x) = E(k_x) + O(k_x)$ is the Fourier transform of $f(x)$ with even and odd parts given by E and O, respectively, then

$$E(k_x) = \mathcal{F}_x[e(x)].$$

Note that $e^{-ik_z x}$ can be written as a sum of an even and an odd function in x.

1.4 Write the convolution theorem for $\frac{1}{2\pi}\int_{-\infty}^{\infty} F(k_x)G(k_x)H(k_x)e^{ik_z x}dk_x$.

1.5 The zero-order Hankel transform and its inverse are given by

$$F(k_\rho) = \mathcal{B}[f(\rho)] \equiv \int_0^{\infty} f(\rho)J_0(k_\rho\rho)\rho d\rho$$

and

$$f(\rho) = \int_0^{\infty} F(k_\rho)J_0(k_\rho\rho)k_\rho dk_\rho.$$

Prove the following Hankel transform relationships:

(a) If $f(\rho) = \delta(\rho - \rho_0)$ and $\rho_0 \geq 0$, then

$$\mathcal{B}[f(\rho)] = \rho_0 J_0(k_\rho\rho_0).$$

(b) If $f(\rho) = 1$ for $0 \leq a \leq \rho \leq 1$ and zero otherwise, then

$$\mathcal{B}[f(\rho)] \propto \frac{J_1(k_\rho) - aJ_1(k_\rho a)}{k_\rho}.$$

(c) If $\mathcal{B}[f(\rho)] = F(k_\rho)$ then

$$\mathcal{B}[f(ak_\rho)] = F(k_\rho/a)/a^2.$$

Chapter 2

Plane Waves

2.1 Introduction

In this chapter we present the foundations of Fourier acoustics–plane wave expansions. This material is presented in depth to provide a firm foundation for the rest of the book, introducing important concepts like wavenumber space and the extrapolation of wavefields from one surface to another. Fourier acoustics is used to derive some famous tools for the radiation from planar sources: the Rayleigh integrals, the Ewald sphere construction of farfield radiation, the first product theorem for arrays, vibrating plate radiation, and radiation classification theory. Finally, a new tool called supersonic intensity is introduced which is useful in locating acoustic sources on vibrating structures. We begin the chapter with a review of some fundamentals: the wave equation, Euler's equation, and the concept of acoustic intensity.

2.2 The Wave Equation and Euler's Equation

Let $p(x, y, z, t)$ be an infinitesimal variation of acoustic pressure from its equilibrium value which satisfies the acoustic wave equation

$$\nabla^2 p - \frac{1}{c^2} \frac{\partial^2 p}{\partial t^2} = 0 \tag{2.1}$$

for a homogeneous fluid with no viscosity. c is a constant and refers to the speed of sound in the medium. At 20°C $c = 343$ m/s in air and $c = 1481$ m/s in water. The right hand side of Eq. (2.1) indicates that there are no sources in the volume in which the equation is valid. In Cartesian coordinates

$$\nabla^2 \equiv \frac{\partial^2}{\partial x^2} + \frac{\partial^2}{\partial y^2} + \frac{\partial^2}{\partial z^2}.$$

A second equation which will be used throughout this book is called Euler's equation,

$$\rho_0 \frac{\partial \vec{v}}{\partial t} = -\vec{\nabla} p. \tag{2.2}$$

where $\vec{\upsilon}$ (Greek letter upsilon) represents the velocity vector with components $\dot{u}$, $\dot{v}$, $\dot{w}$;

$$\vec{\upsilon} = \dot{u}\hat{\imath} + \dot{v}\hat{\jmath} + \dot{w}\hat{k}. \tag{2.3}$$

where $\hat{\imath}$, $\hat{\jmath}$, and $\hat{k}$ are the unit vectors in the x, y, and z directions, respectively, and the gradient is defined in terms of the unit vectors as

$$\vec{\nabla} \equiv \frac{\partial}{\partial x}\hat{\imath} + \frac{\partial}{\partial y}\hat{\jmath} + \frac{\partial}{\partial z}\hat{k}. \tag{2.4}$$

We use the convention of a dot over a displacement quantity to indicate velocity as is done in Junger and Feit.[1] The displacements in the three coordinate directions are given by u, v, and w.

The derivation of Eq. (2.2) is useful in developing some understanding of the physical meaning of p and $\vec{\upsilon}$. Let us proceed in this direction.

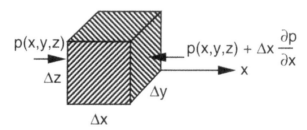

Figure 2.1: Infinitesimal volume element to illustrate Euler's equation.

Figure 2.1 shows an infinitesimal volume element of fluid $\Delta x \Delta y \Delta z$. with the x axis as shown. All six faces experience forces due to the pressure p in the fluid. It is important to realize that pressure is a scalar quantity. There is no direction associated with it. It has units of force per unit area, N/m^2 or Pascals. The following is the convention for pressure,

$$p > 0 \to \text{Compression}$$
$$p < 0 \to \text{Rarefaction}.$$

At a specific point in a fluid, a positive pressure indicates that an infinitesimal volume surrounding the point is under compression, and forces are exerted *outward* from this volume. It follows that if the pressure at the left face of the cube in Fig. 2.1 is positive, then a force will be exerted in the positive x direction of magnitude $p(x, y, z)\Delta y \Delta z$. The pressure at the opposite face $p(x + \Delta x, y, z)$ is exerted in the negative x direction. We expand $p(x + \Delta x, y, z)$ in a Taylor series to first order, as shown in the figure. Note that the force arrows indicate the direction of force for positive pressure. Given the directions of force shown, the total force exerted on the volume in the x direction is

$$(p(x, y, z) - p(x + \Delta x, y, z))\Delta y \Delta z = -\Delta x \Delta y \Delta z \frac{\partial p}{\partial x}.$$

[1]M. C. Junger and D. Feit (1986). *Sound, Structures, and Their Interaction*. 2nd ed. MIT Press, Cambridge, MA.

Now we invoke Newton's equation, $f = ma = m\frac{\partial \dot{u}}{\partial t}$, where f is the force, $m = \rho_0 \Delta x \Delta y \Delta z$ and ρ_0 is the fluid density, yielding

$$\rho_0 \frac{\partial \dot{u}}{\partial t} = -\frac{\partial p}{\partial x}.$$

Carrying out the same analysis in the y and z directions yields the following two equations:

$$\rho_0 \frac{\partial \dot{v}}{\partial t} = -\frac{\partial p}{\partial y}$$

and

$$\rho_0 \frac{\partial \dot{w}}{\partial t} = -\frac{\partial p}{\partial z}.$$

We combine the above three equations into one using vectors yielding Eq. (2.2) above, Euler's equation.

2.3 Instantaneous Acoustic Intensity

It is critical in the study of acoustics to understand certain energy relationships. Most important is the acoustic intensity vector. In the time domain it is called the instantaneous acoustic intensity and is defined as

$$\vec{I}(t) = p(t)\vec{v}(t), \tag{2.5}$$

with units of energy per unit time (power) per unit area, measured as $(\text{joules/s})/\text{m}^2$ or watts/m^2.

The acoustic intensity is related to the energy density e through its divergence,

$$\frac{\partial e}{\partial t} = -\vec{\nabla} \cdot \vec{I}, \tag{2.6}$$

where the divergence is

$$\vec{\nabla} \cdot \vec{I} \equiv \frac{\partial I_x}{\partial x} + \frac{\partial I_y}{\partial y} + \frac{\partial I_z}{\partial z}. \tag{2.7}$$

The energy density is given by

$$e = \frac{1}{2}\rho_0 |\vec{v}(t)|^2 + \frac{1}{2}\kappa p(t)^2 \tag{2.8}$$

where κ is the fluid compressibility,

$$\kappa = \frac{1}{\rho_0 c^2}. \tag{2.9}$$

Equation (2.6) expresses the fact that an increase in the energy density at some point in the fluid is indicated by a negative divergence of the acoustic intensity vector; the intensity vectors are pointing into the region of increase in energy density. Figure 2.2 should make this clear.

If we reverse the arrows in Fig. 2.2, a positive divergence results and the energy density at the center must decrease, that is, $\frac{\partial e}{\partial t} < 0$. This case represents an apparent source of energy at the center.

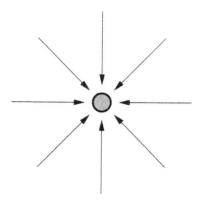

Figure 2.2: Illustration of negative divergence of acoustic intensity. The region at the center has an increasing energy density with time, that is, an apparent sink of energy.

2.4   Steady State

To consider phenomena in the frequency domain, we obtain the steady state solution through Fourier transforms

$$p(t) = \frac{1}{2\pi} \int_{-\infty}^{\infty} \bar{p}(\omega) e^{-i\omega t} d\omega \tag{2.10}$$

leading to the steady state solution

$$\bar{p}(\omega) = \int_{-\infty}^{\infty} p(t) e^{i\omega t} dt. \tag{2.11}$$

Equation (2.10) can be differentiated with respect to time to yield the important relationship

$$\frac{\partial p(t)}{\partial t} = \frac{1}{2\pi} \int_{-\infty}^{\infty} -i\omega \bar{p}(\omega) e^{-i\omega t} d\omega,$$

so that

$$\mathcal{F}(\frac{\partial p(t)}{\partial t}) = -i\omega \bar{p}(\omega), \tag{2.12}$$

where the calligraphic letter $\mathcal{F}$ represents the Fourier transform operation defined in Eq. (2.11).

Equation (2.12) can be used to compute the Fourier transform of the time domain wave equation, Eq. (2.1), yielding the Helmholtz equation

$$\nabla^2 \bar{p} + k^2 \bar{p} = 0, \tag{2.13}$$

where the acoustic wavenumber is $k = \omega/c$, the frequency is given by $2\pi f = \omega$, and $\bar{p}$ is the function $\bar{p}(x, y, z, \omega)$. For simplicity of notation we drop the bar above the variable. It will be clear from the context of the discussion if the quantity is in the frequency or in the time domain. The Fourier transform of Euler's equation, Eq. (2.2), becomes, in the frequency domain

$$i\omega\rho_0\vec{v} = \vec{\nabla}p, \tag{2.14}$$

where Eq. (2.12) has been used again for the time derivative.

2.5   Time Averaged Acoustic Intensity

Now consider the intensity relationship for steady state fields. This is defined as the average of the instantaneous intensity over a period T, where $T = 1/f$ and f is the excitation frequency:

$$\vec{I}(\omega) = \frac{1}{T}\int_0^T p(t)\vec{v}(t)\,dt. \tag{2.15}$$

Using complex variable notation this relationship becomes

$$\vec{I}(\omega) = \frac{1}{2}\mathrm{Re}\big(p(\omega)\vec{v}(\omega)^*\big), \tag{2.16}$$

where * stands for complex conjugate and Re for the real part. The one-half results from the time average process. $\vec{I}$ is the average power over one period passing through unit area. For example, the x component of this flow I_x represents the power passing through an element of area $\Delta y\Delta z$.

Important in this chapter is the radiation from planar radiators. Of particular interest is the power flow crossing an infinite plane. For example, consider the total power crossing the coordinate plane $z = 0$, a quantity expressed in watts or joules per second. We use the symbol $\Pi(\omega)$ to represent the total power in watts crossing the boundary:

$$\Pi(\omega) = \int_{-\infty}^{\infty}\int_{-\infty}^{\infty} I_z(x,y,0)\,dx\,dy. \tag{2.17}$$

If there are no sources in the upper half space, then Π is the total power radiated to the farfield ($z \to \infty$). Besides the plane at $z = 0$, every other plane defined by $z = z_0$ has the same power passing through it, since there is no absorption in the fluid and there are no sources above the boundary.

The equation of continuity, Eq. (2.6), becomes

$$\frac{1}{T}\int_0^T \frac{\partial e}{\partial t}dt = \frac{e(T) - e(0)}{T} = -\vec{\nabla}\cdot\vec{I}(\omega). \tag{2.18}$$

By the definition of steady state the energy density at time T is the same as the density at time 0, so that we have

$$\vec{\nabla}\cdot\vec{I}(\omega) = 0. \tag{2.19}$$

This means that in a source-free field the divergence of the time averaged acoustic intensity must always be zero. The only way the intensity field can have a non-zero divergence is if there are sources or sinks of energy within the medium, or losses in the medium.

2.6 Plane Wave Expansion

We turn now to plane wave solutions of the wave equation, seeking solutions of the Helmholtz equation, Eq. (2.13), in the frequency domain.

2.6.1 Introduction

Our concern in this chapter is with solutions of the wave equation in Cartesian coordinates. These solutions will be useful in the study of sources which are planar (or nearly planar) in geometry such as vibrating plates. We note that Eq. (2.1) is very similar to the equation for a vibrating string:

$$\frac{\partial^2 w}{\partial x^2} - \frac{1}{c_s^2}\frac{\partial^2 w}{\partial t^2} = 0, \tag{2.20}$$

where w is normal displacement of the string, and c_s is the wave speed, a constant. A solution to this equation is given by

$$w(x,t) = A e^{i(k_x x - \omega t)} + B e^{i(-k_x x - \omega t)}, \tag{2.21}$$

where A and B are arbitrary constants. For this solution to satisfy Eq. (2.20) we must have

$$k_x = \omega/c_s. \tag{2.22}$$

We introduce the string solution to understand the plane wave solutions of Eq. (2.1). In Eq. (2.21) k_x is called the wavenumber in the x direction.

Consider the phase term in Eq. (2.21) given by $\phi(x,t) = k_x x - \omega t$. We track the crest of a wave traveling down the string by choosing a constant value of phase and then following it as a function of position and time. The position of the crest, choosing $\phi = 0$ arbitrarily, is given by $x = \omega t/k_x = c_s t$. Thus, c_s is the velocity of the crest in the positive x direction and is called the phase velocity of the wave. The solution corresponding to the second term in Eq. (2.21) is a wave traveling in the negative x direction. At a fixed time the phase repeats over a distance $\Delta x = \lambda_x$. Over this distance the phase term in Eq. (2.21) changes by 2π, giving $2\pi = k_x \Delta x = k_x \lambda_x$, leading to the important relationship

$$k_x = 2\pi/\lambda_x. \tag{2.23}$$

λ_x is the wavelength in the x direction and is the distance over which the phase of the wave changes by 2π when time is held constant.

One final note on the reality of Eq. (2.21). Motions of strings and pressure waves in a fluid are never complex quantities, although we represent them with complex numbers. They must always be real, as no microphone or accelerometer provides a complex signal voltage in the process of measurement. To simulate the actual motion of the string, say with a graphics program on a computer, we must choose either the real or the imaginary part of Eq. (2.21). Choosing the real part we would use $\cos(k_x x - \omega t)$ to represent a wave traveling to the right.

2.6.2 Plane Waves

We now consider the general solution of Eq. (2.13), the Helmholtz equation, in three dimensions. Consider the following solution for $\bar{p}(\omega)$

$$p(\omega) = A(\omega)e^{i(k_x x + k_y y + k_z z)}, \qquad (2.24)$$

where $A(\omega)$ is an arbitrary constant. This equation satisfies Eq. (2.13) as long as

$$k^2 = k_x^2 + k_y^2 + k_z^2. \qquad (2.25)$$

Since k is a constant the three wavenumbers are not independent of one another. We can choose a maximum of two as independent variables, the third being dependent. Throughout this work we will use k_z as the dependent variable so that

$$k_z^2 = k^2 - k_x^2 - k_y^2.$$

Note that there is no restriction on the values of k_x and k_y; they can extend over all real numbers from $-\infty$ to $+\infty$.

Equation (2.24) represents a plane wave solution to the wave equation, although the time dependence is not indicated. To demonstrate this time dependence we must consider what we mean by a plane wave. It has meaning only for a single frequency, that is, in the steady state. Mathematically, writing the arbitrary constant in Eq. (2.24) as

$$A(\omega) = 2\pi B(\omega)\delta(\omega - \omega_0),$$

then

$$p(\omega) = 2\pi B\delta(\omega - \omega_0)e^{i(k_x x + k_y y + k_z z)}, \qquad (2.26)$$

where the delta function expresses the monochromatic nature of the time dependence. To arrive at the plane wave we take the inverse Fourier transform given in Eq. (2.10) above. This yields the final result, a plane wave at the frequency ω_0:

$$p(t) = Ae^{i(k_x x + k_y y + k_z z - \omega_0 t)}, \qquad (2.27)$$

where $k = \omega_0/c$. Now we will study the nature of plane waves in more detail.

Consider the pressure in the infinite x, z coordinate plane at $y = 0$. The phase term is given by (including the time dependent part, but dropping the zero subscript for simplicity)

$$\phi = (k_x x + k_z z - \omega t). \qquad (2.28)$$

This term looks like the string expression given in the introduction above if we plot it along each of the coordinate axes. That is, along the x axis $(z = 0)$ the phase expression is $k_x x - \omega t$ and the wavelength in the x direction is $\lambda_x = 2\pi/k_x$. λ_x is called the trace wavelength in the x direction and k_x the trace wavenumber in the x direction. As the plane wave travels through space, there is a trace wave along the x axis moving with a phase speed $c_x = \omega/k_x$. Similarly along the z axis the phase speed of the trace wave is $c_z = \omega/k_z$. To determine the direction of the plane wave consider Fig. 2.3.

For this example assume that $k_y = 0$, that is, there is no variation of pressure in the y direction. The sinusoidal variation of pressure along the two axes is shown gray scale

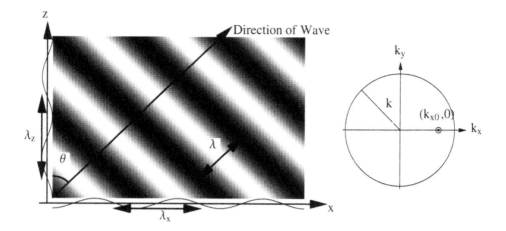

Figure 2.3: Plane wave illustration showing trace matching. Snapshot at $t = 0$. The insert on the right shows the equivalent k-space representation. The trace wave is supersonic, and lies within the radiation circle of radius k.

coded, with white-gray representing positive pressure and black-gray negative pressure. Since pressure must be continuous throughout space (no sources), adjacent regions along the two axes must have the same pressure. These regions are separated by diagonal lines of constant gray scale, as shown, which are called wavefronts. In this example the trace wavelengths in the vertical and horizontal directions are equal; $k_z = k_{z0} = k_x = k_{x0}$. This figure must reflect the relationship $k^2 = k_x^2 + k_z^2$, where the acoustic wavelength λ is shown along the direction of travel of the wave (perpendicular to the wavefronts; the lines of constant phase). The wave propagates in the direction given by θ, as shown in Fig. 2.3. Thus the following must be true:

$$
\begin{aligned}
\lambda_x \sin \theta &= \lambda \\
\lambda_z \cos \theta &= \lambda \\
k_x &= k \sin \theta \\
k_z &= k \cos \theta \\
k_x^2 + k_z^2 &= k^2.
\end{aligned}
\tag{2.29}
$$

The insert on the right in Fig. 2.3 shows a k-space diagram that displays the wavenumbers which exist in the (x, y) plane. In this example only one wavenumber exists, $(k_{x0}, 0)$, corresponding to a single plane wave. The large circle is called the radiation circle and has a radius k. In this example we have $k_{x0} = k_{z0}$ so that $k_{x0} = k/\sqrt{2}$, the latter determining the location of the plane wave wavenumber on the k-space diagram–shown in the small circle in the figure.

If we form the vector $\vec{k} = k_x \hat{i} + k_z \hat{k}$, where $|\vec{k}| = k$; then, since $k_x = k \sin \theta$ and $k_z = k \cos \theta$, $\vec{k}$ points in the direction of propation of the plane wave. This result is completely general; the direction of any plane wave is given by $\vec{k} = k_x \hat{i} + k_y \hat{j} + k_z \hat{k}$.

Since $k = \omega/c$ and $k_x = k \sin\theta$, then

$$c_x \;=\; c/\sin\theta, \tag{2.30}$$
$$c_z \;=\; c/\cos\theta.$$

The trace velocities along the coordinate axes are greater than the speed of sound. In fact, when $\lambda_x \to \infty$ then $c_x \to \infty$ ($\lambda_x = c_x/f$) and the crest of the wave moves at infinite speed along the x axis. Since both c_x and c_z are greater than c, the trace waves are classified as supersonic waves. Note that nothing (especially energy) actually travels at this speed–we are only looking at the projection of the wave on the coordinate axes as it travels through space.

A plane wave can be represented using condensed notation;

$$e^{i(k_x x + k_y y + k_z z)} = e^{i(\vec{k}\cdot\vec{r})}, \tag{2.31}$$

where $\vec{r} = x\hat{\imath} + y\hat{\jmath} + z\hat{k}$ represents the position vector to the observation point in the sound field, and $\vec{k}$ gives the direction of the wave.

Returning to Fig. 2.3 we can see that as the wave turns towards the normal to the (x,y) plane ($\theta = 0$), then $k_x \to 0$ and $k_z \to k$. This case is illustrated in Fig. 2.4. Here the wavelength along the x axis is infinite, and the wavefronts are planes parallel to the (x,y) plane. The k-space diagram on the right indicates that this plane wave is represented by a point at the origin ($k_{x0} = 0$).

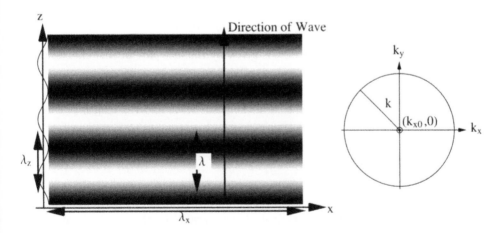

Figure 2.4: Plane wave traveling normal to x axis. In k-space this wave appears as a dot at the origin, as shown on the right.

Figure 2.5 presents the case in which the wavelength in the vertical direction is infinite. In this case we see that wave propagation is in the x direction ($\theta = \pi/2$) along the x axis and $c_x = c$. In k-space this wave is represented by a single dot located on the radiation circle.

Consider now the average intensity vector for plane waves. First we must determine the velocity associated with a plane wave. This is given by inserting Eq. (2.24) into

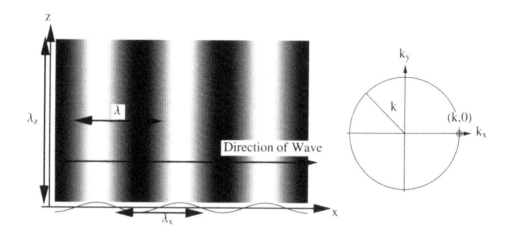

Figure 2.5: Plane wave traveling parallel to x axis, infinite wavelength in the vertical direction. In k-space the wave falls on the radiation circle as shown.

Eq. (2.14):

$$\vec{v}(\omega) = \frac{1}{\omega \rho_0}(k_x \hat{i} + k_y \hat{j} + k_z \hat{k}) p(\omega). \tag{2.32}$$

Thus Eq. (2.16) yields

$$\vec{I} = \frac{|A|^2}{2\omega \rho_0}(k_x \hat{i} + k_y \hat{j} + k_z \hat{k}). \tag{2.33}$$

Clearly the direction of the power flow is given by the last part of this expression, or

$$\vec{k} = (k_x \hat{i} + k_y \hat{j} + k_z \hat{k}),$$

where the direction of $\vec{k}$ is the direction of the plane wave and the length of this vector satisfies Eq. (2.25).

2.6.3 Evanescent Waves

Figures 2.3 and 2.4 illustrate the conditions for wave propagation away from the (x, y) plane. In Fig. 2.5 the wave does not propagate away from the (x, y) plane but travels parallel to it with wavefronts in vertical planes parallel to the (y, z) plane. Looking again at Eq. (2.25) we have

$$k_z = \pm\sqrt{k^2 - k_x^2 - k_y^2}, \tag{2.34}$$

so that Fig. 2.5 corresponds to the case, since $k_y = 0$, where the argument of the square root vanishes ($k_z = 0$). Particle motion is now only in the x direction.

It is important to realize that the plane wave solution still satisfies the wave equation, Eq. (2.13), when k_x or $k_y > k$, a condition under which the plane waves turn into evanescent waves. Note that Eq. (2.34) becomes

$$k_z = \pm i\sqrt{k_x^2 + k_y^2 - k^2} = \pm i k_z'. \tag{2.35}$$

where k_z' is real and the plane wave, turned evanescent, has the form

$$p = Ae^{\mp k_z' z}e^{i(k_x x + k_y y)}. \tag{2.36}$$

If the sources exist in the half space defined by $z < 0$, then the $e^{+k_z' z}$ solution to the wave equation is non-physical, since it blows up at $+\infty$, and we restrict our solution to the decaying term (Sommerfeld radiation condition [see Chapter 8]):

$$p = Ae^{-k_z' z}e^{i(k_x x + k_y y)}. \tag{2.37}$$

This is the form of an evanescent wave, decaying in amplitude in the z direction. We illustrate this case in Fig. 2.6 below, again for the case where $k_y = 0$. Since $k_x > k$ we see that the trace velocity of the wave along the x axis is less than the sound speed, since $c_x < c$. As a result, this wave is called a subsonic wave in contrast to the nonevanescent waves discussed above which resulted in supersonic trace velocities. Also since $k_x > k$, then $\lambda_x < \lambda$ and the trace wavelengths are less than the acoustic wavelength. In k-space, shown on the right in Fig. 2.6, the wavenumber is outside the radiation circle–always the case for evanescent waves.

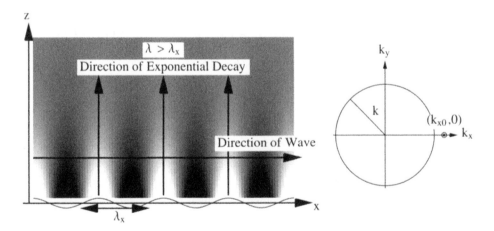

Figure 2.6: Evanescent wave traveling parallel to x axis, decaying exponentially in the vertical direction. In k-space this wave falls outside the radiation circle, as shown. All plane waves outside the radiation circle are subsonic.

From Eq. (2.14) the particle velocity associated with an evanescent wave is

$$\vec{v} = \frac{1}{\omega\rho_0}(k_x\hat{i} + k_y\hat{j} + ik_z'\hat{k})p(\omega) \tag{2.38}$$

so that the intensity is

$$\vec{I} = \frac{|A|^2 e^{-2k_z' z}}{2\omega\rho_0}(k_x\hat{i} + k_y\hat{j}). \tag{2.39}$$

From this it is clear that the power flows parallel to the (x, y) plane in the direction $k_x\hat{i} + k_y\hat{j}$, decaying exponentially in the z direction. The direction of the evanescent wave is $\vec{k}_{ev} = k_x \hat{i} + k_y \hat{j}$. Note that this differs from the plane wave case. where k_z enters into the direction of the wave.

Evanescent waves are also called inhomogeneous waves. They have tremendous relevance in the studies of radiation from plates and wave reflection and transmission between two differing media. Also. they are important for any vibrating structure that supports subsonic waves (wavelength less that the wavelength in the medium) and we will meet them quite frequently throughout this book.

A mathematical particular is the need to select the correct branch of the square root function when the argument is complex. When sources are confined to the lower half space ($z < 0$) and radiation is considered into the space $z > 0$, we must choose for the proper behavior of the evanescent waves

$$k_z = +i\sqrt{k_x^2 + k_y^2 - k^2}. \tag{2.40}$$

2.7 Infinite Plate Vibrating in a Normal Mode

In order to tie the concept of plane waves to a coupled vibration/radiation problem, we next consider the radiation from an infinite plate located in the $z = 0$ plane, which is vibrating in a standing wave mode at a single frequency with normal surface velocity η given by

$$\eta(x, y) = \eta_0 \cos(k_{x0}x) \cos(k_{y0}y). \tag{2.41}$$

The distance between nodal lines on the plate in the x and y directions is given by

$$\begin{aligned} \lambda_{x0}/2 &= \pi/k_{x0}, \\ \lambda_{y0}/2 &= \pi/k_{y0}. \end{aligned}$$

We now specify the conditions needed to solve this boundary value problem, that is, to determine the pressure in the half-space above the plate. There are four conditions for the pressure p and fluid particle velocity $\dot{w}$ which must be satisfied:

(1) p must satisfy the Helmholtz equation. Eq. (2.13), for $z \geq 0$,

(2) $\dot{w}$ and p must satisfy Euler's equation. Eq. (2.14),

(3) $\eta(x, y)$ must equal $\dot{w}(x, y, z)$ at the interface, $z = 0$, and

(4) there are no sources above ($z > 0$) the plate.

The third condition implies that the fluid always stays in contact with the plate. The displacement of the fluid particles near the plate boundary and normal to the plate surface must be continuous. It is important to realize, however, that this continuity requirement is not imposed in the x and y directions. Thus if the plate undergoes motion in-plane, this motion need not be continuous with the fluid particle motion in the x and y directions. This implies that the plate is allowed to slip under the fluid, like a frictionless boundary. Because it is frictionless it cannot drag the fluid particles along with it. This results from our neglect of viscosity in the fluid.

We make an educated guess at the solution to Eq. (2.13), in view of Eq. (2.41):

$$p(x, y, z) = p_0 e^{ik_{z0}z} \cos(k_{x0}x) \cos(k_{y0}y), \tag{2.42}$$

and insert this result into conditions (1) and (2) above. Condition (1) yields

$$k_{z0} = \pm\sqrt{k^2 - k_{x0}^2 - k_{y0}^2}$$

and condition (4) eliminates the negative value of k_{z0}, since $e^{-ik_{z0}z - i\omega t}$ is a wave traveling towards the plate which could only arise from a source above the plate (such as a reflecting boundary). As usual $k = \omega/c$, where ω is the radian frequency of plate oscillation. Thus

$$k_{z0} = \sqrt{k^2 - k_{x0}^2 - k_{y0}^2}. \tag{2.43}$$

Inserting Eq. (2.42) into condition (2),

$$\dot{w}(x, y, 0) = \frac{1}{i\rho_0 ck} \frac{\partial p}{\partial z}\bigg|_{z=0};$$

yields

$$\dot{w}(x, y, 0) = \frac{p_0 k_{z0}}{\rho_0 ck} \cos(k_{x0}x) \cos(k_{y0}y). \tag{2.44}$$

Imposing condition (3), that is, $\dot{w}(x, y, 0) = \eta_0 \cos(k_{x0}x) \cos(k_{y0}y)$, yields

$$p_0 = \frac{\eta_0 \rho_0 ck}{k_{z0}}$$

so that the final result is

$$p(x, y, z) = \frac{\eta_0 \rho_0 ck}{k_{z0}} e^{ik_{z0}z} \cos(k_{x0}x) \cos(k_{y0}y). \tag{2.45}$$

Equation (2.45) is the steady state pressure radiated from a vibrating plate with surface velocity given by Eq. (2.41). k_{x0} and k_{y0} are the given wavenumbers of the modal pattern of the plate, and k_{z0} is a function of them representing the variation of the pressure in the direction normal to the plate.

2.8 Wavenumber Space: k-space

Vibrations due to wave phenomena and resulting radiation can be represented in k-space. Often this presentation is extremely powerful in displaying the physics underlying the phenomena. We have already seen the k-space representations of various plane waves shown in Figs 2.3–2.6 and discussed in the captions. To illustrate further we continue with the infinite plate example of the last section.

We cast Eq. (2.45) in terms of the plane waves discussed earlier noting that

$$\cos(k_{x0}x) = \frac{1}{2}(e^{ik_{x0}x} + e^{-ik_{x0}x})$$

and

$$\cos(k_{y0}y) = \frac{1}{2}(e^{ik_{y0}y} + e^{-ik_{y0}y}),$$

so that the product of the two cosines results in four plane waves given by

$$e^{i(k_{z0}z \pm k_{x0}x \pm k_{y0}y)}.$$

The four plane waves can be illustrated by using a wavenumber space diagram (k-space) as shown in Fig. 2.7.

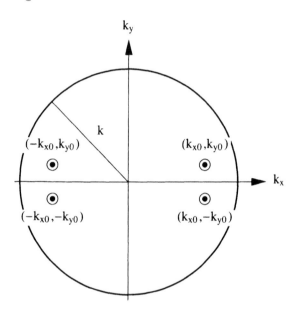

Figure 2.7: k-space diagram showing locations of 4 plane waves radiating from a vibrating plate with standing wave pattern given by k_{x0} and k_{y0}.

The small circles with dots indicate the location of the four plane wave components resulting from a standing wave on the plate. The large circle has a radius given by the acoustic wavenumber k and is called the radiation circle. The case illustrated here is for supersonic wavenumbers in the x and y directions; the components $(\pm k_{x0}, \pm k_{y0})$ are all located within the radiation circle. In this case Eq. (2.43) dictates that k_{z0} is real and less than k in magnitude. The directions of the plane waves are given by $\vec{k} = k_{x0}\hat{\imath} + k_{y0}\hat{\jmath} + k_{z0}\hat{k}$.

The directions of the plane waves can be illustrated using a hemisphere, with the k-space diagram shown in Fig. 2.7 in the equatorial plane, and the k_z axis as the polar axis. This is illustrated in Fig. 2.8.

We use a spherical coordinate system to describe the radiation. The spherical coordinates with polar angle θ and azimuthal angle ϕ are shown. The direction of the plane wave shown in the figure with the arrow labeled k is then, in spherical coordinates,

$$\vec{k} = k\cos\phi\sin\theta\,\hat{\imath} + k\sin\phi\sin\theta\,\hat{\jmath} + k\cos\theta\,\hat{k}, \tag{2.46}$$

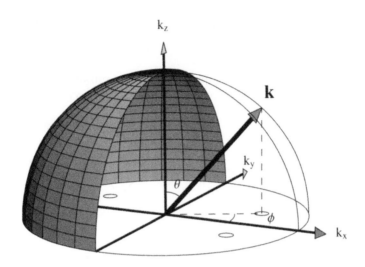

Figure 2.8: Radiation sphere in k-space. The k-space diagram of Fig. 2.7 is the equatorial plane. The direction of the plane wave with positive components (k_{x0}, k_{y0}) is shown by the vector $\vec{k}$, with the spherical angles θ and ϕ. The radius of the hemisphere is k.

where we must have

$$
\begin{aligned}
k_{x0} &= k \cos \phi_0 \sin \theta_0 \\
k_{y0} &= k \sin \phi_0 \sin \theta_0 \\
k_{z0} &= k \cos \theta_0.
\end{aligned}
\tag{2.47}
$$

This equation expresses the important inter-relationship between the wavenumbers on the plate and the direction of propagation of the radiated plane waves. For example, we can see that as the distance between the nodal lines (see Eq. (2.41)) on the plate increases, keeping the frequency fixed (as if we were making the plate stiffer), then $k_{x0} \to 0$ and $k_{y0} \to 0$, then Eq. (2.47) shows that $\theta \to 0$ and the plane wave travels in the direction of the z axis, normal to the plate.

The case illustrated above assumes that both k_{x0} and k_{y0} are supersonic, that is, both are less than or equal to k. However, if either wavenumber is greater than k, one of the components is subsonic (c_{x0} or $c_{y0} < c$) and the square root in Eq. (2.43) becomes

$$
k_{z0} = i\sqrt{k_{x0}^2 + k_{y0}^2 - k^2} = ik'_{z0},
\tag{2.48}
$$

where the argument of the square root is now positive and k'_{z0} is real. The pressure above the plate, Eq. (2.45), now becomes

$$
p(x, y, z) = \frac{-i\eta_0 \rho_0 c k}{k'_{z0}} e^{-k'_{z0}z} \cos(k_{x0}x) \cos(k_{y0}y),
\tag{2.49}
$$

decaying exponentially away from the plate boundary. Again because of the product of
cosines in Eq. (2.49), this pressure is composed of four **evanescent** waves given by

$$e^{-k'_{z0}z}e^{i(\pm k_{x0}x \pm k_{y0}y)}$$

The k-space diagram (in this case c_{x0} and c_{y0} are both subsonic) is shown in Fig. 2.9.
The four plane wave components fall outside the radiation circle. Clearly no radiation

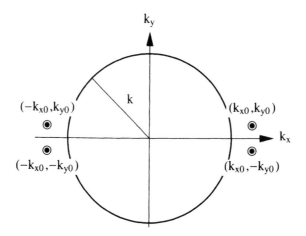

Figure 2.9: k-space diagram showing locations of 4 plane waves radiating
from a vibrating plate with a subsonic standing wave pattern.

reaches the farfield in this case. In other words, an infinite plate with a subsonic wave
component in either (or both) the x or y direction does not radiate to the farfield.
Thus waves with k-space components outside the radiation circle do not radiate to the
farfield.

Consider now the acoustic intensity for the evanescent case. The normal acoustic
intensity (time averaged) at the plate surface is given by

$$I_z(x,y,0) = \frac{1}{2}\mathrm{Re}(p(x,y,0)\eta^*(x,y))$$

which we can compute from Eqs (2.49) and (2.41). We see that since p is purely
imaginary and η is real, the time-averaged, normal acoustic intensity is identically zero;
$I_z(x,y,0) = 0$. This is a very important statement of the fact that there is no average
power transfer from the plate to the fluid. In fact Eq. (2.17) gives the result $\Pi(\omega) = 0$;
there is no power radiated into the half-space $z > 0$, and thus no power radiated to the
farfield. This happens when the nodal lines in either direction on the plate are separated
by less than $\lambda/2$ in the fluid. This condition is often referred to as a hydrodynamic short
circuit, alluding to the fact that adjacent regions of negative and positive velocity tend
to cancel one another as they push against the fluid in an effort to launch radiation into
the farfield. This condition is illustrated in Fig. 2.10 below.

Equation (2.49) also indicates that as the nodal line separation gets smaller and
smaller, then $k'_{z0} \to \infty$ and the pressure at the surface of the plate diminishes to zero;

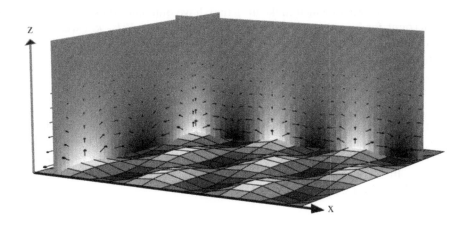

Figure 2.10: Hydrodynamic short circuit occurring when adjacent regions on the plate push fluid into one another, canceling any radiation from the plate. The arrows show the magnitude and direction of the velocity of the fluid above the plate, revealing the circulation of energy in the nearfield. The vertical planes provide a gray scale mapping of the fluid pressure, illustrating the exponential decay of the sound field resulting from the hydrodynamic short circuit.

another indication of the effect of the hydrodynamic short circuit. Notice there appears to be a problem when $k_{z0} = 0$, when the plate turns from supersonic to subsonic (called coincidence), and the plane waves excited travel parallel to the surface of the plate extending without decay to infinity. The zero in the denominator of Eq. (2.49) then implies an infinite pressure above the plate. This impossible condition results from a violation of condition (3) above, which requires the normal velocity of the plate to be continuous with the fluid velocity in contact with it. However, a plane wave traveling parallel to the plate has only a fluid velocity parallel to the plate (in the direction of travel) and zero normal velocity. The contradiction in our problem is expressed in the mathematics by the appearance of an infinity.

This completes our discussion of the radiation from an infinite plate with a given modal pattern of vibration. We included this example to build up our understanding of plane and evanescent waves, critical to the analysis which we are about to present. In a more general way than was presented in the plate problem above, we now show how any arbitrary pressure distribution in a source-free half space can be decomposed into plane and evanescent waves.

2.9 The Angular Spectrum: Fourier Acoustics

Consider a general unknown, steady state pressure distribution $p(x, y, z)$ in a source-free half space, $z > 0$. This pressure can be expressed uniquely and completely by a sum of plane and evanescent waves of the form discussed above. We must keep in mind that k_x and k_y are independent variables, and that k_z depends upon them. Each of the plane/evanescent waves which make up p may have different amplitudes and phases which we account for by using a multiplying coefficient term $P(k_x, k_y)$ which depends on the two wavenumbers. That is, we expect any pressure distribution in a source-free region to be expressible as a wave sum such as

$$p(x, y, z) = \sum_{k_x} \sum_{k_y} P(k_x, k_y) e^{i(k_x x + k_y y + k_z z)}$$

where we recognize the exponential term as a plane/evanescent wave. This is similar in concept to the result from the plate example presented above in which four plane or evanescent wave components were needed to solve for the pressure field for a given plate velocity. In this case, the sums above would only contain two terms each ($\pm k_{x0}$ and $\pm k_{y0}$) and, as Eq. (2.45) shows,

$$P(k_x, k_y) = \frac{\eta_0 \rho_0 c k}{4 k_{z0}}.$$

For a general problem, because of the infinite extent in the x and y directions, we expect a continuum of possible wavenumbers so that the sums above have to be represented by integrals to accommodate the continuum of values. Thus the pressure field can be written in general as

$$p(x, y, z) = \frac{1}{4\pi^2} \int_{-\infty}^{\infty} dk_x \int_{-\infty}^{\infty} dk_y P(k_x, k_y) e^{i(k_x x + k_y y + k_z z)}. \tag{2.50}$$

The introduction of the arbitrary constant $1/4\pi^2$ is for purposes which will become clear later. This equation is extremely important and central to this book. The integrals are over all values, supersonic and subsonic, of the wavenumbers and, as before,

$$k_z = \sqrt{k^2 - k_x^2 - k_y^2}.$$

Note that only positive k_z values are taken from the square root. This expresses the fact that we are dealing with a half-space problem, that is, the sources are confined to $z \leq 0$ and thus no plane waves can travel in the negative z direction.

Now consider the interpretation of the complex quantity $P(k_x, k_y)$. If $z = 0$ in Eq. (2.50) we have

$$p(x, y, 0) = \frac{1}{4\pi^2} \int_{-\infty}^{\infty} dk_x \int_{-\infty}^{\infty} dk_y P(k_x, k_y) e^{i(k_x x + k_y y)}. \tag{2.51}$$

This equation is an expression for the pressure in the infinite plane at $z = 0$. Comparing this to Eq. (1.17) reveals that the integrals represent two inverse Fourier transforms in k_x and k_y, respectively. Thus, in view of Eq. (1.17), the complex amplitude $P(k_x, k_y)$ is given by the corresponding two-dimensional Fourier transform:

$$P(k_x, k_y) = \int_{-\infty}^{\infty} dx \int_{-\infty}^{\infty} dy \, p(x, y, 0) e^{-i(k_x x + k_y y)}. \tag{2.52}$$

The Fourier transform guarantees that *any* pressure distribution, $p(x, y, 0)$ can be represented by Eq. (2.51). $P(k_x, k_y)$ is called the angular spectrum.[2]

---

[2] J. Goodman (1968). *Introduction to Fourier Optics.* McGraw-Hill, New York.

2.9.1 Wave Field Extrapolation

This analysis leads us to a significant result which forms the backbone of Fourier acoustics (as well as Fourier optics). Once $P(k_x, k_y)$ is known, computed from the pressure in the plane $z = 0$, then we can use Eq. (2.50) to compute the pressure field over the three-dimensional volume from $z = 0$ to infinity, without any more information. We will see that this ability to extrapolate fields from one plane to another in such simple fashion will provide powerful tools for numerical and experimental applications, and forms the foundation of nearfield acoustical holography.

Another way to express this extrapolation is to note that the Fourier transform of the pressure in a plane $z = $ constant is related to the transform of the pressure in the plane $z = 0$ by

$$\mathcal{F}_x \mathcal{F}_y [p(x, y, z)] \equiv P(k_x, k_y, z) = P(k_x, k_y) e^{ik_z z}, \tag{2.53}$$

so that the plane wave amplitudes only undergo a phase change (for k_z real) from one horizontal plane to another. For example, if $k_x = k_y = 0$, then $k_z = k$ and $P(0,0) e^{ikz} = P(0, 0, z)$ and e^{ikz} is the phase change of a plane wave traveling in the z direction (normal). The general expression then to extrapolate the angular spectrum in the plane $z = z'$ to a plane $z = z$ is

$$P(k_x, k_y, z) = P(k_x, k_y, z') e^{ik_z(z-z')}. \tag{2.54}$$

Note that the evanescent waves are also included in this expression. When k_z is purely imaginary then

$$P(k_x, k_y, z) = P(k_x, k_y, z') e^{-|k_z|(z-z')}. \tag{2.55}$$

Note the following definition has been assumed,

$$P(k_x, k_y) \equiv P(k_x, k_y, 0). \tag{2.56}$$

Very simple algebraic equations also exist relating the k-space velocity and the k-space pressure. These are derived from a two-dimensional Fourier transform of Euler's equation, Eq. (2.14). Let the transform of the normal velocity be

$$\dot{W}(k_x, k_y, z) \equiv \mathcal{F}_x \mathcal{F}_y [\dot{w}(x, y, z)]. \tag{2.57}$$

Similarly, the Fourier components of the other velocity components are $\dot{U}(k_x, k_y, z)$ and $\dot{V}(k_x, k_y, z)$. From Eq. (2.54)

$$\mathcal{F}_x \mathcal{F}_y \left[\frac{\partial p(x, y, z)}{\partial z} \right] = \frac{\partial P(k_x, k_y, z)}{\partial z} = ik_z P(k_x, k_y, z') e^{ik_z(z-z')}, \tag{2.58}$$

and, given Eq. (1.10), the Fourier transform of Euler's equation is

$$\dot{U}(k_x, k_y, z)\hat{\imath} + \dot{V}(k_x, k_y, z)\hat{\jmath} + \dot{W}(k_x, k_y, z)\hat{k} = \frac{1}{\rho_0 ck} (k_x\hat{\imath} + k_y\hat{\jmath} + k_z\hat{k}) P(k_x, k_y, z). \tag{2.59}$$

Note that the partial differentiation operations of the gradient have been replaced with a multiplication by ik_x and ik_y. This formula makes the rather significant statement that from a knowledge of the angular spectrum of the pressure in a plane, one can easily determine the three angular spectrum components of the vector velocity.

In particular, we can use Eq. (2.54) to relate the velocity to the pressure in a different plane;

$$
\dot{U}(k_x, k_y, z)\hat{\imath} + \dot{V}(k_x, k_y, z)\hat{\jmath} + \dot{W}(k_x, k_y, z)\hat{k}
$$
$$
= \frac{1}{\rho_0 ck}(k_x\hat{\imath} + k_y\hat{\jmath} + k_z\hat{k})P(k_x, k_y, z')e^{ik_z(z-z')}. \tag{2.60}
$$

Thus, the normal component of velocity is

$$
\dot{W}(k_x, k_y, z) = \frac{k_z}{\rho_0 ck}P(k_x, k_y, z')e^{ik_z(z-z')}. \tag{2.61}
$$

This important equation relates the angular spectrum components of normal velocity in one plane to components of pressure in a different plane.

2.10 Derivation of Rayleigh's Integrals

We will now use the angular spectrum to derive some rather famous integrals. It is a tribute to the angular spectrum approach that we are able to proceed in this direction. The following integrals were developed using a different approach by Rayleigh.[3] First, we will derive his second integral formula (also called the second diffraction formula). Start with the inverse Fourier transform of Eq. (2.54):

$$
p(x, y, z) = \frac{1}{4\pi^2}\int_{-\infty}^{\infty}\int_{-\infty}^{\infty} P(k_x, k_y, z')e^{ik_z(z-z')}e^{i(k_x x + k_y y)}dk_x dk_y.
$$

Note that this integral is the product of two Fourier transforms. In shorthand notation this equation is,

$$
p(x, y, z) = \mathcal{F}_x^{-1}\mathcal{F}_y^{-1}\left[P(k_x, k_y, z')e^{ik_z(z-z')}\right].
$$

If we let $F(k_x, k_y) = P(k_x, k_y, z')$ and $G_p(k_x, k_y, z - z') \equiv e^{ik_z(z-z')}$ then we can use the convolution theorem, Eq. (1.20), in x and y:

$$
f(x, y) * *g_p(x, y, z - z') = \mathcal{F}_x^{-1}\mathcal{F}_y^{-1}\left[F(k_x, k_y)G_p(k_x, k_y, z - z')\right],
$$

to arrive at

$$
p(x, y, z) = \iint p(x', y', z')g_p(x - x', y - y', z - z')dx'dy', \tag{2.62}
$$

where the inverse transform of $P(k_x, k_y, z')$ is $p(x', y', z')$ and the inverse transform of $G_p(k_x, k_y, z - z')$ is $g_p(x, y, z)$. To determine $g_p(x - x', y - y', z - z')$ we note that

$$
g_p(x, y, z - z') = \mathcal{F}_x^{-1}\mathcal{F}_y^{-1}[e^{ik_z(z-z')}],
$$

---

[3]J. W. S. Rayleigh (1897). "On the passage of waves through apertures in plane screens, and allied problems", Philosophical Magazine, **43**, pp. 259–272.

so that, using the shift theorem, $g_p(x - x', y - y', z - z')$ is given by the integral

$$g_p(x - x', y - y', z - z') = \frac{1}{4\pi^2} \int_{-\infty}^{\infty} \int_{-\infty}^{\infty} e^{i[k_x(x-x')+k_y(y-y')]} e^{i(z-z')\sqrt{k^2 - k_x^2 - k_y^2}} \, dk_x dk_y.$$

(2.63)

Note k_z has been written out in terms of k_x and k_y.

Now we will evaluate the integrals on the right. They can be cast in the form of a well known integral called Weyl's integral[4] as was pointed out by Lalor.[5] Weyl's integral provides the expansion of the free space Green function, which will be discussed in more detail in Section 6.5.1, in terms of plane waves:

$$\frac{e^{ik|\vec{r} - \vec{r}'|}}{|\vec{r} - \vec{r}'|} = \frac{i}{2\pi} \int_{-\infty}^{\infty} \int_{-\infty}^{\infty} e^{i[k_x(x-x')+k_y(y-y')]} \frac{e^{ik_z|z-z'|}}{k_z} \, dk_x dk_y,$$

(2.64)

where $\vec{r} = (x, y, z)$, $\vec{r}' = (x', y', z')$, and

$$|\vec{r} - \vec{r}'| = \sqrt{(x - x')^2 + (y - y')^2 + (z - z')^2}.$$

Differentiating both sides of this integral with respect to z' and restricting $z > z'$ yields,

$$\frac{\partial}{\partial z'} \left[\frac{e^{ik|\vec{r} - \vec{r}'|}}{|\vec{r} - \vec{r}'|} \right] = -\frac{1}{2\pi} \int_{-\infty}^{\infty} \int_{-\infty}^{\infty} e^{i[k_x(x-x')+k_y(y-y')]} e^{ik_z(z-z')} \, dk_x dk_y.$$

(2.65)

Comparison with Eq. (2.63) gives

$$g_p(x - x', y - y', z - z') = -\frac{1}{2\pi} \frac{\partial}{\partial z'} \left[\frac{e^{ik|\vec{r} - \vec{r}'|}}{|\vec{r} - \vec{r}'|} \right].$$

(2.66)

Finally we insert this result into Eq. (2.62) to yield **Rayleigh's second integral formula**:

$$p(x, y, z) = -\frac{1}{2\pi} \int_{-\infty}^{\infty} \int_{-\infty}^{\infty} p(x', y', z') \frac{\partial}{\partial z'} \left[\frac{e^{ik|\vec{r} - \vec{r}'|}}{|\vec{r} - \vec{r}'|} \right] dx' dy',$$

(2.67)

with $z \geq z'$. This formula relates the spatial pressure in one plane to the spatial pressure in another plane. It gives a forward propagation formula by convolving the pressure in the plane $z' =$ constant with a propagator g_p projecting the field to a more distant plane ($z > z'$). Later we will look at the inverse of this equation so that we can backpropagate the spatial acoustic pressure field.

When $z = z'$ in Eq. (2.67) an identity arises and one can not solve for the pressure field. That is, when $z = z'$ Eq. (2.63) yields, using the delta function relation Eq. (1.36),

$$g_p(x - x', y - y', 0) = \delta(x - x')\delta(y - y'),$$

so that Eq. (2.67) is simply

$$p(x, y, z') = \iint p(x', y', z')\delta(x - x')\delta(y - y') \, dx' dy',$$

---

[4]H. Weyl (1919). Ann. Physik, **60**, p. 481.
[5]E. Lalor (1968). "Inverse Wave Propagator", J. Math. Phys., **9**, p. 2001.

which reduces to an identity.

We will discuss Eq. (2.67) in more detail after we derive Rayleigh's first integral formula, which is better known than his second and is extensively used in the study of radiation from finite plates. Again we use the angular spectrum approach for the derivation. This formula relates the velocity on the surface to the pressure radiated and thus we look for the angular spectrum representation which relates normal velocity to pressure. This is given by Eq. (2.61), resulting from Euler's equation, which we rewrite as

$$P(k_x, k_y, z) = \rho_0 ck \dot{W}(k_x, k_y, z') \frac{e^{ik_z(z-z')}}{k_z}, \tag{2.68}$$

where we have interchanged z and z', and assumed $z \geq z'$. We can now take the inverse Fourier transforms in k_x and k_y of both sides of this equation, using the convolution theorem, Eq. (1.20), again:

$$p(x, y, z) = \mathcal{F}_x^{-1} \mathcal{F}_y^{-1} \left[\dot{W}(k_x, k_y, z') \right] * * \mathcal{F}_x^{-1} \mathcal{F}_y^{-1} \left[\frac{\rho_0 ck e^{ik_z(z-z')}}{k_z} \right]. \tag{2.69}$$

Defining the function $g_v(x, y, z)$ as

$$g_v(x, y, z) \equiv \mathcal{F}_x^{-1} \mathcal{F}_y^{-1} \left[\frac{\rho_0 ck e^{ik_z z}}{k_z} \right], \tag{2.70}$$

then by definition of convolution Eq. (2.69) becomes

$$p(x, y, z) = \iint \dot{w}(x', y', z') g_v(x - x', y - y', z - z') dx' dy'. \tag{2.71}$$

By the shift theorem,

$$g_v(x - x', y - y', z - z') = \mathcal{F}_x^{-1} \mathcal{F}_y^{-1} \left[\rho_0 ck \frac{e^{-ik_x x'} e^{-ik_y y'} e^{ik_z(z-z')}}{k_z} \right], \tag{2.72}$$

that is,

$$g_v(x - x', y - y', z - z') = \frac{\rho_0 ck}{4\pi^2} \iint e^{i[k_x(x-x') + k_y(y-y')]} \frac{e^{ik_z(z-z')}}{k_z} dk_x dk_y. \tag{2.73}$$

The integral is recognized as Weyl's integral, Eq. (2.64), as long as $z \geq z'$; thus

$$g_v(x - x', y - y', z - z') = -i\rho_0 ck \frac{e^{ik|\vec{r}-\vec{r}'|}}{2\pi|\vec{r}-\vec{r}'|}. \tag{2.74}$$

Substitution into Eq. (2.71) yields the final result, again for $z \geq z'$,

$$p(x, y, z) = \frac{-i\rho_0 ck}{2\pi} \int_{-\infty}^{\infty} \int_{-\infty}^{\infty} \dot{w}(x', y', z') \frac{e^{ik|\vec{r}-\vec{r}'|}}{|\vec{r}-\vec{r}'|} dx' dy'. \tag{2.75}$$

This very important equation is **Rayleigh's first integral formula** and is used extensively in the literature in solving radiation problems from plates. Both of Rayleigh's formulas provide a means to compute the radiation into a half space ($z \geq z'$) given either the pressure on a surface $z = z'$, Eq. (2.67), or the normal velocity on a surface $z = z'$, Eq. (2.75). These surface fields are convolved with a propagator, a process which implies that all the source points (x', y', z') contribute to the radiation at a single field point (x, y, z).

The following figure should make clear the geometrical ramifications of Rayleigh's integral.

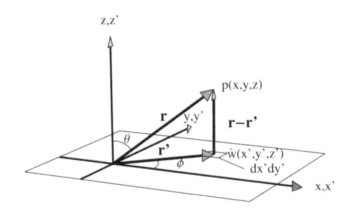

Figure 2.11: Geometric interpretation of Rayleigh's first integral formula. The small box represents an element of area $dx'dy'$ with normal velocity $\dot{w}$. The integral indicates that for a fixed field point (x, y, z) this area sweeps over the complete (x', y') plane.

The normal velocity of the element of area $dx'dy'$ on the surface radiates to the field point at (x, y, z) with an amplitude and phase given by the propagator. One must add the contributions of all of the area elements in the infinite plane to determine the pressure at a single field point.

2.10.1   The Velocity Propagator

The propagator g_v defined in Eq. (2.74) is called the velocity propagator since it determines the pressure radiated to an outward plane through convolution with the normal velocity distribution on a surface. This propagator is proportional to the pressure from a *baffled* point source located at (x', y', z'), as we will now show.

We can represent a baffled point source located at $(x_0, y_0, 0)$ using delta functions:

$$\dot{w}(x', y', 0) = Q_h \delta(x' - x_0)\delta(y' - y_0). \tag{2.76}$$

Q_h represents the strength of the source in the units of meters per second times area, or volume per unit time (cubic meters per second). This is the amount of fluid injected into the medium per unit time. The angular spectrum is found by taking the Fourier transform of Eq. (2.76),

$$\dot{W}(k_x, k_y, 0) = Q_h e^{-ik_x x_0} e^{-ik_y y_0}.$$

The pressure spectrum associated with this is. from Eq. (2.61),

$$P(k_x, k_y, 0) = \frac{Q_h \rho_0 ck}{k_z} e^{-ik_x x_0} e^{-ik_y y_0}.$$

which can be extended to a different value of z by multiplication by $\exp(ik_z z)$:

$$P(k_x, k_y, z) = \frac{Q_h \rho_0 ck}{k_z} e^{-ik_x x_0} e^{-ik_y y_0} e^{ik_z z}.$$

Review of Eq. (2.72) and Eq. (2.74) reveals that the inverse transform of this expression is

$$p(x, y, z) = \mathcal{F}_x^{-1} \mathcal{F}_y^{-1}[P(k_x, k_y, z)] = \frac{-iQ_h \rho_0 ck}{2\pi} \frac{e^{ik|\vec{r} - \vec{r}_0|}}{|\vec{r} - \vec{r}_0|}, \qquad (2.77)$$

with $z' = z_0 = 0$ and $\vec{r}_0 = (x_0, y_0, 0)$. This equation gives the pressure field radiated by a point source in a baffle[6] with source strength Q_h. This source is like a tiny dome loudspeaker (radius a and surface area $2\pi a^2$) in an infinite baffle moving with a radial velocity $\dot{w}_r$ so that

$$Q_h = 2\pi a^2 \dot{w}_r.$$

We will now show how we can use the Rayleigh integral to calculate the farfield radiation from planar sources.

2.11 Farfield Radiation: Planar Sources

Rayleigh's integral is the springboard for a very powerful formula which relates the farfield radiation from planar sources to the Fourier transform of the surface velocity. The reader with a strong background in Fourier transforms will find that he has a complementary knowledge of the farfield patterns for many kinds of planar sources.

To begin the derivation, we let the field point move far from the source plane in the z direction. We will assume that any given source distribution will always be confined to a finite area in the (x, y) plane, and that outside this area there is a rigid baffle extending to infinity. A rigid baffle is defined by vanishing normal velocity on its surface, that is, when (x', y') is on the baffle then $\dot{w}(x', y', 0) = 0$. If the surface velocity is contained within an area S then Rayleigh's integral has finite limits and becomes,

$$p(x, y, z) = \frac{-i\rho_0 ck}{2\pi} \iint_S \dot{w}(x', y', 0) \frac{e^{ik|\vec{r} - \vec{r}'|}}{|\vec{r} - \vec{r}'|} dx' dy'. \qquad (2.78)$$

For $\vec{r}'$ in S ($z' = 0$) the definition of the farfield (see Fig. 2.11) is $r >> r'$, where

$$r \equiv |\vec{r}| \text{ and } r' \equiv |\vec{r}'|.$$

[6]L. E. Kinsler and A. R. Frey (1962). *Fundamentals of Acoustics*. Wiley & Sons, New York, 2nd ed., p. 165.

Under these conditions,

$$
\begin{aligned}
|\vec{r} - \vec{r}'| &= ((x - x')^2 + (y - y')^2 + z^2)^{1/2} \\
&\approx r\left(1 - \frac{x}{r}x' - \frac{y}{r}y'\right).
\end{aligned}
\tag{2.79}
$$

Define a vector $\vec{k} = k_x\hat{\imath} + k_y\hat{\jmath} + k_z\hat{k}$ in the same direction as $\vec{r}$ so that

$$
\frac{\vec{k}}{k} = \frac{\vec{r}}{r},
$$

that is, the unit vectors point in the same direction. Thus $\frac{x}{r} = \frac{k_x}{k}$ and $\frac{y}{r} = \frac{k_y}{k}$, and

$$
\frac{e^{ik|\vec{r} - \vec{r}'|}}{|\vec{r} - \vec{r}'|} \approx \frac{e^{ikr}}{r} e^{-i(k_x x' + k_y y')}.
\tag{2.80}
$$

Note that whereas one can replace $|\vec{r} - \vec{r}'|$ with r in the denominator, we can not do so in the phase term of the exponential, since the latter is an oscillating function with range. Since $\vec{k}$ is in the same direction as $\vec{r}$ then the same spherical angles describe them both. Thus in spherical coordinates we have

$$
\begin{array}{ll}
x = r\sin\theta\cos\phi, & k_x = k\sin\theta\cos\phi, \\
y = r\sin\theta\sin\phi, & k_y = k\sin\theta\sin\phi, \\
z = r\cos\theta, & k_z = k\cos\theta.
\end{array}
\tag{2.81}
$$

With these results Rayleigh's integral becomes

$$
p(r,\theta,\phi) = -i\rho_0 ck \frac{e^{ikr}}{2\pi r} \iint_S \dot{w}(x',y',0) e^{-i(k_x x' + k_y y')} dx' dy'
\tag{2.82}
$$

or, noting that the integrals here are Fourier transforms,

$$
p(r,\theta,\phi) = -i\rho_0 ck \frac{e^{ikr}}{2\pi r} \mathcal{F}_{x'}\mathcal{F}_{y'}\left[\dot{w}(x',y',0)\right].
\tag{2.83}
$$

The final result is

$$
p(r,\theta,\phi) = -i\rho_0 ck \frac{e^{ikr}}{2\pi r} \dot{W}(k_x,k_y,0),
\tag{2.84}
$$

where k_x and k_y are given above in terms of spherical coordinates. This powerful formula states that the farfield of any planar source is determined from the two-dimensional Fourier transform of its normal velocity distribution as long as the direction of the farfield point is taken to be that of $\vec{k}$.

Most often what is plotted, when one asks for the farfield, is the directivity function. This is defined so as to remove the $\exp(ikr)/r$ factor:

$$
p(r,\theta,\phi) = \frac{e^{ikr}}{r} D(\theta,\phi)
\tag{2.85}
$$

with

$$
D(\theta,\phi) = \frac{-i\rho_0 ck}{2\pi} \dot{W}(k_x,k_y,0).
\tag{2.86}
$$

The directivity function has the units of Pascal-meters.

Equations (2.84) and (2.86) are quite significant. As we will demonstrate in the examples below, once the Fourier transform is computed for a given source distribution in a plane, these equations provide the directivity patterns for any frequency. Furthermore, these patterns can be constructed almost trivially using a procedure called the Ewald sphere construction.

Before we develop this construction, we present the farfield formula for vibrators with circular symmetry.

2.11.1 Vibrators with Circular Symmetry

For circular vibrators, such as a baffled drum head, the farfield radiation can be expressed in terms of a Hankel transform. If the vibration pattern of the vibrator is expressed as a function of polar coordinates, $\dot{w}(x, y, z = 0) \to \dot{w}(\rho, \phi)$, then we can use a Fourier series, Eq. (1.24), to represent the surface velocity:

$$\dot{w}(\rho, \phi') = \sum_{n=-\infty}^{\infty} \dot{w}_n(\rho) e^{in\phi'}. \tag{2.87}$$

The two-dimensional Fourier transform in rectangular coordinates in Eq. (2.83) can now be translated into polar coordinates with $k_x = k_\rho \cos\phi$, $k_y = k_\rho \sin\phi$, $k_\rho \equiv k \sin\theta$, $x' = \rho \cos\phi'$, and $y' = \rho \sin\phi'$:

$$\mathcal{F}_{x'}\mathcal{F}_{y'}\left[\dot{w}(x', y', 0)\right] = \sum_n \int_0^\infty \rho\, d\rho\, \dot{w}_n(\rho) \oint d\phi'\, e^{in\phi'} e^{-ik_\rho\rho(\cos\phi\cos\phi' + \sin\phi\sin\phi')}$$

$$= \sum_n \int_0^\infty \rho\, d\rho\, \dot{w}_n(\rho) \oint d\phi'\, e^{-ik_\rho\rho\cos(\phi-\phi') + in\phi'},$$

where the integral over ϕ' is written as $\oint$ since it is circular spanning an interval of 2π. To reduce this further we need an integral representation of the Bessel function,

$$J_n(z) = \frac{1}{2\pi} \int_{-\pi}^{\pi} e^{-iz\sin\xi + in\xi} d\xi, \tag{2.88}$$

and the substitution $\phi - \phi' = \pi/2 - \xi$ to arrive at

$$\oint d\phi'\, e^{-ik_\rho\rho\cos(\phi-\phi') + in\phi'} = 2\pi e^{-in\pi/2} e^{in\phi} J_n(k_\rho\rho). \tag{2.89}$$

Finally, since the integral over ρ is an nth order Hankel transform given in Eq. (1.29),

$$\int_0^\infty \dot{w}_n(\rho) J_n(k_\rho\rho) \rho\, d\rho \equiv \mathcal{B}_n[\dot{w}_n(\rho)],$$

and we arrive at the final result:

$$\mathcal{F}_{x'}\mathcal{F}_{y'}\left[\dot{w}(x', y', 0)\right] = 2\pi \sum_n (-i)^n e^{in\phi} \mathcal{B}_n[\dot{w}_n(\rho)]. \tag{2.90}$$

Returning to Eq. (2.83), the farfield pressure for a vibrator with surface velocity expressed in polar coordinates becomes, with $k_\rho = k \sin \theta$,

$$p(r, \theta, \phi) = \rho_0 c k \frac{e^{ikr}}{r} \sum_{n=-\infty}^{\infty} (-i)^{n+1} e^{in\phi} \mathcal{B}_n[\dot{w}_n(\rho)], \qquad (2.91)$$

and the directivity pattern is given by

$$D(\theta, \phi) = \rho_0 c k \sum_{n=-\infty}^{\infty} (-i)^{n+1} e^{in\phi} \mathcal{B}_n[\dot{w}_n(\rho)]. \qquad (2.92)$$

2.11.2 Ewald Sphere Construction

As a first example we will investigate the steady state radiation from a traveling wave on an infinite plate located in the plane $z = 0$. Let k_{x0} and k_{y0} be the supersonic wavenumbers (contained in the radiation circle) of the wave which has the form,

$$\dot{w}(x, y) = \dot{w}_0 e^{ik_{x0}x} e^{ik_{y0}y}. \qquad (2.93)$$

Noting that $\mathcal{F}_x[\exp(ik_{x0}x)] = 2\pi\delta(k_x - k_{x0})$ then

$$\dot{W}(k_x, k_y, 0) = 4\pi^2 \dot{w}_0 \delta(k_x - k_{x0})\delta(k_y - k_{y0}), \qquad (2.94)$$

and Eq. (2.86) yields the directivity function

$$D(\theta, \phi) = -2\pi i \dot{w}_0 \rho_0 c k \delta(k_x - k_{x0})\delta(k_y - k_{y0}), \qquad (2.95)$$

under the condition that

$$k_{x0} = k \sin \theta_0 \cos \phi_0, \qquad (2.96)$$
$$k_{y0} = k \sin \theta_0 \sin \phi_0. \qquad (2.97)$$

These conditions provide two equations to solve for the two unknown angles:

$$\sin \theta_0 = \sqrt{k_{x0}^2 + k_{y0}^2}/k,$$
$$\tan \phi_0 = k_{y0}/k_{x0}.$$

Now a simple mapping procedure called the Ewald sphere construction procedure allows us to plot the directivity function on a hemisphere for any given source whose Fourier transform we know. The term Ewald sphere is borrowed from X-ray diffraction theory.[7] Figure 2.12 illustrates the concept for the example problem. In the figure k is the radius of the hemisphere so that the equatorial plane contains the radiation circle.

The Fourier transform of the source is plotted in the radiation circle making up the base of the hemisphere. In this case only a single delta function is plotted there located at $(k_x, k_y) = (k_{x0}, k_{y0})$. The amplitude (and phase) of the transform is then assigned

---

[7]A. Guiner (1963). *X-ray Diffraction.* Translated by Paul Lorrain and Dorothée Sainte-Marie Lorrain, SanFranscisco & London: W. H. Freeman & Co.

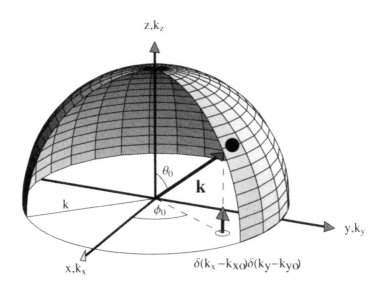

Figure 2.12: Construction of farfield in spherical coordinates using Ewald sphere construction.

to a point on the hemisphere determined by the vertical (upward) projection through the point to the hemisphere. This vertical projection satisfies the condition set up in Eq. (2.96) and Eq. (2.97) above. In this way the directivity pattern of the traveling wave is seen to be a delta function in the farfield at the spherical angles (θ_0, ϕ_0), and zero at all other angles.

Another simple example is a point source located at the origin surrounded by an infinite rigid baffle. Thus,

$$\dot{w}(x, y, 0) = \delta(x)\delta(y),$$

and

$$\dot{W}(k_x, k_y, 0) = 1,$$

The projection of this covers the full Ewald sphere with a constant, unit amplitude. This constant directivity is verified by Eq. (2.86) since

$$D(\theta, \phi) = \frac{-i\rho_0 ck}{2\pi};\tag{2.98}$$

a constant over all angles. As we expect the point source is omnidirectional.

2.11.3 A Baffled Square Piston

For the next example of the Ewald sphere construction we compute the farfield from a square piston vibrator with surface velocity

$$\dot{w}(x,y,0) = \begin{cases} 1 & \text{if } -L/2 < x < L/2, \\ & \quad -L/2 < y < L/2 \\ 0 & \text{otherwise.} \end{cases}$$

These conditions define the rectangle function, so that $\dot{w}(x,y,0) = \Pi(x/L)\Pi(y/L)$.

The transform of this can be found using Eq. (1.41),

$$\dot{W}(k_x, k_y, 0) = L^2 \text{sinc}(k_x L/2) \, \text{sinc}(k_y L/2), \tag{2.99}$$

and the directivity function is, from Eq. (2.86),

$$D(\theta, \phi) = \frac{-i\rho_0 ck L^2}{2\pi} \text{sinc}(\tfrac{kL}{2} \sin\theta \cos\phi) \, \text{sinc}(\tfrac{kL}{2} \sin\theta \sin\phi). \tag{2.100}$$

The following figure, Fig. 2.13, provides a linear plot of the Fourier transform of the square piston using a three-dimensional plot. The value at the center is unity ($\dot{W}/L^2$ is plotted) and the levels are greatest along the coordinate axes, decaying rapidly from the center. Along the diagonals the decay is most rapid.

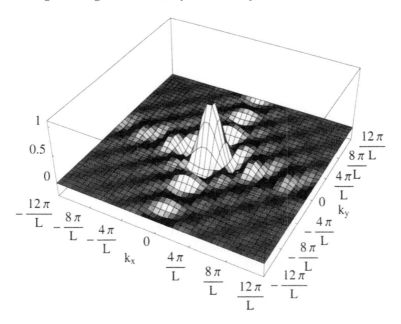

Figure 2.13: Surface plot of $\text{sinc}(k_x L/2) \, \text{sinc}(k_y L/2)$.

It is often desirable to plot the farfield on a dB scale, to show more dynamic range in the plot. Figure 2.14 is a contour plot of Eq. (2.99) using a logarithmic scale ($20\log_{10}[|\dot{W}|/L^2]$), cutoff at −40 dB. Ten contours are used to represent the region

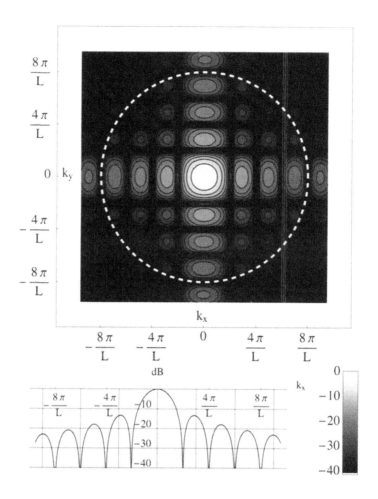

Figure 2.14: Logarithmic contour plot of the Fourier transform of square piston. The dashed circle has a radius $k = 8\pi/L$.

from –40 to 0 dB, with a contour line every 4 dB. The gray scale plot indicates levels as shown in the key. The levels are distributed equally on the log scale. The bottom plot on the figure shows a horizontal slice through the center of the contour plot. The first sidelobe is at $k_x L/2 = 3\pi/2$ and the level has dropped by –13.5 dB from the maximum of the main lobe.

We can now use the Ewald sphere construction procedure to map out the farfield at any frequency using Fig. 2.14. In the following example we have chosen the frequency such that $k = 8\pi/L$. We draw a circle of this radius, $k = 8\pi/L$, on the k-space contour plot (Fig. 2.14). This circle is indicated by the dashed white circle. This circle serves as the equatorial base of the Ewald hemisphere shown in Fig. 2.15. A great semi-circle of that hemisphere is shown in the $k_y = 0$ plane. We continue by projecting, as indicated by the arrows, the values of the function, plotted on the base plane, vertically upwards to the hemisphere to determine the farfield directivity pattern. The values of the projections are plotted as a polar plot with a dB scale with the semi-circle as a base as shown in the figure. This polar plot above provides the levels of the farfield pressure in the corresponding direction (radially outward from the tips of the arrows). The conventional polar plot of the directivity is shown on the right in the figure for

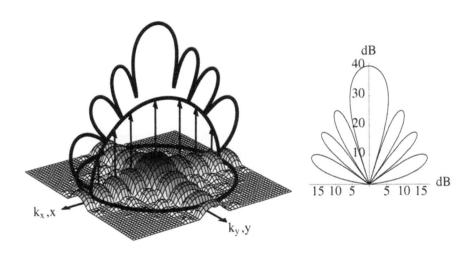

Figure 2.15: The Ewald sphere construction of the farfield from projection of Fig. 2.14 using a radius of $k = 8\pi/L$.

comparison. One can see that the main lobe, centered at the north pole, covers a fairly wide angle (about 35 degrees), and that the lowest levels correspond to the projected nulls in the surface plot below.

It is easy now to sketch directivity patterns for other frequencies by returning to Fig. 2.14 and drawing a circle for the desired frequency. We can see that at very low frequencies (small radius) there is very little variation in the transform in the circle so that the farfield would be nearly omnidirectional. Whereas at $k = 8\pi/L$ the farfield will have three side lobes, as already shown in Fig. 2.15.

It is interesting to consider the limit as the square plate becomes infinite, since it should provide the same result as before; we expect a delta function directivity pattern to occur. This fact is illustrated in Fig. 2.14 if we make L large keeping k fixed. The function contracts towards the center, concentrating to a spot at the center in the limit. Indeed, as a result of Eq. (1.36),

$$\lim_{L \to \infty} L \operatorname{sinc}(k_x L/2) = \pi\delta(k_x). \tag{2.101}$$

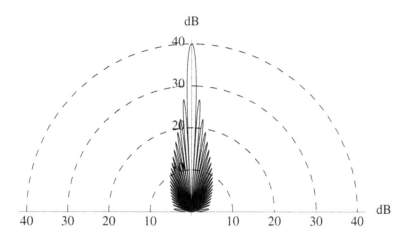

Figure 2.16: Directivity pattern of square plate along x axis for 20 wavelengths across length, $k = 40\pi/L$.

Thus the directivity pattern becomes a delta function located at $k_x = 0$, with infinite pressure at normal $(\theta = 0)$ and zero pressure at any other angle.

The same delta function directivity occurs if, instead of increasing L, we increase frequency. Figure 2.16 plots a high frequency case, $k = 40\pi/L$. It shows that the piston is very directive, approaching a delta function in directivity. In the Ewald sphere construction this case corresponds to a radiation circle five times the radius shown in Fig. 2.14.

2.11.4   Baffled Square Plate with Traveling Wave

The example above was for a vibrator with no variation in velocity across its face. We will now see that vibrators with complicated vibration patterns can be studied easily from a knowledge of the piston results. To illustrate this we choose the simplest example, a square plate with a traveling wave vibration pattern in the x direction. The velocity on the surface is given by

$$\dot{w}(x,y,0) = \begin{cases} \dot{w}_0 e^{ik_{x0}x} & \text{if } -L/2 < x < L/2, \\ & \quad -L/2 < y < L/2 \\ 0 & \text{otherwise.} \end{cases}$$

This is a somewhat unrealistic vibrator, since there are no waves traveling in the negative x direction, but mathematically simple.

We construct the Fourier transform by writing the surface velocity using the rectangle function of Section 1.6. This introduces the effect of the baffle. Thus

$$\dot{W}(k_x, k_y, 0) = \dot{w}_0 \int_{-\infty}^{\infty} \Pi(y/L) e^{-ik_y y} dy \int_{-\infty}^{\infty} \Pi(x/L) e^{ik_{x0}x} e^{-ik_x x} dx.$$

The first integral is just $L \operatorname{sinc}(k_y L/2)$ as before. The second integral is recognized as a product of two spatial functions, and thus we can use the k-space convolution theorem,

Eq. (1.15). The Fourier transform of $e^{ik_{x0}x}$ is $2\pi\delta(k_x - k_{x0})$, given by Eq. (1.36), so that

$$\dot{W}(k_x, k_y, 0) = \dot{w}_0 L^2 \operatorname{sinc}(k_y L/2) \big(\operatorname{sinc}(k_x L/2) * \delta(k_x - k_{x0}) \big).$$

The convolution with a delta function, Eq. (1.37), yields

$$\dot{W}(k_x, k_y, 0) = \dot{w}_0 L^2 \operatorname{sinc}(k_y L/2) \operatorname{sinc}\big((k_x - k_{x0})L/2 \big). \tag{2.102}$$

From Eq. (2.96) then

$$\operatorname{sinc}\big[(k_x - k_{x0})L/2 \big] = \operatorname{sinc}\big[kL/2(\sin\theta \cos\phi - \sin\theta_0 \cos\phi_0) \big]. \tag{2.103}$$

This result is similar to the square piston except that the main lobe in the x direction has been shifted off the normal to an angle given by $k_{x0} = k \sin\theta_0$. As long as k_{x0} is supersonic we can see that the Ewald construction provides a shifted main lobe and side lobes, with the main lobe pointing in the direction of θ_0. Figure 2.17 presents the case for $k_{x0} = k/2$, or $\theta_0 = 30$ degrees and $\phi_0 = 0$. Note that $\phi = \pi$ is in the second quadrant of the plot. The phase velocity of the piston wave is $c_{x0} = 2c$, twice the velocity of sound in the fluid.

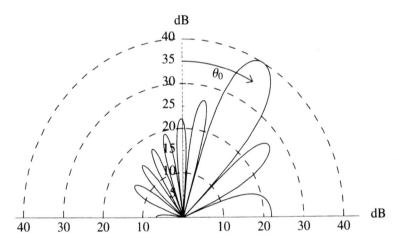

Figure 2.17: Traveling wave on baffled plate, for $k_{x0} = k/2$ and $k = 30/L$.

If the traveling wave is just sonic so that $k_{x0} = k$ (the phase velocity of the wave, $c_{x0} = c$), we expect that the main lobe will now lie in the (x, y) plane. This condition represents beaming parallel to the plate. Figure 2.18 illustrates this case for $k = 30/L$.

The progression of the main lobe towards the horizon, as shown in Figs 2.17 and 2.18 indicates that as the wave traveling on the plate becomes subsonic, this main lobe will drop below the horizon. However, one can see that the side lobes remain above, radiating into the farfield. Thus, the subsonic wave is able to radiate to the farfield unlike the case of the infinite plate discussed in Section 2.7. This is an important phenomenon related to the truncation of the traveling wave at the baffle. We will discuss this in much more detail later when we encounter edge and corner modes of plate radiators.

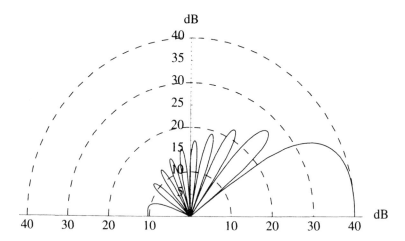

Figure 2.18: Traveling wave on baffled plate illustrating beaming in the horizontal plane, $k_{x0} = k$ and $k = 30/L$.

2.11.5 Baffled Circular Piston

The vibration of a baffled circular piston of radius a located in the $z = 0$ plane is defined by

$$\dot{w}(r, \phi) = \dot{w}(r) = \begin{cases} \dot{w}_0 & \text{if } 0 \leq r \leq a \\ 0 & \text{otherwise.} \end{cases}$$

The farfield radiation follows directly from Eq. (2.92) with $n = 0$:

$$\begin{aligned} D(\theta, \phi) &= -i\rho_0 ck \int_0^\infty \dot{w}(r) J_0(k_r r) r \, dr \\ &= -i\rho_0 ck \dot{w}_0 \int_0^a J_0(k_r r) r \, dr, \end{aligned}$$

where the link to the farfield is given by $k_r = k \sin \theta$. The indefinite integral over the Bessel function is well known,

$$\int J_0(x) x \, dx = x J_1(x), \tag{2.104}$$

so that

$$\int_0^a J_0(k_r r) r \, dr = \frac{a}{k_r} J_1(k_r a), \tag{2.105}$$

and thus

$$D(\theta, \phi) = -i\rho_0 cka^2 \dot{w}_0 \frac{J_1(k_r a)}{k_r a}.$$

Given $\pi a^2 \dot{w}_0 = Q_h$ (the volume velocity of the piston) then

$$D(\theta, \phi) = \frac{-i\rho_0 ckQ_h}{\pi} \frac{J_1(ka \sin \theta)}{ka \sin \theta}. \tag{2.106}$$

An example of the directivity pattern for a circular piston is shown in Fig. 2.19 where the logarithm of D is plotted for $ka = 10$. The maximum on the polar axis is set arbitrarily at 40 dB. Two side lobes arise at this value of ka. Using the Ewald sphere construction one can visualize $D(\theta, \phi)$ at any frequency from the logarithmic plot of $J_1(x)/x$ shown in Fig. 2.20.

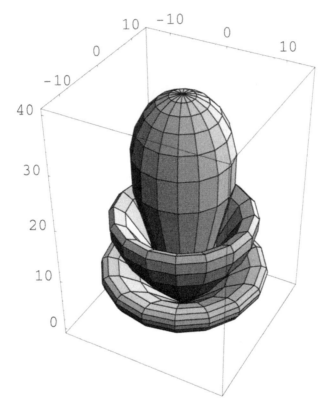

Figure 2.19: Logarithm of the directivity function for a circular piston, $20\log_{10}(D)$ for $ka = 10$.

2.11.6　First Product Theorem for Arrays

We now derive the first product theorem which states: *the directivity pattern of an array of N identical (size and shape) radiators is equal to the product of the directivity pattern of one of the radiators times the transform of an array of N baffled point sources positioned at the centers of the original radiators (now removed) with the same relative amplitude and phases as the original radiators.*

To prove the first product theorem, let $D_0(\theta, \phi)$ be the directivity pattern of one of the radiators in the array located, however, at the origin. Equation (2.86) shows that this pattern is proportional to $Q_1 \ddot{W}(k_x, k_y, 0)$, the Fourier transform of the normal velocity distribution $Q_1 \dot{w}(x, y, 0)$ across one radiator and its baffle. The other radiators

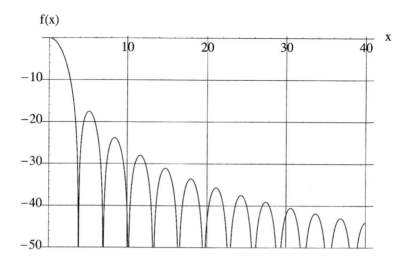

Figure 2.20: Logarithmic plot of $f(x) = 20\log_{10}(|2J_1(x)/x|)$ which can be used to visualize the directivity function for $ka < 40$.

are replaced by the baffle. Q_1 is the complex amplitude of one radiator. If we shift this radiator in the x and y directions by an amount (x_n, y_n), and let it have a different amplitude of vibration, Q_n, then the velocity of the shifted radiator is $Q_n \dot{w}(x-x_n, y-y_n, 0)$ and the directivity is (by the shift theorem) $Q_n \ddot{W}(k_x, k_y, 0)e^{-ik_x x_n}e^{-ik_y y_n}$. If we consider a collection of N of these radiators, all identical in geometry then by superposition the farfield pressure would be

$$D(\theta, \phi) = \frac{-i\rho_0 ck}{2\pi} \ddot{W}(k_x, k_y, 0) \sum_{n=1}^{N} Q_n e^{-ik_x x_n} e^{-ik_y y_n}. \tag{2.107}$$

Referring back to Section 2.10.1 on page 37. we recognize $Q_n e^{-ik_x x_n}e^{-ik_y y_n}$ as the transform of a point source with strength Q_n located at the point (x_n, y_n), that is, $\delta(x - x_n)\delta(y - y_n)$. Thus,

$$\sum_{n=1}^{N} Q_n e^{-ik_x x_n} e^{-ik_y y_n}.$$

represents the transform of a sum of point sources of complex strengths given by Q_n located at the positions of the original, physically identical radiators. Equation (2.107) represents a statement of the **first product theorem**: the directivity pattern of an array of like vibrators is the product of the directivity pattern of one of the radiators (with unit amplitude) with the sum of N point sources located at the positions of the original N radiators. This completes the proof of the first product theorem.

Example: Two Baffled Square Pistons

Consider two identical square pistons located on the x axis and separated by a distance d, as shown in Fig. 2.21. Assume that they are vibrating 180 degrees out of phase with respect to one another but with equal amplitudes. Thus we set $Q_1 = 1$ and $Q_2 = -1$. f D_0 is the directivity pattern of a square piston of width L, centered on the origin.

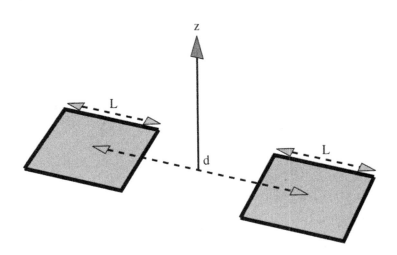

Figure 2.21: Two baffled square pistons on the x axis, each of length L and separated by a distance d.

The first product theorem takes the form

$$D(\theta, \phi) = D_0(\theta. \phi)(e^{ik_z d/2} - e^{-ik_z d/2}),$$

since the point sources are located at $x = \pm d/2$, $y = 0$. Thus

$$D(\theta, \phi) = 2i D_0(\theta, \phi) \sin(k_x d/2)$$

and since the directivity pattern of a square piston centered at the origin, Eq. (2.99), is

$$D_0(\theta, \phi) = \frac{-i\rho_0 ck L^2}{2\pi} \operatorname{sinc}(k_x L/2) \operatorname{sinc}(k_y L/2),$$

then

$$
\begin{aligned}
D(\theta, \phi) &= \frac{\rho_0 ck L^2}{\pi} \operatorname{sinc}(k_x L/2) \operatorname{sinc}(k_y L/2) \sin(k_x d/2) & (2.108) \\
&= \frac{\rho_0 ck L^2}{\pi} \operatorname{sinc}(kL/2 \sin\theta \cos\phi) \sin(kd/2 \sin\theta \cos\phi) \operatorname{sinc}(kL/2 \sin\theta \sin\phi).
\end{aligned}
$$

The directivity pattern of the single piston is modulated by the sine term with the amount of modulation depending on the distance between the pistons. When d is small (pistons overlapping) the opposing motions of the pistons cancel and the radiation is nearly extinguished. At the angle of maximum radiation from the single piston ($\theta = 0$), the dipole pair has a null for any d and k.

2.12   Radiated Power

The total power radiated into a half-space from planar radiators is given by the normal acoustic intensity integrated over the area of the vibrating region. For baffled radiators this region, S, is finite, covering only the non-baffled area, otherwise the area S is infinite. From Eq. (2.17), with $ds = dx\,dy$ we have

$$\Pi(\omega) = 1/2 \iint_S \mathrm{Re}\Big[p(x,y,0)\dot{w}^*(x,y,0)\Big]ds. \tag{2.109}$$

Inserting the angular spectrum representations (see Eq. (2.51)),

$$p(x,y,0) = \frac{1}{4\pi^2} \int_{-\infty}^{\infty}\int_{-\infty}^{\infty} P(k_x,k_y,0)e^{i(k_x x + k_y y)}dk_x dk_y$$

and

$$\dot{w}^*(x,y,0) = \frac{1}{4\pi^2} \int_{-\infty}^{\infty}\int_{-\infty}^{\infty} \dot{W}^*(k_x',k_y',0)e^{-i(k_x' x + k_y' y)}dk_x' dk_y',$$

and the delta function relation (see Eq. (1.5))

$$\frac{1}{4\pi^2}\int_{-\infty}^{\infty}\int_{-\infty}^{\infty} e^{i(k_x - k_x')x}e^{i(k_y - k_y')y}dx\,dy = \delta(k_x - k_x')\delta(k_y - k_y') \tag{2.110}$$

into Eq. (2.109), yields

$$\Pi(\omega) = \frac{1}{8\pi^2}\mathrm{Re}\bigg[\int_{-\infty}^{\infty}\int_{-\infty}^{\infty} P(k_x,k_y,0)\dot{W}^*(k_x,k_y,0)dk_x dk_y\bigg].$$

Using Eq. (2.61) with $z = z' = 0$,

$$P(k_x,k_y,0) = \frac{\rho_0 c k}{k_z}\dot{W}(k_x,k_y,0),$$

the equation for power becomes,

$$\Pi(\omega) = \frac{\rho_0 c k}{8\pi^2}\mathrm{Re}\bigg(\int_{-\infty}^{\infty}\int_{-\infty}^{\infty} \frac{|\dot{W}(k_x,k_y,0)|^2}{\sqrt{k^2 - k_x^2 - k_y^2}}dk_x dk_y\bigg). \tag{2.111}$$

If S_r is the area inside and including the radiation circle defined by

$$\int_{S_r} dk_x dk_y \equiv \int_{-k}^{k} dk_y \int_{-\sqrt{k^2 - k_y^2}}^{\sqrt{k^2 - k_y^2}} dk_x, \tag{2.112}$$

then, since the integrand is imaginary outside of S_r, we can rewrite this integral restricting the limits of the integration:

$$\Pi(\omega) = \frac{\rho_0 c k}{8\pi^2}\iint_{S_r} \frac{|\dot{W}(k_x,k_y,0)|^2}{\sqrt{k^2 - k_x^2 - k_y^2}}dk_x dk_y. \tag{2.113}$$

Equation (2.113) provides a means of computing the power radiated by a source from a knowledge of the angular spectrum of its normal surface velocity. As we have seen before, only the part of the angular spectrum within the radiation circle radiates to the farfield, and thus only this part contributes to the integrand in Eq. (2.113).

There is a very interesting counterpart to Eq. (2.113) in the space domain, but is less well known and seems to have been first provided by Bouwkamp:[8]

$$\Pi(\omega) = \frac{\rho_0 ck}{4\pi} \iint_S \iint_{S'} \dot{w}(x', y', 0) \frac{\sin(kR)}{R} \dot{w}^*(x, y, 0) ds\, ds',\qquad (2.114)$$

where $R \equiv |\vec{r} - \vec{r}^{\,*}|$ is defined in the same way as in Rayleigh's integrals (Section 2.10) and $ds' = dx' dy'$ is integrated over the infinite surface S'.

To prove this relationship consider the following. Equation (2.111) can be written as

$$\Pi(\omega) = \frac{\rho_0 ck}{8\pi^2} \int_{-\infty}^{\infty} \int_{-\infty}^{\infty} |\dot{W}(k_x, k_y, 0)|^2 \mathrm{Re}\left(\frac{1}{k_z}\right) dk_x dk_y.$$

Inserting $\left. e^{i(k_x x + k_y y)} \right|_{x=y=0}$ into the integrand,

$$\Pi(\omega) = \frac{\rho_0 ck}{8\pi^2} \int_{-\infty}^{\infty} \int_{-\infty}^{\infty} \dot{W}(k_x, k_y, 0) \dot{W}^*(k_x, k_y, 0) \mathrm{Re}\left(\frac{1}{k_z}\right) e^{i(k_x x + k_y y)} dk_x dk_y \Big|_{x=y=0},$$

where the exponential has been added to make the integral look like an inverse Fourier transform (evaluated at the origin). This integral has the form

$$\mathcal{I} = \mathcal{F}_x^{-1} \mathcal{F}_y^{-1} [\dot{W} \{\dot{W}^* \mathrm{Re}(1/k_z)\}] \Big|_{x=y=0},$$

where

$$\Pi(\omega) = \frac{\rho_0 ck}{2} \mathcal{I}.$$

Using the inverse Fourier transform of the convolution theorem, Eq. (1.19), we have

$$\mathcal{I} = \dot{w}(x, y, 0) ** \left(w^*(-x, -y, 0) ** \mathcal{F}_x^{-1} \mathcal{F}_y^{-1} [\mathrm{Re}(1/k_z)] \right),$$

where we have used the following to derive the inverse transform of $\dot{W}^*$:

$$\frac{1}{4\pi^2} \iint \dot{W}^*(k_x, k_y) e^{i(k_x x + k_y y)} dk_x dk_y = \left(\frac{1}{4\pi^2} \iint \dot{W}(k_x, k_y) e^{ik_x(-x) + ik_y(-y)} dk_x dk_y \right)^*$$
$$= \dot{w}^*(-x, -y, 0).$$

The inverse transform of $1/k_z$ is found from Weyl's integral with $z = r' = 0$ (Eq. (2.74) and Eq. (2.70)):

$$-i\frac{e^{ikr}}{2\pi r} = \mathcal{F}_x^{-1} \mathcal{F}_y^{-1} [\frac{1}{k_z}].\qquad (2.115)$$

---

[8]C. J. Bouwkamp (1970). "Theoretical and Numerical Treatment of Diffraction Through a Circular Aperture," IEEE Trans. Antennnas Propag., **AP-18**, pp. 152–176.

To determine the inverse transform of $\mathrm{Re}[1/k_z]$ we need to draw upon a theorem about Fourier transforms.[9] $f(x)$ can be split unambiguously into an even and odd function,

$$f(x) = e(x) + o(x) \tag{2.116}$$

where e and o are even and odd functions, respectively. Its transform $F(k_x)$ is then

$$F(k_x) = E(k_x) + O(k_x), \tag{2.117}$$

where E and O are even and odd functions in transform space. It is straightforward to show, using

$$\mathrm{Re}\left[\frac{-ie^{ikr}}{2\pi r}\right] = \frac{\sin kr}{2\pi r},$$

that

$$\mathrm{Re}[E(k_x)] = \mathcal{F}_x[\mathrm{Re}(e(x))], \tag{2.118}$$

$$\mathrm{Im}[E(k_x)] = \mathcal{F}_x[\mathrm{Im}(e(x))]. \tag{2.119}$$

The same result can be applied to two-dimensional even functions and for the inverse Fourier transform. Thus, since k_z is even in k_x and k_y, Eq. (2.115) can be written as

$$\mathrm{Re}\left(-i\frac{e^{ikr}}{2\pi r}\right) = \mathcal{F}_x^{-1}\mathcal{F}_y^{-1}\left[\mathrm{Re}(\frac{1}{k_z})\right], \tag{2.120}$$

so that

$$\mathcal{I} = \dot{w}(x,y,0) * * \left(\dot{w}^*(-x,-y,0) * * \frac{\sin(kr)}{2\pi r}\right)\bigg|_{x=y=0}, \tag{2.121}$$

where $r = \sqrt{x^2 + y^2}$. This is essentially the final result, although the convolutions need to be written out in order to compare with Eq. (2.114). To help in the untangling of the multiple convolutions we write the one-dimensional equivalent in general terms,

$$f(x) * \{g(x) * h(x)\} \equiv f(x) * \int dx' g(x')h(x - x') \equiv f(x) * q(x),$$

where

$$q(x) \equiv \int dx' g(x')h(x - x').$$

Thus, remembering that the result of the convolutions is a function of x only,

$$f(x) * \{g(x) * h(x)\} \equiv f(x) * q(x) \equiv \int dx'' f(x'')q(x - x'')$$

$$= \int dx'' f(x'') \int dx' g(x')h(x - x'' - x'). \tag{2.122}$$

Finally, returning to the expression for $\mathcal{I}$ we have

$$\mathcal{I} = \int_{-\infty}^{\infty}\int_{-\infty}^{\infty} dx'' dy'' \int_{-\infty}^{\infty}\int_{-\infty}^{\infty} dx' dy' w(x'', y'', 0)w^*(-x', -y', 0)\frac{\sin(k\bar{R})}{2\pi\bar{R}}\bigg|_{x=y=0}$$

[9]R. N. Bracewell (1978). *The Fourier Transform and Its Application*. McGraw-Hill, New York, 2nd ed., p. 14.

where $\bar{R} = \sqrt{(x - x'' - x')^2 + (y - y'' - y')^2}$. Therefore, noting that substituting x' for $-x'$ and y' for $-y'$ does not change the value of the integrals, the final result for the radiated power is

$$\Pi(\omega) = \frac{\rho_0 c k}{4\pi} \int_{-\infty}^{\infty} \int_{-\infty}^{\infty} dx'' dy'' \int_{-\infty}^{\infty} \int_{-\infty}^{\infty} dx' dy' w(x'', y'', 0) w^*(x', y', 0) \frac{\sin(kR')}{R'},$$
(2.123)

where $R' = \sqrt{(x' - x'')^2 + (y' - y'')^2}$. This completes the proof of Eq. (2.114).

2.12.1 Low Frequency Expansion

This formula, Eq. (2.114), is extremely useful in deriving a low frequency series expression for the power radiated from a planar vibrator.[10] We can expand $\sin(kR)/R$ in a MacLaurin series

$$\sin(kR)/R = \sum_{m=0}^{\infty} \frac{k^{2m+1}}{(2m+1)!} (-1)^m [(x - x')^2 + (y - y')^2]^m.$$
(2.124)

Furthermore, we can expand the term $[(x-x')^2 + (y-y')^2]^m$ using the binomial theorem,

$$[(x - x')^2 + (y - y')^2]^m = \sum_{l=0}^{m} \binom{m}{l} (x - x')^{2m-2l} (y - y')^{2l},$$
(2.125)

where

$$\binom{m}{l} \equiv \frac{m!}{l!(m-l)!}.$$
(2.126)

With a little effort we finally arrive at

$$\Pi(\omega) = \frac{\rho_0 c}{4\pi} \sum_{m=0}^{\infty} \sum_{l=0}^{m} \sum_{p=0}^{2m-2l} \sum_{q=0}^{2l} \frac{k^{2m+2}}{(2m+1)!} \binom{m}{l} \binom{2m-2l}{p} \binom{2l}{q}$$

$$\times [(\frac{\partial}{\partial k_x})^{2m-2l-p} (\frac{\partial}{\partial k_y})^{2l-q} \dot{W}^*(k_x, k_y, 0)]$$
(2.127)

$$\times [(\frac{\partial}{\partial k_x})^{p} (\frac{\partial}{\partial k_y})^{q} \dot{W}(k_x, k_y, 0)] \Big|_{k_x = k_y = 0}.$$

In particular, for example, the low frequency limit has one term ($m = 0$) yielding

$$\Pi_0 = \frac{\rho_0 c k^2}{4\pi} |\dot{W}(0, 0, 0)|^2.$$
(2.128)

The low frequency power can be related to the volume flow of the radiator, Q_h, by recognizing that the forward transform is

$$\dot{W}(0, 0, 0) = \iint \dot{w}(x, y, 0) dx\, dy \equiv Q_h,$$
(2.129)

---

[10]E. G. Williams (1983). "A series expansion of the acoustic power radiated from planar sources", J. Acoust. Soc. Am., **73**, pp. 1520–1524.

so that Eq. (2.128) becomes

$$\Pi_0 = \frac{\rho_0 c k^2}{4\pi}|Q_h|^2. \tag{2.130}$$

2.13 Vibration and Radiation from an Infinite Point-driven Plate

Mathematical models of some simple vibrating structures are invaluable in the application of NAH to practical vibration/radiation problems. These models help in the understanding of k-space and its physical significance. Outside of NAH these models are essential for understanding the vibration and radiation from more complex physical structures. Point driven structures occur quite often in practice. To gain some understanding of the radiation fields we study the infinite plate excited at a point with a known force. We present the equation of motion of a vibrating plate without derivation since this is not within the scope of this text.

Let $w(x, y, 0)$ be the normal displacement of the plate and let a point force excite the plate at the origin. The differential equation for the plate motion which must be solved is of fourth order. Let the plate be driven by a point force of magnitude F at the point $(x, y, z) = (0, 0, 0^-)$ and be loaded with the acoustic fluid on the top side $z = 0^+$, with a vacuum on the bottom side. If $p_a(x, y, t)$ is the pressure of the fluid acting on the plate, then

$$D\left(\frac{\partial^4 w}{\partial x^4} + 2\frac{\partial^4 w}{\partial x^2 \partial y^2} + \frac{\partial^4 w}{\partial y^4}\right) + \rho_s h\frac{\partial^2 w}{\partial t^2} = F(t)\delta(x)\delta(y) - p_a(x, y, t), \tag{2.131}$$

where D is the flexural rigidity of the plate given by

$$D = \frac{Eh^3}{12(1 - \nu^2)}, \tag{2.132}$$

E is Young's modulus, ν is Poisson's ratio, h is the thickness of the plate and ρ_s is the density of the plate. For convenience we will need Skudrzyk's plate constant[11] α which we will use later:

$$\alpha \equiv \left(\frac{D}{\rho_s h}\right)^{1/4} = \left(\frac{Eh^2}{12\rho_s(1 - \nu^2)}\right)^{1/4}. \tag{2.133}$$

The plate equation of motion is almost never solved in the time domain. If one assumes that the time dependence is given as usual by $e^{-i\omega t}$ then Eq. (2.131) becomes (where w and p_a are functions of ω now)

$$D\left(\frac{\partial^4 w}{\partial x^4} + 2\frac{\partial^4 w}{\partial x^2 \partial y^2} + \frac{\partial^4 w}{\partial y^4}\right) - \rho_s h\omega^2 w = F(\omega)\delta(x)\delta(y) - p_a(x, y). \tag{2.134}$$

---

[11] Eugen Skudrzyk (1968). *Simple and Complex Vibratory Systems*. The Pennsylvania State University Press, University Park, PA.

Since the point drive at the origin dictates circular symmetry we can write the equation of motion in polar coordinates with the transformation

$$x = r\cos\phi$$

$$y = r\sin\phi$$

$$x^2 + y^2 = r^2.$$

Note that we are using r instead of ρ here, and the context of the development should prevent any confusion with the definition of r used up to now. Given that the normal displacement $w(x,y) \to w(r)$ the equation of motion becomes

$$D\left(\frac{d^2}{dr^2} + \frac{d}{r\,dr}\right)^2 w - \rho_s h\omega^2 w = F(\omega)\frac{\delta(r)}{2\pi r} - p_a(r), \qquad (2.135)$$

where the shorthand notation is used for the derivatives:

$$\left(\frac{d^2}{dr^2} + \frac{d}{r\,dr}\right)^2 \equiv \left(\frac{d^2}{dr^2} + \frac{d}{r\,dr}\right)\left(\frac{d^2}{dr^2} + \frac{d}{r\,dr}\right).$$

To simplify the solution we assume light fluid loading, $p_a(r) \approx 0$, so that the pressure term in Eq. (2.135) can be ignored. The solution to the resulting equation is well known. It is obtained by use of the Hankel transform pair, Section 1.4. That is,

$$W(k_r) \equiv \int_0^\infty w(r)J_0(k_r r)r\,dr, \qquad (2.136)$$

and

$$w(r) = \int_0^\infty W(k_r)J_0(k_r r)k_r\,dk_r \equiv \mathcal{B}^{-1}[W(k_r)]. \qquad (2.137)$$

We will find in Chapter 4 (from Eqs (4.13) and (4.17)) that

$$\left(\frac{d^2}{dr^2} + \frac{d}{r\,dr}\right)J_0(k_r r) = -k_r^2 J_0(k_r r).$$

Thus applying the differential operators to Eq. (2.137) yields

$$\begin{aligned}
\left(\frac{d^2}{dr^2} + \frac{d}{r\,dr}\right)w(r) &= \int_0^\infty W(k_r)\left(\frac{d^2}{dr^2} + \frac{d}{r\,dr}\right)J_0(k_r r)k_r\,dk_r \\
&= \int_0^\infty W(k_r)(-k_r^2)J_0(k_r r)k_r\,dk_r
\end{aligned}$$

which, written in shorthand, is

$$\mathcal{B}^{-1}[-k_r^2 W(k_r)] = \frac{d^2 w(r)}{dr^2} + \frac{1}{r}\frac{dw(r)}{dr}.$$

Inverting this equation (writing as a forward transform) yields

$$\int_0^\infty \left(\frac{d^2 w(r)}{dr^2} + \frac{1}{r}\frac{dw(r)}{dr}\right)J_0(k_r r)r\,dr = -k_r^2 W(k_r),$$

so that the Hankel transform of Eq. (2.135) leads to the simple result:

$$W(k_r) = \frac{F(\omega)}{2\pi D(k_r^4 - k_f^4)}, \tag{2.138}$$

where k_f is the free wavenumber of the plate,

$$k_f = (m_s \omega^2 / D)^{1/4}, \tag{2.139}$$

and $m_s = \rho_s h$ is the mass per unit area of the plate. The normal velocity is

$$\dot{W}(k_r) = \frac{-i\omega F}{2\pi D(k_r^4 - k_f^4)}. \tag{2.140}$$

Note that the free wavenumber of the plate is dispersive, depending on $\sqrt{\omega}$. The dispersion equation is simply, using Eq. (2.133),

$$k_f = \frac{\sqrt{\omega}}{\alpha}. \tag{2.141}$$

The inverse Hankel transform of Eq. (2.138), provides the solution for the displacement:

$$w(r) = \frac{F}{2\pi D} \int_0^\infty \frac{J_0(k_r r)}{(k_r^4 - k_f^4)} k_r dk_r. \tag{2.142}$$

This integral can be solved by using contour integration and residue evaluation. Towards that end we extend the integral to $-\infty$ by noting that (Eq. (4.20))

$$\begin{aligned} J_0(x) &= \frac{1}{2} H_0^{(1)}(x) + \frac{1}{2} H_0^{(2)}(x) \\ &= \frac{1}{2} H_0^{(1)}(x) - \frac{1}{2} H_0^{(1)}(-x), \end{aligned} \tag{2.143}$$

since $H_0^{(1)}(-x) = -H_0^{(2)}(x)$. Equation (2.142) becomes

$$w(r) = \frac{F}{4\pi D} \int_{-\infty}^\infty \frac{H_0^{(1)}(k_r r)}{(k_r^4 - k_f^4)} k_r dk_r. \tag{2.144}$$

The important point regarding the contour integration is the fact that the residues are determined from the zeros of the denominator of Eq. (2.144), that is, the poles of the k-space velocity (Eq. (2.140)). Generally the poles are related to the wave types which are free to travel on the structure, being independent of the forcing function driving the plate. In this case the poles are given by $k_r^4 - k_f^4 = 0$ which has the four solutions $k_r = \pm k_f$ and $k_r = \pm i k_f$.

Recall that from residue theory at each of the simple poles x_0

$$\oint F(x) dx = 2\pi i R(x_0),$$

where the residue $R(x_0) = \lim_{x \to x_0}[(x - x_0)F(x)]$. Using this fact with the poles $k_r = k_f$ and $k_r = ik_f$ within the closed contour, and $\dot{w} = -i\omega w$, Eq. (2.144) becomes,

$$\dot{w}(r) = \frac{F}{8\alpha^2 m_s}\left[H_0^{(1)}(k_f r) - H_0^{(1)}(ik_f r)\right],$$

where α was given in Eq. (2.133). The second solution, $H_0^{(1)}(ik_r r)$ is equivalent to a MacDonald function (Eq. (4.34)):

$$H_0^{(1)}(ik_r r) = \frac{-2i}{\pi} K_0(k_f r) \tag{2.145}$$

so that the complete solution in physical space has the form

$$\dot{w}(r) = \frac{F}{8\alpha^2 m_s}\left[H_0^{(1)}(k_f r) + \frac{2i}{\pi} K_0(k_f r)\right]. \tag{2.146}$$

Equation (2.143) indicates that at $r = 0$ the term in the square bracket above is unity (since $J_0(0) = 1$) so that

$$\dot{w}(0) = F/(8\alpha^2 m_s). \tag{2.147}$$

The Hankel function provides the free wave solution, a cylindrical wave traveling outward from the origin, and the decaying MacDonald function provides the flexural nearfield. Whereas the Hankel function is complex, the MacDonald function is purely real. Figure 2.22 is a plot of the imaginary part of the traveling wave solution, $H_0^{(1)}(k_f r)$,

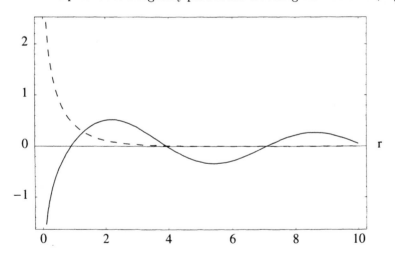

Figure 2.22: Traveling wave solution, $\text{Im}[H_0(k_f r)]$ (solid line) compared to flexural wave nearfield, $K_0(k_f r)$ (dashed line) for $k_f = 1$. The nearfield exists only close to the drive-point. Both curves blow up at $r = 0$.

compared with the flexural nearfield, $K_0(k_f r)$. The asymptotic versions of these two functions (Eqs (4.22) and (4.39)) clearly indicate the wave types:

$$H_0^{(1)}(k_f r) \rightarrow \sqrt{\frac{2}{\pi k_f r}} e^{i(k_f r - \pi/4)}, \tag{2.148}$$

and

$$K_0(k_f r) \to \sqrt{\frac{\pi}{2k_f r}} e^{-k_f r}, \tag{2.149}$$

both valid for large values of $k_f r$.

The plate solution also provides a very useful structural formula, given in Eq. (2.147); the drive-point impedance of a plate is

$$Z_p \equiv F/\dot{w}(0) = 8\alpha^2 m_s. \tag{2.150}$$

Associated with the free wavenumber is the phase velocity of the flexural (bending) wave defined through

$$k_f = \omega/c_b. \tag{2.151}$$

Thus Eq. (2.141) leads to

$$c_b = \alpha\sqrt{\omega}. \tag{2.152}$$

Note that the bending wave phase velocity c_b is dispersive since it depends on frequency.

2.13.1   Farfield Radiation

The farfield directivity pattern is determined from Eq. (2.140) and the Ewald sphere construction process, Section 2.11.2. To obtain the farfield we saw in Section 2.11.1 that we must make the replacement, $k_r = k \sin\theta$. Thus the pressure in the farfield is

$$p(r,\theta,\phi) = \frac{-i\rho_0 ck}{2\pi} \frac{e^{ikr}}{r} \dot{W}(k_r) = \frac{-i\rho_0 ck}{2\pi} \frac{e^{ikr}}{r} \dot{W}(k \sin\theta). \tag{2.153}$$

The k-space velocity $\dot{W}(k_r)$ is plotted in Fig. 2.23. It is interesting to note that the k-space velocity is infinite when the radial wavenumber equals the free wavenumber in the plate. This forms a circle in k-space, representing a locus of possible values for the free, flexural waves traveling in the plate. If the radiation circle, $k_r = k$, is located outside of this free wavenumber circle, $k > k_f$ (supersonic free wavenumber), then we can see that the dominant plate radiation is at a polar angle θ_0 given by

$$\sin\theta_0 = k_f/k.$$

The farfield pressure is infinite at $\theta = \theta_0$. The radiation pattern is circumferentially symmetric, since $\dot{W}$ is independent of angle.

We can write the k-space normal velocity in partial fraction form,

$$\dot{W}(k_r) = \frac{-i\omega F}{2\pi D(k_r^4 - k_f^4)} = \frac{-i\omega F}{2\pi D k_f^2} \left[\frac{1}{(k_r^2 - k_f^2)} - \frac{1}{(k_r^2 + k_f^2)} \right]. \tag{2.154}$$

The radiation from the plate consists of two similar terms, differing by a minus sign. Each of these terms represents a wave type. The first term corresponds to a diverging wave from the origin of the plate, and the second to an exponentially decaying wave; as we have seen, the inverse Hankel transform results in $H_0^{(1)}(k_r r)$ and $H_0^{(1)}(ik_r r)$ for each of these terms, respectively. Thus returning to Fig. 2.23 we see that a traveling

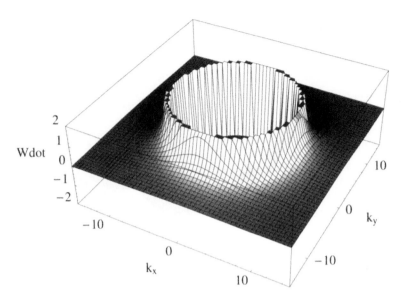

Figure 2.23: k-space diagram for an infinite, point-driven plate. The k-space velocity is infinite when $k_r = k_f$ as indicated by the circle.

wave solution produces more than just a delta function in k-space, as was the case with plane waves (Eq. (2.95)). The circle in the figure is not a delta function due to the fact that it is spread in k_r with a drop off in amplitude given by $1/(k_r^4 - k_f^4)$. This spread is due partly to the outgoing wave and partly due to the flexural nearfield of the plate located around the driver. This flexural nearfield is a local distortion of the plate at the drive-point; as though the driver were punching through the plate. As we can see by referring back to Fig. 2.22, this local distortion is much like a spread delta function in physical space. Thus it must have a broad spectrum in k-space with a corresponding broad directivity pattern in the farfield.

For the case $k_f > k$ (subsonic free wave on the plate) the radiation circle is located inside the free wavenumber circle in Fig. 2.23; we have the situation shown in Fig. 2.24. Drawing the radiation circle in this figure, say at $k = 10$, indicates that the plate still radiates to the farfield and that the directivity pattern of the plate has a maximum amplitude at $\theta = \pi/2$. The level of the pressure radiated to the farfield is small, however, compared to the supersonic wave case. As $k \to 0$ we can see from Eq. (2.154), since $k_r = k \sin \theta$, that the contributions of the traveling and nearfield waves are about equal, each varying as $1/k_f^2$.

The frequency at which $k = k_f$, when the radiation circle falls on the peak of the k-space velocity, is called the coincidence frequency. In this case the bending wave speed c_b just equals the speed of sound in the fluid. The peak in directivity occurs on the horizon. As has been discussed above, at frequencies below the coincidence frequency the plate is a poor radiator (k_f is subsonic). At frequencies above coincidence the plate radiates very efficiently (k_r is supersonic).

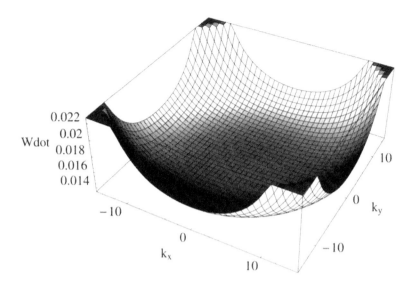

Figure 2.24: k-space diagram of the magnitude of the k-space surface velocity for an infinite, point-driven plate with radiation circle inside of free wavenumber circle.

The coincidence frequency f_c is given by $\alpha\sqrt{\omega_c} = c$ or

$$f = c^2/(2\pi\alpha^2). \tag{2.155}$$

As an example of the coincidence frequency consider a 2 inch steel plate with $E = 19.5 \times 10^{10}$ newtons/m^2, $\nu = 0.28$ and $\rho_s = 7700$ kg/m^3. Figure 2.25 is a plot of c_b as a function of frequency. The coincidence frequencies for the 2 inch plate in water and in air are 4.5 kHz and 240 Hz, respectively. As can be seen, the coincidence frequency is higher for the thinner plate.

2.14   Vibration and Radiation of a Finite, Simply Supported Plate

The finite vibrating plate represents a more realistic source than the infinite one. The vibration problem is of interest because it provides an introduction to normal modes, and the resulting expansion using orthogonal functions (normal modes) to solve the vibration problem for arbitrary excitation. The corresponding radiation problem provides an introduction to edge and corner mode radiation which occurs due to the finite boundaries in the problem. Furthermore, we introduce concepts of radiation impedance and radiation efficiency and write them in terms of normal modes.

The literature on finite plates is vast, and is reviewed extensively in a book by Leissa.[12] The way the plate is supported along its boundaries is crucial to its vibration

---

[12] A. W. Leissa (1969). *Vibration of Plates.* NASA SP-160. Office of Technology Utilization, National Aeronautics and Space Administration, Washington, D.C.

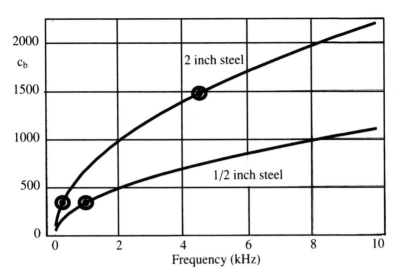

Figure 2.25: Bending wave speed in 2 inch and 1/2 inch steel plates. The circles indicate the coincidence frequencies for sound speed in water (1481 m/s) and air (343 m/s).

and radiation. To determine its radiation the plate is almost always surrounded by an infinite, rigid plane baffle, otherwise the problem is intractable.

We will consider here only one particular type of boundary support, called the simply supported boundary condition, because it leads to a simple solution which can be used to illuminate important concepts.

First we present the homogeneous equation of motion in the frequency domain of the plate along with the boundary condition at the edges. The plate is rectangular of length L_x and width L_y. The origin of the coordinate system is located at the lower left corner of the plate. The equation of motion is the same as the infinite plate:

$$D\left(\frac{\partial^4 w}{\partial x^4} + 2\frac{\partial^4 w}{\partial x^2 \partial y^2} + \frac{\partial^4 w}{\partial y^4}\right) - \rho_s h \omega^2 w = 0, \tag{2.156}$$

which is often written as

$$\nabla^4 w(x, y, \omega) - k_f^4 w(x, y, \omega) = 0 \tag{2.157}$$

where $\nabla^4 = (\frac{\partial^2}{\partial x^2} + \frac{\partial^2}{\partial y^2})^2$ and the bending wavenumber is $k_f = (\rho_s h \omega^2/D)^{1/4}$, the same as the infinite plate. The simply supported boundary condition implies a knife edge support where the plate can not move in the z direction but is free to rotate about this support. The first boundary condition implies that $w(x, y) = 0$ along the boundary and the second that the bending moment vanishes along the edges. The moments anywhere in the plate are related to the plate displacement:[13]

$$M_x = -D\left(\frac{\partial^2 w}{\partial x^2} + \nu\frac{\partial^2 w}{\partial y^2}\right), \tag{2.158}$$

---

[13]E. Skudrzyk (1968). *Simple and Complex Vibratory Systems*. The Pennsylvania State University Press, University Park, PA.

$$M_y = -D(\frac{\partial^2 w}{\partial y^2} + \nu \frac{\partial^2 w}{\partial x^2}). \tag{2.159}$$

The simply supported boundary condition is

$$\begin{aligned} w(x,y) = M_x(x,y) = 0, && x = && 0, \text{ and } L_x, \\ w(x,y) = M_y(x,y) = 0, && y = && 0, \text{ and } L_y, \end{aligned} \tag{2.160}$$

which leads to the conclusion,

$$\begin{aligned} \frac{\partial^2 w(x,y)}{\partial x^2} = 0, && x = && 0, \text{ and } L_x, \\ \frac{\partial^2 w(x,y)}{\partial y^2} = 0. && y = && 0, \text{ and } L_y. \end{aligned} \tag{2.161}$$

In proceeding with the solution to Eq. (2.157) we can not use Fourier transforms as we did with the infinite plate because the equation of motion is specified over a limited area. Consider the following set of modes:

$$\Phi_{mn}(x,y) = \frac{2}{\sqrt{L_x L_y}} \sin(m\pi x/L_x) \sin(n\pi y/L_y), \ n = 1,2,3\cdots, \ m = 1,2,3\cdots.$$
$$\tag{2.162}$$

They satisfy the boundary conditions, Eq. (2.161), and also the equation of motion, Eq. (2.157), as long as

$$(m\pi/L_x)^2 + (n\pi/L_y)^2 = k_f^2. \tag{2.163}$$

The infinite set of functions given by Eq. (2.162). the modes of the plate, represent a set of orthonormal functions; they satisfy the very important relations,

$$\begin{aligned} \int_0^{L_x} \int_0^{L_y} \Phi_{mn}(x,y)\Phi_{pq}(x,y)dx\,dy &= 0 \text{ if } m \neq p, \text{ or } n \neq q \\ &= 1 \text{ if } m = p, \text{ and } n = q. \end{aligned} \tag{2.164}$$

Figure 2.26 shows the spatial variation of the first nine modes for a simply supported square plate. The orthogonality of the modes is actually guaranteed by the theory of partial differential equations, and similar sets of orthogonal modes exist for other boundary conditions, although it is more difficult to write down expressions for them. This same theory[14] guarantees that the modes form a complete set. We will use Eq. (2.164) to solve for the response of the plate when it is driven by external forces.

The eigenvalue equation given by Eq. (2.163) relates the modes to the frequency of vibration since, by Eq. (2.141), $k_f = \sqrt{\omega}/\alpha$. Thus for each mode (m,n) there corresponds an eigenfrequency ω_{mn} given by the equation,

$$\omega_{mn} = \alpha^2[(m\pi/L_x)^2 + (n\pi/L_y)^2]. \tag{2.165}$$

Only at this eigenfrequency, however, is this mode a solution to the equation of motion.

---

[14]Discovered by Sturm and Liouville in 1836 and 1837. See B. Diprima (1977). *Elementary Differential Equations and Boundary Value Problems*. Wiley and Sons, New York, 3rd ed., pp. 531–541.

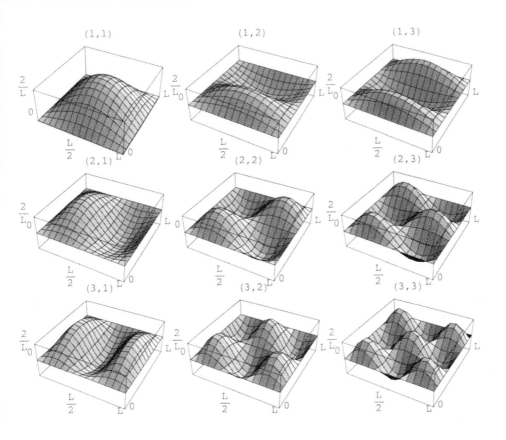

Figure 2.26: Some of the orthonormal modes, $\Phi_{mn}(x,y)$, of a simply supported plate. At the top of each mode is the (m,n) index corresponding to it.

Now we are in a position to formulate the solution for the point excited, simply-supported rectangular plate. If the point source is of magnitude F and is located at the point (x_0, y_0), then the equation of motion is (see Eq. (2.131))

$$\nabla^4 w(x,y,\omega) - k_f^4 w(x,y,\omega) = \frac{1}{D}[F(\omega)\delta(x - x_0)\delta(y - y_0) - p_a(x,y,\omega)], \qquad (2.166)$$

where p_a is the loading (forces) due to the fluid. Note that, if we assume that the plate is in an infinite rigid baffle, p_a can be expressed in terms of the displacement by using Rayleigh's integral, Eq. (2.75):

$$p(x,y,\omega) = \frac{-\omega^2 \rho_0}{2\pi} \int_{-\infty}^{\infty} \int_{-\infty}^{\infty} w(x',y',\omega) \frac{e^{ik|\vec{r}-\vec{r}'|}}{|\vec{r} - \vec{r}'|} dx' dy'.$$

This makes Eq. (2.166) an integral-differential equation. When the fluid is light, such as air, p_a can be ignored unless $\rho_s h$ is very small (as in a timpanic membrane). We proceed to study the rectangular plate with negligible fluid loading.

One of the most common tools for solving problems of this type is to use the fact that the orthonormal modes form a complete set for the specified boundary condition, which means that any displacement on the plate can be expressed as a sum of these modes with appropriate weighting coefficients. Thus we postulate a solution of Eq. (2.166) of the form

$$w(x, y, \omega) = \sum_{m=1}^{\infty} \sum_{n=1}^{\infty} A_{mn}(\omega)\Phi_{mn}(x, y). \tag{2.167}$$

The solution requires three steps:

(1) We insert this solution into Eq. (2.166),

(2) multiply both sides by $\Phi_{pq}(x, y)$ and

(3) integrate over the surface of the plate and use Eq. (2.164).

This will result in an equation for the unknown coefficients $A_{mn}(\omega)$, which together with Eq. (2.167) provide the final solution.

Implementing these three steps leads to the solution

$$w(x, y, \omega) = -\frac{F}{\rho_s h} \sum_{m=1}^{\infty} \sum_{n=1}^{\infty} \frac{\Phi_{mn}(x_0, y_0)\Phi_{mn}(x, y)}{\omega^2 - \omega_{mn}^2}. \tag{2.168}$$

In the process of obtaining this equation we used

$$D\nabla^4 \Phi_{mn}(x, y) = \rho_s h \omega_{mn}^2 \Phi_{mn}(x, y),$$

which resulted from the fact that the modes are solutions of the homogeneous equation, Eq. (2.157), when $\omega^2 = \omega_{mn}^2$.

Actually the procedure leading to Eq. (2.168) is very general, and can be used for vibrating systems in general, whenever the modes of the system can be identified. Note that the amplitude of each mode in Eq. (2.168) is given by that mode evaluated at the excitation point (x_0, y_0). This leads to the familiar notion that when the force is located on a nodal line for a mode, that mode is not excited. For example, if the force were located at the center of the plate only odd-odd (m and n odd) modes would be excited. In general, however, almost all the modes of the plate are excited at any frequency.

When the driving frequency of the force, which we recall is given by $F(\omega)e^{-i\omega t}$, equals the eigenfrequency (resonance frequency) of a mode, the denominator vanishes and the response becomes infinite. These frequencies are the resonances of the plate. The response is finite, however, if damping is added to the plate by making Young's modulus E complex. As a result ω_{mn} becomes complex, since it depends upon $\sqrt{E}$, and the denominator of Eq. (2.168) no longer has any real zeros.

Equation (2.168) leads to an equation for the transfer mobility (ratio of velocity over excitation force) $Y(\omega)$ of the plate:

$$Y(\omega) \equiv \frac{-i\omega w}{F} = \frac{i\omega}{\rho_s h} \sum_{m=1}^{\infty} \sum_{n=1}^{\infty} \frac{\Phi_{mn}(x_0, y_0)\Phi_{mn}(x, y)}{\omega^2 - \omega_{mn}^2}. \tag{2.169}$$

When $x = x_0$ and $y = y_0$ this equation leads to the drive-point mobility. The transfer mobility provides the Green function for the plate when we put $F = 1$.

2.14.1 Rectangular Plate with Fluid Loading

Before we consider the radiation from the simply-supported plate, we look briefly at the vibration when the plate is fluid loaded. The integral-differential equation was provided by Eq. (2.166). One can still attempt to solve this equation using the in vacuo modes, a technique that is sometimes used in the literature. Of course the in vacuo modes are not the real modes of the system, and the theory of partial differential equations does not provide us with a formulation which even proves that normal modes exist in this system. This is a subject of continuing debate.

One can proceed with the three step solution provided above and obtain a result cast in terms of coupled in vacuo modes, as has been done by Davies.[15] Another solution technique was provided by Lax[16] which is quite a bit easier to use on a computer. In the latter case the solution is obtained without fluid loading, using the sum of the in vacuo modes as outlined above. This is the zero order solution, $w^{(0)}(x, y)$. Using this solution p_a can be approximated using Rayleigh's integral, and Eq. (2.166) solved again to obtain a first order solution, $w^{(1)}(x, y)$. This process in continued until convergence is obtained.

2.14.2 Radiation from Rectangular Plates: Radiation Impedance and Efficiency

Since the vibration of a point-driven plate can be expressed as summations of the in vacuo normal modes, we will study the radiation from a single mode first. This is particularly important when the plate is driven near a resonance frequency corresponding to one of the modes, since the vibration is then dominated by a single normal mode. Also we will present some very basic concepts of plate radiation; radiation efficiency, radiation impedance and the concept of edge and corner mode radiation. We will assume throughout that the plate is in an infinite, rigid baffle.

The real part of the radiation impedance is called the radiation resistance and is defined for the (m, n)th mode of the plate as

$$R_{mn}(\omega) \equiv \Pi(\omega)/\frac{1}{2}\langle|\dot{w}_{mn}|^2\rangle, \tag{2.170}$$

where the spatial average of the square of surface velocity is defined by

$$\langle|\dot{w}_{mn}|^2\rangle \equiv \frac{1}{L_x L_y} \int_0^{L_x} \int_0^{L_y} |\dot{w}_{mn}(x, y)|^2 dx\, dy. \tag{2.171}$$

The radiation efficiency is defined in general by

$$S \equiv \frac{\Pi}{\Pi_0} \equiv \frac{\Pi}{\frac{1}{2}\rho_0 c L_x L_y \langle|\dot{w}|^2\rangle}, \tag{2.172}$$

---

[15] H. G. Davies (1969). "Acoustic Radiation from Fluid Loaded Rectangular Plates," MIT, TR71476-1, December.

[16] M. Lax (1944). "The Effect of Radiation on the Vibrations of a Circular Diaphragm," J. Acoust. Soc. Am., **16**, pp. 5–13.

where $\Pi_0 \equiv \frac{1}{2}\rho_0 c L_x L_y \langle |\dot{w}|^2 \rangle$ is the power radiated by the area $L_x L_y$ of an infinite plate vibrating as an infinite rigid piston with a velocity amplitude given by the average square velocity $\langle |\dot{w}_{mn}|^2 \rangle$. This follows from the fact that for a plane wave, $p = \rho_0 c \dot{w}$, and thus $I = \frac{1}{2}\rho_0 c |\dot{w}|^2$ and $\Pi_0 = \int I dA = \frac{1}{2}\rho_0 c L_x L_y \langle |\dot{w}|^2 \rangle$. Thus it follows, if S_{mn} is the radiation efficiency of mode (m,n), that

$$S_{mn} = \frac{R_{mn}}{\rho_0 c L_x L_y} = \frac{\Pi}{\frac{1}{2}\rho_0 c L_x L_y \langle |\dot{w}_{mn}|^2 \rangle}. \tag{2.173}$$

Radiation efficiency eliminates the dependence on the panel size. A radiation efficiency of unity is considered 100% efficient (compared with a plane wave).

For a single normal mode of plate vibration $\Phi_{mn}(x,y)$ given in Eq. (2.162) the velocity of the plate is, from Eq. (2.167),

$$\dot{w}_{mn}(x,y) = -i\omega A_{mn}\Phi_{mn}(x,y), \tag{2.174}$$

and the spatial average of the square of the velocity is

$$\left\langle |\dot{w}_{mn}|^2 \right\rangle = \omega^2 \left\langle |A_{mn}\Phi_{mn}|^2 \right\rangle = \frac{\omega^2 |A_{mn}|^2}{L_x L_y},$$

since the mode is orthonormal. The radiation efficiency is then

$$S_{mn} = \frac{\Pi}{\frac{1}{2}\omega^2 |A_{mn}|^2 \rho_0 c}, \tag{2.175}$$

where Π is the power radiated from the mode.

We now compute the power radiated by a normal mode. We will approach this computation by determining the farfield pressure of a mode and using the following formula for the power:

$$\Pi(\omega) = \int_0^{2\pi} \int_0^{\pi/2} \frac{|p(r,\theta,\phi,\omega)|^2}{2\rho_0 c} r^2 \sin\theta \, d\theta \, d\phi. \tag{2.176}$$

Note the integrand is simply the power per unit area through a hemisphere of radius r.

To determine the farfield pressure we return to Eq. (2.84), with $\dot{W}(k_x, k_y)$ the Fourier transform of the modal velocity, $k_x = k \sin\theta \cos\phi$, and $k_y = k \sin\theta \sin\phi$. Thus we must evaluate the Fourier transform of Eq. (2.174):

$$\dot{W}(k_x, k_y) = -i\omega \mathcal{F}_x \mathcal{F}_y [A_{mn}\Phi_{mn}(x,y)].$$

Using Eq. (2.162) and taking Fourier transforms yields

$$\dot{W}(k_x, k_y) = -i\omega 4mn\pi^2 \sqrt{L_x L_y} A_{mn} \left[\frac{(-1)^m e^{-ik_x L_x} - 1}{(k_x L_x)^2 - (m\pi)^2} \right] \left[\frac{(-1)^n e^{-ik_y L_y} - 1}{(k_y L_y)^2 - (n\pi)^2} \right], \tag{2.177}$$

so that the farfield pressure becomes (remember that k_x and k_y are functions of θ and ϕ)

$$p(r,\theta,\phi) = -\omega^2 \rho_0 \frac{e^{ikr}}{r} 2mn\pi \sqrt{L_x L_y} A_{mn} \left[\frac{(-1)^m e^{-ik_x L_x} - 1}{(k_x L_x)^2 - (m\pi)^2}\right] \left[\frac{(-1)^n e^{-ik_y L_y} - 1}{(k_y L_y)^2 - (n\pi)^2}\right].$$
(2.178)

With a bit of algebra we obtain the radiation efficiency of a mode in an infinite baffle:

$$S_{mn} = \frac{64k^2 L_x L_y}{\pi^6 m^2 n^2} \int_0^{2\pi} \int_0^{\pi/2} \left(\frac{\{^{\cos}_{\sin}\}(\frac{k_x L_x}{2})\{^{\cos}_{\sin}\}(\frac{k_y L_y}{2})}{[(k_x L_x/m\pi)^2 - 1][(k_y L_y/n\pi)^2 - 1]}\right)^2 \sin\theta\, d\theta\, d\phi.$$
(2.179)

In this expression cosine is used when the integer m or n is odd, and sine when it is even. This expression can not be simplified any further and we must turn to computer evaluation. Calculations by Wallace[17] are shown in Figs 2.27 and 2.28 for the radiation efficiency of various normal modes.

First consider the radiation efficiency of the $(m,n) = (1,1)$ mode whose shape is shown on the top left of Fig. 2.26. Its radiation efficiency is shown in Fig. 2.27 plotted against γ where, using Eq. (2.163) (k_f is the free bending wavenumber),

$$\gamma = k/\sqrt{(m\pi/L_x)^2 + (n\pi/L_y)^2} = k/k_f.$$
(2.180)

Thus γ is a measure of the coincidence frequency: when $\gamma < 1$ the mode is below coincidence and when $\gamma > 1$ it is above. The latter condition implies strong radiation. The radiation efficiency in Fig. 2.27 confirms this. We can see that all the modes reach 100% efficiency at and above coincidence. In fact, close to $\gamma = 1$ the efficiency is greater than 1, reaching values of 2 and sometimes 3.

To understand how the plate reaches efficiencies greater than one, we consider the radiation impedance of an infinite plate. Equation (2.45) in Section 2.7 indicates that the pressure above an infinite plate vibrating in a normal mode is infinite at the coincidence frequency, $k_{z0} = 0$. An infinite pressure implies that the radiation resistance must also be infinite at coincidence; this results from the fact that the surface velocity is finite (and non-zero), so that when the plate pushes against this infinite resistance an infinite pressure is produced. Of course, in a physical experiment the plate would be loaded by this resistance and its velocity would be diminished as a result. Under the same conditions the radiation efficiency of a *finite* plate is no longer infinite, but still reaches values greater than one as reflected in Fig. 2.27. Put in other terms, the radiation impedance used in the denominator of Eq. (2.175) is finite, always given by $\rho_0 c$, so that when the radiation impedance of the actual mode is greater than this, efficiencies larger than unity occur.

The frequency region below coincidence is very important and an extremely interesting region. This is especially true for plates in water, because the coincidence frequency is much higher and thus this region (see Fig. 2.25) covers a broader range of frequencies. It becomes clear from studying Figs 2.27 and 2.28 (from Wallace) that at low frequencies

[17]C. E. Wallace (1972). "Radiation Resistance of a Rectangular Panel", J. Acoust. Soc. Am., **51**, pp. 946–952.

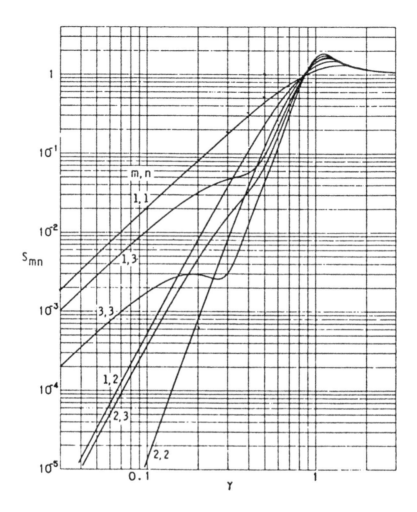

Figure 2.27: Radiation efficiency for the low-numbered modes of a square plate (from Wallace).

the efficiencies are quite different for different modes. We now attempt to explain in some depth the nature of the radiation efficiencies in this region.

We turn to some pioneering work by Gideon Maidanik.[18] In this work he classified the modes of a simply supported baffled panel as edge, corner and surface modes, depending on where the wavenumbers of the modes fall in k-space. The surface mode (supersonic condition) occurs when the wavenumbers of the mode fall inside the radiation circle, $k_x = m\pi/L_x < k$ and $k_y = n\pi/L_y < k$; an edge mode when one of these wavenumbers is greater than k; and a corner mode when both these wavenumbers are greater than k. The latter two represent subsonic wave cases. Figure 2.29 illustrates

---

[18]G. Maidanik (1962). "Response of Ribbed Panels to Reverberant Acoustic Fields", J. Acoust. Soc. Am., **34**, pp. 809–826.

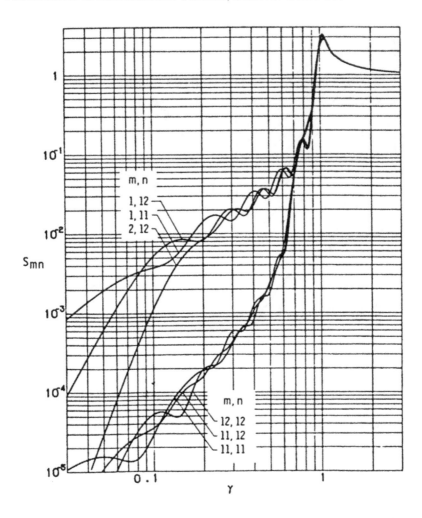

Figure 2.28: Radiation efficiency for the typical high-numbered modes of a square plate (from Wallace).

the definitions.

To understand the physical significance of this classification scheme we must consider the mutual interaction of adjacent regions of surface velocity for a normal mode. Consider Fig. 2.30. This figure represents the Φ_{36} mode. We want to determine the mutual effects of two adjacent cells as shown on the right in the figure. Assume that these two cells are isolated in an infinite baffle and that they are small in dimensions compared to an acoustic wavelength. By the first product theorem from Section 2.11.6 we can replace these cells by two point sources at the centers; the farfield is the product of the farfield of the point sources and the directivity pattern of one of the cells. Since the cell is assumed small compared with a wavelength, the directivity pattern of a cell is nearly omnidirectional and can be ignored.

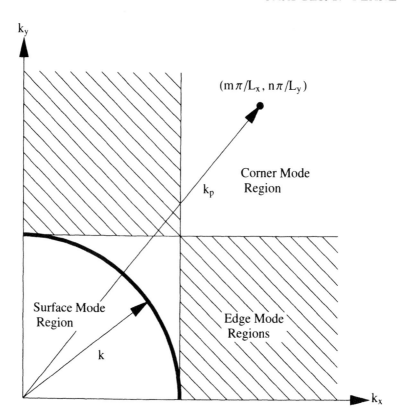

Figure 2.29: Radiation classification of normal modes of a simply supported plate.

The power radiated into the half-space by one of these point sources from Eq. (2.130) $(Q_h = 1)$ is just

$$\Pi_1 = \frac{\rho_0 c k^2}{4\pi}.$$

The power radiated by both can easily be computed using Eq. (2.114) with

$$\dot{w}(x,y,0) = \delta(x)\delta(y-b) \pm \delta(x)\delta(y), \tag{2.181}$$

where the positive cell is located a distance b above the negative cell (minus sign in the second term). The case where both cells are positive is included for completeness and is represented by the positive instead of the negative sign in the second term. The double integral over S' in Eq. (2.114) is

$$\iint_{S'} \dot{w}(x',y',0)\frac{\sin(kR)}{kR}dx'dy' = \frac{\sin(k\sqrt{x^2+(y-b)^2})}{\sqrt{x^2+(y-b)^2}} \pm \frac{\sin(k\sqrt{x^2+y^2})}{\sqrt{x^2+y^2}}.$$

The double integral over S in Eq. (2.114), using the results above, becomes

$$\Pi(\omega) = \frac{\rho_0 c k^2}{4\pi}\left[2(1 \pm \frac{\sin(kb)}{kb})\right] = 2\Pi_1(1 \pm \frac{\sin(kb)}{kb}). \tag{2.182}$$

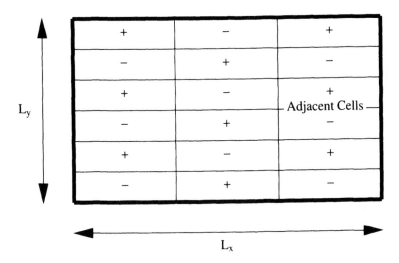

Figure 2.30: Surface velocity for a $m = 3$, $n = 6$ normal mode. Adjacent cells shown on the right used to compute the mutual radiation impedance between them.

Although not relevant to this discussion we note that when the two point sources are in phase with each other the power radiated is four times that of one of them, if they are located very close to one another ($kb << 1$). This is a well known result and is due to the increase in the radiation resistance from the presence of the second source. However, if the two sources are separated by more than an acoustic wavelength the sinc term is small and the total power radiated is increased only by a factor of 2. The two sources do not "see" each other, and radiate independently.

We see a similar circumstance with the adjoining cells of opposing sign (using the minus sign in the second term of Eq. (2.182)); when they are far apart, the power is also increased by a factor of 2. When the sources are $\lambda/4$ apart the power radiated, however, is *reduced* by 36%. At $\lambda/8$ the power is down to 10%. As b tends towards zero the power radiated goes to zero. This is a statement of the hydrodynamic short circuit which occurs when two sources vibrate out of phase close to one another. We now apply this argument to various normal modes of the simply supported plate to derive the radiation classification scheme.

Consider first the effect of adjacent cell cancellation in a mode vibrating at a frequency so that the wavenumbers of the mode are within the region labeled as a corner mode in Fig. 2.29. In this case the separation distance between the centers of adjacent cells in both the x and y directions is less than $\lambda/2$ since $k_\rho > k$. In the limit (separations much less than a wavelength) the result of the cancellation of adjacent cells is to leave four regions at the corners which have no corresponding sources to cancel. This is shown in Fig. 2.31. Of course, as the frequency increases for the same mode shape the cancellation across nodal lines becomes less complete. The power radiated by this mode is similar to the power radiated from four point sources located at the corners of the plate. Due to cancellations the mode radiates inefficiently.

In the next figure, Fig. 2.32, we illustrate the case of an edge mode. Reference to

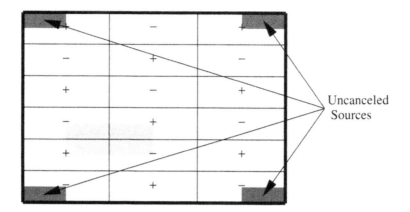

Figure 2.31: Example of a corner mode. All adjacent regions (one shown in light gray) cancel. Only the four corner regions are left uncancelled.

Fig. 2.29 indicates that this is a mode with adjacent cells in one direction separated by less than a half wavelength, whereas the cells in the other direction are separated by more than a half wavelength. Thus adjacent cells in the latter direction do not cancel one another. In this example the wavelength in the vertical direction is much larger

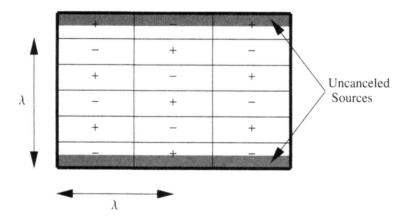

Figure 2.32: Example of an edge mode. Adjacent vertical regions cancel one another, but the horizontal regions on the edge no longer cancel leaving two strips uncancelled.

than the separation between nodal lines. This is not the case in the horizontal direction, where $\lambda < \lambda_m$, $\lambda_m = 2\pi/k_m$. Thus cancellation occurs vertically but not horizontally. The power radiated from this mode is similar to the power radiated by two baffled horizontal strips. This radiator is more efficient than the edge mode due to the larger radiating area of the strips versus the four corners.

The third kind of radiation classification is called the surface mode. In this case no adjacent regions cancel and the whole area of the plate radiates to the farfield. The mode wavelengths in each direction are each greater than the fluid wavelength, and each cell radiates independently.

Now we return to the corner mode and consider the low frequency limit. When the largest dimension of the plate is less than $\lambda/2$, all eigenmodes are corner modes. We must consider the fact that the distance between the corner regions, in the low frequency limit, is much less than a wavelength, and thus they are coupled through their mutual radiation resistances, an effect we studied above in reference to Eq. (2.182). There are three cases to consider which depend on the evenness or oddness of the mode. These cases are illustrated in Fig. 2.33. The odd-odd mode combination, illustrated

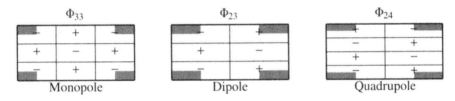

Figure 2.33: Low frequency limits for the radiation classification of normal modes.

with a (3,3) mode in the figure results in four positive corner regions which interfere constructively to produce monopole-like radiation which has

$$\Pi(\omega) \propto k^2. \tag{2.183}$$

The even-odd (or odd-even) mode combination shown in the center figure is two horizontal dipoles adding constructively in the vertical direction and thus have

$$\Pi(\omega) \propto k^4. \tag{2.184}$$

In this case note that $1 - \sin(kb)/kb \approx (kb)^2/6$ so that Eq. (2.182) leads to a k^4 dependency. The last case, with an even-even mode combination, is two dipoles placed in opposition to one another, destructively interfering in both directions. This combination is known as a quadrupole and its frequency dependence is

$$\Pi(\omega) \propto k^6 \tag{2.185}$$

and thus is the least efficient of the three combinations. In view of Eq. (2.182), the product of two factors like $1 - \sin(kb)/kb$ occurs and results in the product of two k^2 dependencies, one for each coordinate direction leading to a k^6 dependency.

We can now return to Figs 2.27 and 2.28 and view the radiation efficiencies in the light of what we have just learned. For example, in Fig. 2.27 we can clearly see that for $\gamma < 0.1$, the radiation efficiencies of the (1,1), (1,3) and (3,3) monopole modes are the highest; the (1,2) and (2,3) dipole modes are next; and the (2,2) quadrupole mode is the least efficient. The slopes of the curves indicate the k dependence.

In the middle frequency region below the coincidence frequency we see that since the panel is square, then the (1,3) will correspond to an edge mode, whereas the (3,3) mode must be a corner mode. The efficiency curves verify the fact that the edge mode is more efficient than the corner mode (by about a factor of ten in this case). The same holds true for the modes shown in Fig. 2.28. Here the (11,11), (11,12) and (12,12) modes are corner modes (since the panel is square), and the (1,11), (1,12) and (2,12) modes are edge modes. In the region $0.1 < \gamma < 0.6$ the edge modes are nearly 100 times more efficient. In the region $\gamma < 0.1$ the (1,11),(1,12) and (2,12) modes reach their low frequency limits of monopole, dipole and quadrupole, respectively.

Returning to the point-driven, simply supported plate and having gained an understanding of how the individual normal modes radiate, we apply these ideas to the general solution, Eq. (2.168), which demands a sum over all the normal modes. Figure 2.34 illustrates the spectrum of modes for the point-driven case. The array of dots

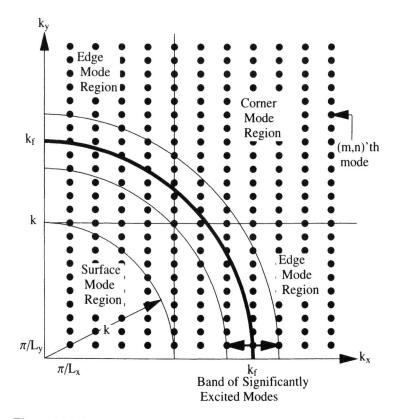

Figure 2.34: k-space diagram with eigenmodes displayed as dots. The circular ring is a region in which the amplitudes of the excited modes are the largest. The plate is excited at the frequency ω, $k = \omega/c$ and $k < k_f$.

represent the individual modes. A quarter circle representing the resonance condition, $\omega = \omega_{mn}$, is shown which has a radius given by the flexural wavenumber, $k_f = \sqrt{\omega}/\alpha$. This radius corresponding to the condition in which the denominator of Eq. (2.168) vanishes. A second quarter circle with radius $k = \omega/c$ is shown. The frequency is chosen so that the resonance modes of the plate are subsonic. The annulus region shown, drawn with somewhat arbitrary width, represents the modes which are strongly excited, since their resonance frequencies are close to the resonance circle. As Eq. (2.168) implies, all modes are excited as long as $\Phi_{mn}(x_0, y_0)$ is non-zero (which can be obtained if the point drive is very close to a corner of the plate). Figure 2.34 indicates that all three radiation classes exist at the same time. Each dot in the figure represents an excited mode. Thus corner, edge and surface modes are all excited. It is a curious, and perhaps remarkable, fact that the low order modes (the surface modes) are excited above their individual coincidence frequencies ($\gamma > 1$ so they are very efficient radiators). However, their amplitudes of excitation are very small due to $\omega^2 - \omega_{mn}^2$ in the denominator. On the other hand the resonant modes are strongly excited since $\omega^2 - \omega_{mn}^2$ is close to zero, but are very inefficient radiators.

Finally, we note that if damping is added to the plate (ω_{mn} becomes complex) the resonance amplitudes are no longer infinite and the relative amplitudes of the low order surface modes are greater. How much of a role they actually play in the radiation from the plate remains an interesting and unanswered question.

2.15 Supersonic Intensity

Supersonic intensity is an outgrowth of Fourier acoustics with tremendous utility for sound source localization. It was first introduced in 1996 in applications to source localization on cylinders.[19] Supersonic intensity is a powerful tool, yet simple in its derivation, which we present in this section.

The supersonic intensity is built from the supersonic plane wave components of the velocity and pressure, the latter defined by

$$p^{(s)}(x, y, z) \equiv \frac{1}{4\pi^2} \iint_{S_r} P(k_x, k_y, z) e^{ik_x x} e^{ik_y y} dk_x dk_y, \qquad (2.186)$$

where S_r is the area in the radiation circle; the integration is over values of k_x and k_y such that $k_x^2 + k_y^2 \leq k^2$. The superscript s indicates a supersonic quantity. Similarly the supersonic normal surface velocity is

$$\dot{w}^{(s)}(x, y, z) \equiv \frac{1}{4\pi^2} \iint_{S_r} \dot{W}(k_x, k_y, z) e^{ik_x x} e^{ik_y y} dk_x dk_y. \qquad (2.187)$$

The supersonic intensity (normal component) is defined in the same way as the total normal intensity, Eq. (2.16):

$$\Pi^{(s)} \equiv I^{(s)}(x, y, z) \equiv \frac{1}{2} \text{Re}[p^{(s)}(x, y, z) \dot{w}^{(s)}(x, y, z)^*]. \qquad (2.188)$$

Eliminating the subsonic plane wave components from the intensity, eliminates the circulation of power flow which arises from the beating of the subsonic and supersonic plane wave components near vibrating structures with subsonic flexural waves.

The credibility of the concept of supersonic intensity lies in the fact that power is conserved. That is the total (real) power passing through the plane at $z =$constant is

---

[19] Earl G. Williams (1995). "Supersonic acoustic intensity", J. Acoust. Soc. Am., **97**, pp. 121–127.

identical to the supersonic power which passes through that plane. That is,

$$\iint_{-\infty}^{\infty} I^{(s)}(x, y, z_0)\, dx\, dy = \iint_{-\infty}^{\infty} \frac{1}{2}\mathrm{Re}[p(x, y, z_0)\dot{w}(x, y, z_0)]\, dx\, dy, \qquad (2.189)$$

where p and $\dot{w}$ are the nonfiltered fields.

The proof of this is quite simple. Expanding $p^{(s)}$ and $\dot{w}^{(s)}$ in their Fourier transforms, the left hand side of Eq. (2.189) becomes

$$\Pi^{(s)}(\omega) = \frac{1}{2}\mathrm{Re}\left[\frac{1}{(4\pi^2)^2}\iint_{-\infty}^{\infty}\iint_{S_r}\iint_{S'_r} P(k_x, k_y, z)\dot{W}(k'_x, k'_y, z)^* \right.$$
$$\left. \times e^{i(k_x - k'_x)x}e^{i(k_y - k'_y)y}dx\, dy\, dk_x dk_y dk'_x dk'_y\right].$$

The integral over x and y yields $4\pi^2\delta(k_x - k'_x)\delta(k_y - k'_y)$ (see Eq. (1.36)) so that the right hand side collapses to a double integral. From Eq. (2.61) we have

$$P(k_x, k_y, z) = \frac{\rho_0 c k}{k_z}\dot{W}(k_x, k_y, z),$$

so that

$$\Pi^{(s)}(\omega) = \frac{1}{8\pi^2}\iint_{S_r}\mathrm{Re}\left[\frac{\rho_0 c k}{k_z}\right]|\dot{W}(k_x, k_y, z)|^2 dk_x dk_y.$$

This result is identical to Eq. (2.113), an expression for the total real power passing through the (x, y) plane. Thus $\Pi^{(s)}(\omega) = \Pi(\omega)$ and power is conserved. To show the utility of the supersonic intensity for source localization, we turn to an example.

2.15.1   Supersonic Intensity for a Point Source

The concept of supersonic intensity, and its ability to locate the regions on a structure which radiate to the farfield, is clarified by considering a point source in an infinite baffle. Let the source plane be located at $z = 0$. The source has a strength given by its volume flow Q_h of Eq. (2.76) which in polar coordinates must be

$$\dot{w}(\rho) = \frac{Q_h}{2\pi}\frac{\delta(\rho)}{\rho}, \qquad (2.190)$$

since $Q_h \equiv \iint \dot{w}(x, y)\, dx\, dy$. From Eq. (2.85) for an axisymmetric source we have

$$\dot{W}(k_x, k_y) = 2\pi\mathcal{B}[\dot{w}(\rho)] = Q_h. \qquad (2.191)$$

Following the definition, Eq. (2.187), the supersonic surface velocity is

$$\dot{w}^{(s)}(x, y) = \frac{Q_h}{4\pi^2}\iint_{S_r} e^{ik_x x}e^{ik_y y}dk_x dk_y.$$

Transforming to polar coordinates and using Eq. (2.89) yields

$$\dot{w}^{(s)}(x, y) = \frac{Q_h}{2\pi}\int_0^k J_0(k_\rho\rho)k_\rho dk_\rho = \frac{kQ_h}{2\pi}\frac{J_1(k\rho)}{\rho}, \qquad (2.192)$$

where we have used the indefinite integral relation

$$\int J_0(x)x\,dx = xJ_1(x). \tag{2.193}$$

Similarly, following the definition of the supersonic pressure, Eq. (2.186), in the plane $z = 0$, along with Eq. (2.61) on page 34 with $z = z'$, we have

$$p^{(s)}(x,y) = \frac{1}{4\pi^2}\iint_{S_r} P(k_x,k_y)e^{ik_x x}e^{ik_y y}dk_x dk_y$$

$$= \frac{\rho_0 ck}{4\pi^2}\iint_{S_r} \frac{\dot{W}(k_x,k_y)}{k_z}e^{ik_x x}e^{ik_y y}dk_x dk_y.$$

Again transforming to polar coordinates and using Eq. (2.89) we find

$$p^{(s)}(x,y) = \frac{Q_h\rho_0 ck}{2\pi}\int_0^k \frac{J_0(k_\rho\rho)}{\sqrt{k^2 - k_\rho^2}}k_\rho dk_\rho.$$

The integral is given in tables:[20]

$$\int_0^1 \frac{J_0(\gamma x)}{\sqrt{1-x^2}}x\,dx = \frac{\sin\gamma x}{\gamma x},$$

so that finally

$$p^{(s)}(x,y) = \frac{Q_h\rho_0 ck^2}{2\pi}\frac{\sin k\rho}{k\rho}. \tag{2.194}$$

The supersonic pressure follows a sinc function. Following Eq. (2.188) we have for the normal supersonic intensity,

$$I^{(s)}(x,y) = \frac{Q_h^2\rho_0 ck^2}{8\pi^2}\frac{J_1(k\rho)\sin(k\rho)}{\rho^2}. \tag{2.195}$$

Figure 2.35 is a plot of $J_1(k\rho)\sin(k\rho)/(k\rho)^2$, the spatial variation of the supersonic intensity over the plane. Note that there are almost no negative values and the side lobes are small in level. To show the side lobe levels better, Fig. 2.36 is a dB plot. The first major side lobe is 16 dB below the main peak. The small sidelobes between the larger ones are the small regions where the intensity is negative.

Figure 2.36 clarifies the assertion that the supersonic intensity localizes the sources on a vibrating structure, identifying the location of the "hot spots" which radiate to the farfield. Even though $I^{(s)}$ spreads over the whole plane, it is mainly confined to an area $\lambda/2$ on either side of the actual location of the point source. The side lobes are an inevitable phenomena of the sharp cutoff at $k_\rho = k$, which produces ringing in real space. Certainly, using a taper on this cutoff, a technique which we will study in detail

---

[20] I. S. Gradshteyn and I. M. Ryzhik (1965). *Tables of integrals, series and products*, Academic Press, New York and London.

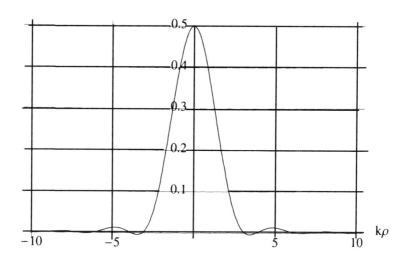

Figure 2.35: Plot of $\frac{J_1(k\rho)\sin(k\rho)}{(k\rho)^2}$. Note the regions of negative intensity are almost nonexistent, and the side lobes are small.

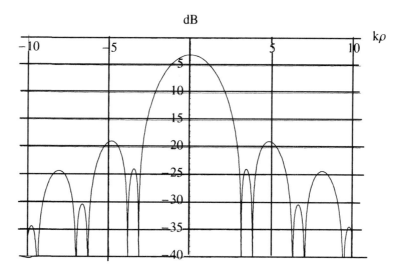

Figure 2.36: Logarithmic plot of Fig. 2.35: $10\log_{10}|\frac{J_1(k\rho)\sin(k\rho)}{(k\rho)^2}|$ to show more clearly the side lobe levels.

in Chapter 3, would suppress the side lobe levels even more. We will not, however, pursue this any further here.

To demonstrate that the total power is conserved we calculate

$$\Pi^{(s)} = \frac{Q_h^2\rho_0 ck^2}{8\pi^2}\int_0^\infty\int_0^{2\pi}\frac{J_1(k\rho)\sin(k\rho)}{\rho^2}\rho\,d\rho\,d\phi = \frac{Q_h^2\rho_0 ck^2}{4\pi},$$

having used the relation[21]

$$\int_0^\infty \frac{J_1(k\rho)\sin(k\rho)}{\rho} d\rho = 1.$$ (2.196)

This is identical to Eq. (2.130) and thus power is conserved. We now consider an example for a vibrating plate.

2.15.2 Supersonic Intensity of a Mode of a Simply Supported Plate

Consider a normal mode of a simply-supported, baffled square plate of dimensions (L_x, L_y). The modes are given by Eq. (2.162) and for this example we choose $m = 11$, $n = 9$, and $L_x = L_y = 2$. Figure 2.37 shows the mode shape using a density plot, with white and black indicating maximum positive and negative values, respectively. The region of the baffle is shown surrounding the plate. The legend indicates the cor-

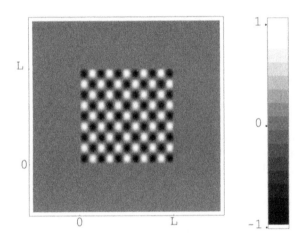

Figure 2.37: Mode shape shown in grey scale for a $m = 11$, $n = 9$ mode of a square plate, $\Psi_{mn}(x,y) = \sin(11\pi x/L_x)\sin(9\pi y/L_y)$. $k = 6$ and $L = L_x = L_y = 2$. The baffle is shown surrounding the plate.

responding levels of vibration. We assume that this mode is forced into excitation at a frequency such that $k = 6$. The actual eigenfrequency for this mode is irrelevant since we are not considering any of the elastic details and we are interested only in the radiation. However, this mode corresponds to a free wavenumber, $k_f = 22.3$ (Eq. (2.163)), and $\gamma = k/k_f = 0.27$ (Eq. (2.180)). Thus the plate is excited below coincidence and according to the radiation classification of Fig. 2.29 a corner mode exists.

The pressure on the plate and baffle is computed using Rayleigh's integral, Eq. (2.61). Fourier transforms of the resulting pressure and specified velocity (mode shape) provide the integrands for Eqs (2.186) and (2.187). These equations are then used to compute the supersonic intensity, using Eq. (2.188). for this mode. The result is shown in

---

[21] Gradshteyn and Ryzhik, *Tables of integrals, series and products.*

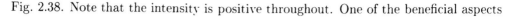

Fig. 2.38. Note that the intensity is positive throughout. One of the beneficial aspects

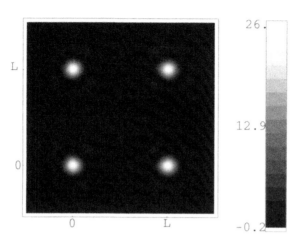

Figure 2.38: Supersonic intensity for the normal mode shown in Fig. 2.37. The intensity is only positive, with black near zero level. White indicates maximum level and locates the regions of the plate which radiate to the farfield.

of the supersonic intensity is the removal of the circulating power flow. Furthermore, the largest levels of intensity (watts/m^2), shown in white, localize the regions on the plate which radiate to the farfield. Note that the sources are the four corner regions of the plate, perfectly consistent with the theory of mode classification presented in Section 2.14.2, in which this mode is classified as a corner mode, depicted in Fig. 2.31.

The power radiated by each identified source is easily obtained by computing the surface integral over the source region, so that if S_s is the area identified as a source (such as one of the corner regions in Fig. 2.38, then the power radiated from the source is

$$\Pi_s = \iint_{S_s} I^{(s)}(x,y)dx\,dy.$$

One can divide the plate into radiating regions. The conservation of power, Eq. (2.191), guarantees that the sum of all the powers from all the regions must equal the actual power radiated.

As a point of comparison, Fig. 2.39 is a plot of the actual normal intensity on the surface, which includes all the subsonic waves. We note that it is positive and negative throughout, which arises from the beating of subsonic and supersonic plane waves and indicates circulation of the intensity vector. The actual source regions which radiate to the farfield are not as evident, although the four corners have the largest levels.

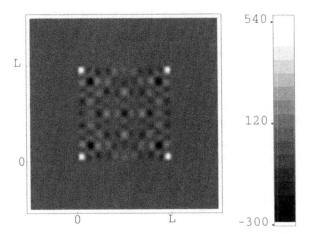

Figure 2.39: Normal acoustic intensity for the normal mode shown in Fig. 2.37. The intensity is both positive (white) and negative (black), indicating circulating intensity flow. The source regions are not evident.

Problems

2.1 A baffled ring is vibrating with constant velocity w_0 over its surface. Given that the inner radius is a and the outer radius is b find the formula for the farfield directivity pattern, $D(\theta, \phi)$.

2.2 An infinite plate in the $z = 0$ plane in an infinite half-space is vibrating with a normal surface velocity, $\eta(x, y) = \eta_0 \sin(3\pi x/L) \sin(6\pi y/L)$.

    (a) In terms of L, what is the distance between the nodal lines in the x and y directions.

    (b) Solve for the pressure field in the half space $z \geq 0$.

    (c) What does the farfield pressure look like if $k = 8\pi/L$?

    (d) What does the farfield pressure look like if $k = 3\pi/L$?

2.3 Consider a baffled, rectangular piston of dimensions L_x and L_y in the x and y directions, respectively. The velocity on its surface is $\dot{w}_0$.

    (a) Write the equation for the k-space surface velocity, $\dot{W}(k_x, k_y, 0)$.

    (b) Write the equation for the directivity function, $D(\theta, \phi)$.

    (c) Use the Ewald construction along with Fig. 2.8 to sketch out the directivity function for the case $L_x = 3L_y$ at the normalized frequency $kL_x/2 = 15$. Your sketch should make clear the difference in directivity in the x and y directions.

2.4 Let n baffled point sources all of equal strength be spaced at intervals of d along the *negative* x axis with the first source located at the origin.

(a) Find the directivity function, $D(\theta, \phi)$. Hint: Note that

$$(1 + e^{ia} + e^{2ia} + \cdots + e^{(n-1)ia}) = \frac{1 - e^{nia}}{1 - e^{ia}},$$

and your final form should be written mostly in terms of sin functions.

(b) Write an equation for the locations of the maxima of the directivity pattern. At what frequency is the first sidelobe maximum in the x, y plane?

(c) Keeping the distance between the first and last elements constant ($nd = L =$ constant), take the limit as $n \to \infty$ and write the expression for the directivity pattern. The is called a continuous line array.

2.5 The formula for radiated power was given by Eq. (2.113)

$$\Pi(\omega) = \frac{\rho_0 c k}{8 \pi^2} \iint_{S_r} \frac{|\dot{W}(k_x, k_y, 0)|^2}{\sqrt{k^2 - k_x^2 - k_y^2}} dk_x dk_y.$$

Convert this integral to polar coordinates, $k_x, k_y \to k_\rho, \psi$ and use the relationship between the farfield and the k-space velocity to show that

$$\Pi(\omega) = \frac{1}{2\rho_0 c} \iint |rp(r, \theta, \phi)|^2 \sin \theta \, d\theta \, d\phi.$$

2.6 Evaluate the differentiation with respect to z in

$$\frac{\partial}{\partial z} \left[\frac{e^{ik|\vec{r} - \vec{r}'|}}{|\vec{r} - \vec{r}'|} \right]$$

where

$$|\vec{r} - \vec{r}'| = \sqrt{(x - x')^2 + (y - y')^2 + (z - z')^2}.$$

What difference do you obtain if you differentiate with respect to z' instead of z?

2.7 Equation (2.67) presented Rayleigh's second integral formula which provides a means of computing the pressure given the pressure in a different plane. That is

$$p(x, y, z) = -\frac{1}{2\pi} \int_{-\infty}^{\infty} \int_{-\infty}^{\infty} p(x', y', 0) \frac{\partial}{\partial z} \left[\frac{e^{ik|\vec{r} - \vec{r}'|}}{|\vec{r} - \vec{r}'|} \right] dx' dy'. \tag{2.197}$$

His first formula, Eq. (2.75) provided a means to compute the pressure given the normal velocity. However, he did not present a formula to compute the normal velocity. Using Eq. (2.197) and Euler's equation, derive an integral formula for $\dot{w}(x, y, z)$.

2.8 Given the following formula,

$$-2\pi \frac{\partial p(x, y, z)}{\partial z} = \int_{-\infty}^{\infty} \int_{-\infty}^{\infty} p(x', y', 0) \frac{\partial^2}{\partial z^2} \left[\frac{e^{ik|\vec{r} - \vec{r}'|}}{|\vec{r} - \vec{r}'|} \right] dx' dy',$$

and use the fact that $\frac{e^{ik|\vec{r}-\vec{r}'|}}{|\vec{r}-\vec{r}'|}$ satisfies the *homogeneous* wave equation (Eq. (2.13)) of the notes) when $r \neq r'$, to derive the differential-integral equation given by Bouwkamp:[22]

$$\frac{\partial p(x,y,z)}{\partial z} = \frac{1}{2\pi}\left[k^2 + \frac{\partial^2}{\partial x^2} + \frac{\partial^2}{\partial y^2}\right]\int_{-\infty}^{\infty}\int_{-\infty}^{\infty} p(x',y',0)\left[\frac{e^{ik|\vec{r}-\vec{r}'|}}{|\vec{r}-\vec{r}'|}\right]dx'dy'.$$

2.9 The phase gradient method of acoustic intensity measurement was presented in 1968.[23] The formula for the steady-state intensity presented was

$$\vec{I} = \frac{1}{2\rho_0 ck}|p|^2\vec{\nabla}\Theta_p,$$

where

$$p(x,y,z) = |p|e^{i\Theta_p(x,y,z)}.$$

Derive his formula starting with the definition of intensity

$$\vec{I} = \frac{1}{2}\mathrm{Re}[p\vec{u}^*]$$

using the fact that

$$\vec{\nabla}p(x,y,z) = p(x,y,z)\vec{\nabla}[\log_e(p(x,y,z))].$$

Mechel's formula provides an important fact about the phase of the pressure field. It must increase in the direction of energy flow.

2.10 The wavelengths of a plane/evanescent wave in the x,y,z directions, respectively, are

$$\lambda_x = 1/3, \quad \lambda_y = \infty, \quad \lambda_z = 1/2.$$

(a) What are the corresponding wavenumbers, k_x, k_y, k_z, in x,y,z directions?

(b) What are the trace velocities c_x, c_y, c_z?

(c) Determine the direction of the wave in spherical coordinates.

(d) What is the frequency, f, for this wave in terms of the speed of sound, c?

2.11 Two plane/evanescent waves traveling in an infinite half-space ($z \geq 0$) are given with $k_x = \pm 2k$, and $k_y = 0$, that is,

$$p(x,y,0) = P_0(e^{i2kx} + e^{-i2kx})e^{-i\omega t}.$$

Write down the expression for $p(x,y,z)$.

2.12 Given a baffled point source at (x_0, y_0), that is,

$$\dot{w}(x,y,0) = Q_0\delta(x-x_0)\delta(y-y_0).$$

Write the expression for the pressure in the $z = 0$ plane, $p(x,y,0)$.

---

[22] C. J. Bouwkamp (1954), "Diffraction Theory", Rep. Progr. Phys., **17**, pp. 35–100.
[23] F. P. Mechel (1968). Proceedings of the ICA, Tokyo.

2.13 Given a point source at the origin with pressure field

$$p(x, y, z) = p_0 e^{ikR}/R,$$

where $R = \sqrt{x^2 + y^2 + z^2}$, compute the vector velocity, $\vec{v}(x, y, z)$ on the x axis $(y = z = 0)$.

2.14 The normal intensity in the $z = 0$ plane is given by

$$I_z(x, y, 0) = \frac{\Pi(x/L_x)\Pi(y/L_y)}{x^2 y^2},$$

where $\Pi(x/L_x)$ and $\Pi(y/L_y)$ are rectangular window functions. Find the total power, $\Pi(\omega)$ (unfortunately the same symbol as the rectangular window), crossing the $z = 0$ plane in the $+z$ direction.

2.15 Let the pressure in the $z = 0$ plane be given by

$$p(x, y, 0) = p_0 \delta(x - x_0)\delta(y - y_0).$$

(a) Compute the angular spectrum, $P(k_x, k_y, 0)$.

(b) Find $P(k_x, k_y, z)$, in the infinite, source-free half-space, $z \geq 0$,

(c) Find the angular spectrum of the normal velocity, $\dot{W}(k_x, k_y, z)$.

(d) Make a rough sketch of $|\dot{W}(k_x, k_y, z)|$ in terms of k along the k_x axis $(k_y = 0)$, where $-\infty < k_x < +\infty$. Include rough sketches when $z \to \infty$ and when $z \to 0$.

(e) Compute the power, $\Pi(\omega)$, radiated into the infinite half-space in terms of k.

2.16 Sketch the farfield over a complete hemisphere given the Fourier transform on the right of the function on the left. Note that $\rho = \sqrt{x^2 + y^2}$ and $k_\rho = \sqrt{k_x^2 + k_y^2}$ as usual and κ is a constant. Consider two cases: (a) when the acoustic wavenumber $k = 2\kappa$ and (b) $k = \kappa/2$. Considering the fact that the function on the left represents the normal velocity of an infinite membrane, discuss the meaning of the answer obtained for case (b).

2.17 In the following problem we are going to simulate a nearfield holography measurement of a baffled point source, located at the origin in the $z = 0$ plane and given by $\dot{w}(x, y, 0) = Q_0 \delta(x)\delta(y)$. The simulated pressure measurement is made in the plane $z = z_h$. Assume that the "measurement" is perfect, over an infinite aperture with infinitesimally close measurement points. The measured pressure is

$$p(x, y, z_h) = \frac{-iQ_0\rho_0 ck}{2\pi} \frac{e^{ikR}}{R},$$

with $R = \sqrt{x^2 + y^2 + z_h^2}$. Following the holographic reconstruction equation, Eq. (3.4),

(a) determine the Fourier transform of the pressure, $P(k_x, k_y, z_h)$.

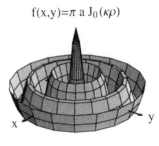

f(x,y)=π a J₀(κρ)

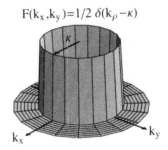

F(k$_x$,k$_y$)=1/2 δ(k$_ρ$−κ)

Figure 2.40: A Bessel function and its Fourier transform.

(b) multiply by the inverse velocity propagator, G, and

(c) apply a rectangular k-space window to the "data" given by

$$\Pi(k_x/(2k_c))\Pi(k_y/(2k_c))$$

and do the inverse Fourier transforms to arrive at an algebraic expression for the reconstructed velocity, $\tilde{w}(x,y,0)$. You should be able to do the integrations.

(d) Discuss the difference between the reconstructed velocity, $\tilde{w}(x,y,0)$, and the actual velocity, $\dot{w}(x,y,0) = Q_0\delta(x)\delta(y)$.

2.18 (Challenge problem) Starting with the Rayleigh integral,

$$p(\vec{r},\omega) = \frac{-i\omega\rho_0}{2\pi} \int_{-\infty}^{\infty}\int_{-\infty}^{\infty} \dot{w}(\vec{r'},\omega)\frac{e^{i\omega|\vec{r}-\vec{r'}|/c}}{|\vec{r}-\vec{r'}|}dx'dy',$$

where $\vec{r} = (x,y,z)$ and the dependence upon ω has been written explicitly, use Fourier transforms and their associated relationships given in Chapter 1 to derive the time-domain version of Rayleigh's integral given by

$$p(\vec{r},t) = \frac{\rho_0}{2\pi}\frac{\partial}{\partial t}\int_{-\infty}^{\infty}\int_{-\infty}^{\infty}dx'dy'\int_{-\infty}^{\infty}dt'\,\dot{w}(x',y',0,t')\frac{\delta(t-t'-|\vec{r}-\vec{r'}|/c)}{|\vec{r}-\vec{r'}|}.$$

2.19 Show that Eq. (2.162) satisfies Eq. (2.156) and that Eq. (2.163) results.

2.20 Using Eq. (2.164), verify that the normal modes given in Eq. (2.162) are orthonormal.

2.21 Derive the result shown in Eq. (2.177).

2.22 Derive Eq. (2.168) following the three steps specified in the text.

Chapter 3

The Inverse Problem: Planar Nearfield Acoustical Holography

3.1   Introduction

Acoustical holography, the predecessor of NAH, appeared in the mid 1960s.[1] Acoustical holography, however, is only an approximation to the inverse problem of reconstructing sound fields. This inverse problem backtracks the pressure field in space and time towards the sources. Only source details greater that the acoustic wavelength can be retrieved in this procedure. Nearfield acoustical holography, which appeared in 1980,[2] provides a rigorous solution, however, to the inversion resulting in an almost unlimited resolution in the reconstruction. When evanescent waves are present (such as plate vibrator sources which contain subsonic waves), an essential requirement for this increased resolution is the measurement of the sound field very close to the sources of interest. This latter fact leads to prefixture of the term "nearfield" to acoustical holography to arrive at the name NAH.

NAH reconstructs not only the pressure but also the three components of the fluid velocity as well as the acoustic intensity vector. The practical implementation of the theory requires the materials presented in the previous chapters. The regularization of the inverse problem is naturally based in k-space analysis, which we have taken care to present in sufficient detail in Chapter 2 so that the reader will be well positioned to understand NAH.

This chapter deals with planar NAH, the stepping stone to other geometries. Because of its great speed, the implementation of the theory is based on the DFT (discrete Fourier transform) and the FFT (fast Fourier transform).

---

[1] B. P. Hildebrand and B. B. Brenden (1974). *An Introduction to Acoustical Holography.* Plenum Press, New York.

[2] E. G. Williams and J. D. Maynard (1980), "Holographic Imaging without the wavelength resolution limit," Phys. Rev. Lett., **45**, pp. 554–557.

3.2 Overview of the Theory

In Section 2.9 on the angular spectrum we found that given knowledge of the pressure on a plane we could determine the pressure and vector velocity on any other plane in a source-free medium. In other words, if the acoustic sources are confined to the half space $z \leq z_s$, and if the pressure is known on a plane $z = z_h \geq z_s$ then the pressure on any other plane is given (through the angular spectrum) as

$$P(k_x, k_y, z) = P(k_x, k_y, z_h)e^{ik_z(z-z_h)}, \tag{3.1}$$

as was presented in Eq. (2.54). Similarly, the vector velocity was determined by Eq. (2.59) and, in particular, the normal velocity was (Eq. (2.61))

$$\dot{W}(k_x, k_y, z) = \frac{k_z}{\rho_0 ck}e^{ik_z(z-z_h)}P(k_x, k_y, z_h) = G(k_x, k_y, z - z_h)P(k_x, k_y, z_h), \tag{3.2}$$

where

$$G(k_x, k_y, z - z_h) \equiv \frac{k_z}{\rho_0 ck}e^{ik_z(z-z_h)}. \tag{3.3}$$

G is called the velocity propagator. In these formulas z and z_h play critical roles. When $z \geq z_h$ the solution is a forward problem, as provided by the Rayleigh integrals, and G is a forward propagator. However, when $z < z_h$ the solution is an inverse problem. In other words, if the field is measured in the plane $z = z_h$, then the solution for the pressure in a plane closer to the sources determines what the pressure must have been there before it reached the measurement plane. G is then the inverse velocity propagator. This is an inverse problem.

The Rayleigh integrals do not provide an inverse solution. They can only provide the pressure *radiated* from the sources. Nearfield acoustical holography provides a solution to the inverse problem, as we will see below.

Nearfield acoustical holography (NAH) was first proposed by Williams and Maynard in 1980.[3] Its major attraction is its solution to the inverse problem, that is the reconstruction of the surface velocity field on plate radiators from a measurement of the pressure in a parallel plane at a small distance from the plate. Let the plane $z = z_h$ be the measurement plane, and $z = z_s$ be the surface of the vibrator. NAH provides the relationship

$$p(x, y, z_h) \Rightarrow \dot{w}(x, y, z_s),$$

where $z_h \geq z_s$. From what we have learned about the angular spectrum we can formulate the solution very easily. The mathematics behind NAH is summarized in the single statement:

$$\dot{w}(x, y, z_s) = \mathcal{F}_x^{-1}\mathcal{F}_y^{-1}\left[\mathcal{F}_x\mathcal{F}_y[p(x, y, z_h)]G(k_x, k_y, z_s - z_h)\right]. \tag{3.4}$$

Using the convolution theorem, Eq. (1.20) on page 4, then the statement is rewritten as

$$\dot{w}(x, y, z_s) = p(x, y, z_h) * *g_v^{-1}(x, y, z_s - z_h) \tag{3.5}$$

---

[3]Williams and Maynard, "Holographic Imaging without the Wavelength Resolution Limit".

where we have defined the inverse velocity propagator,

$$g_v^{-1}(x, y, z_s - z_h) \equiv \mathcal{F}_x^{-1} \mathcal{F}_y^{-1} \left[G(k_x, k_y, z_s - z_h) \right]$$

$$= \mathcal{F}_x^{-1} \mathcal{F}_y^{-1} \left[\frac{k_z}{\rho_0 c k} e^{ik_z(z_s - z_h)} \right], \tag{3.6}$$

consistent with the definition given in Eq. (2.70).

In words, Eq. (3.4) states:

(1) measure the pressure $\rightarrow p(x, y, z_h)$.

(2) compute its angular spectrum $\rightarrow P(k_x, k_y, z_h)$,

(3) multiply by the inverse propagator $G(k_x, k_y, z_s - z_h) \rightarrow \dot{W}(k_x, k_y, z_s)$,

(4) compute the inverse transforms $\rightarrow \dot{w}(x, y, z_s)$.

Because NAH solves an inverse problem, the mathematical solution must be approached with some caution to assure that the solution is unique and stable.[4] That is, straightforward implementation of Eq. (3.4) will lead to disaster! How this equation must be modified to avoid disaster is the springboard for the NAH technique.

3.3 Presentation of Theory for a One-Dimensional Radiator

To understand all the details for the implementation of NAH for the general planar geometry, we will present the theory assuming that the measured field only depends on x and z and is independent of y. This is called a one-dimensional planar radiator. In no way does this compromise our understanding of the full two-dimensional problem, for we will see that it is trivial to make the extrapolation to this case once we have gained the knowledge from the one-dimensional case.

Given a one-dimensional plate vibrator located in the plane $z = z_s$, the measured pressure in the plane $z = z_h$ where $z_h \geq z_s$ is of the form

$$p(x, y, z_h) \rightarrow p(x, z_h),$$

that is, the pressure is independent of y. Given this pressure measurement, we develop the solution for the normal surface velocity in the source plane at $z = z_s$. First we obtain the angular spectrum so that we can backpropagate the field to the plate surface at $z = z_s$:

$$\mathcal{F}_x \mathcal{F}_y [p(x, z_h)] = 2\pi P(k_x, k_y, z_h) \delta(k_y), \tag{3.7}$$

where the Dirac delta function arises from the transform over y. Because of the delta function the following occurs in the inverse Fourier transform in k_y (carrying out the

---

[4]Some precise mathematical details are given in G. C. Sherman (1967). "Integral-transform formulation of Diffraction", J. Opt. Soc. Am. **55**, pp. 1490–1498.

$\mathcal{F}_y^{-1}$ operation in Eq. (3.4)):

$$\frac{1}{2\pi} \int_{-\infty}^{\infty} P(k_x, k_y, z_h) G(k_x, k_y, z_s - z_h) 2\pi \delta(k_y) e^{ik_y y} dk_y =$$

$$= P(k_x, 0, z_h) G(k_x, 0, z_s - z_h). \tag{3.8}$$

In other words, the one-dimensional vibrator is equivalent to evaluating the angular spectrum and the velocity propagator at $k_y = 0$. This is not surprising since this case corresponds to an infinite wavelength in the y direction. Thus the holographic process given in Eq. (3.4) does not depend on transforms in y and the reconstruction equation becomes

$$\dot{w}(x, z_s) = \mathcal{F}_x^{-1} \Big[\mathcal{F}_x[p(x, z_h)] G(k_x, 0, z_s - z_h) \Big]. \tag{3.9}$$

Of particular importance in Eq. (3.9) is the inverse velocity propagator G (from Eq. (3.3)):

$$G(k_x, 0, z_s - z_h) = \frac{k_z}{\rho_0 ck} e^{-ik_z(z_h - z_s)}. \tag{3.10}$$

Here we are presented with our first difficultly, caused by the exponential in this equation. Since $k_y = 0$ for the one-dimensional problem, k_z is defined by $k_z = \sqrt{k^2 - k_x^2}$ within the radiation circle and by $k_z = i\sqrt{k_x^2 - k^2} = ik_z'$ outside of it. In the latter case k_x is subsonic and we have seen in Section 2.8 (Eq. (2.49)) that this leads to exponentially decaying sound fields. However, since we are solving the inverse problem, then we are backpropagating the field, a process which reverses the sign of the exponent producing a rising exponential. In this case the exponential in Eq. (3.10) leads to

$$e^{-ik_z(z_h - z_s)} = e^{k_z'(z_h - z_s)},$$

which multiplies the measured pressure angular spectrum $P(k_x, 0, z_h)$. As k_x increases to infinity the inverse propagator G becomes infinite.

In order for the product $G(k_x, 0, z_s - z_h) P(k_x, 0, z_h)$ to remain finite we must assume that the angular spectrum of the pressure drops off at a faster rate with k_x compared with the rising rate of the propagator. This is in fact guaranteed by the angular spectrum relationship for subsonic waves

$$P(k_x, 0, z_h) = P(k_x, 0, z_s) e^{-k_z'(z_h - z_s)}. \tag{3.11}$$

The exponential decay in this equation balances the increase in the inverse propagator, Eq. (3.10), so that the backpropagation of the sound field is well behaved.

For the experimental problem, however, $p(x, y, z_h)$ is a measured quantity and its angular spectrum is not computed analytically, but through the application of a numerical Fourier transform. As one might imagine, spatial noise in the measurement has a strong likelihood of completely destroying this delicate cancellation process.

3.4 Ill Conditioning Due to Measurement Noise

In order to consider rigorously the effects of noise we will use random process theory applied to spatial random processes.[5] Let $\epsilon(x)$ represent uncorrelated noise introduced from experimental circumstances at the position x. We assume that this noise is spatially uncorrelated (wavenumber white). Let $p(x, 0, z_h)$ be the noiseless pressure signal, so that at location x

$$\tilde{p}(x, 0, z_h) = p(x, 0, z_h) + \epsilon(x), \tag{3.12}$$

where $\tilde{p}$ represents the measured pressure, corrupted by noise.

Assume that an ensemble of N separate array measurements $\tilde{p}_i(x, 0, z_h)$ where $i = 1, \cdots, N$, are made, which we average for each position x in the array to determine the expectation value. Define $\mathcal{E}$ as the expected value across the N experiments,

$$\mathcal{E}[\tilde{p}] \equiv \frac{1}{N} \sum_{i=1}^{N} \tilde{p}_i. \tag{3.13}$$

Assuming $\epsilon(x)$ has zero mean then

$$\mathcal{E}[\tilde{p}] = \mathcal{E}[p] = p,$$

expressing the essence of the technique of signal averaging. The autocorrelation function (covariance) is[6]

$$\mathcal{E}[\epsilon(x)\epsilon^*(x')] = \gamma\delta(x - x'). \tag{3.14}$$

We want to determine how this noise corrupts the angular spectrum of the reconstructed velocity, $\dot{W}(k_x, 0, z_s)$. Let $\tilde{\dot{W}}$ represent the noise corrupted reconstruction. We need to investigate the expected value of its square:

$$\mathcal{E}[|\tilde{\dot{W}}(k_x, 0, z_s)|^2] = \mathcal{E}[|G(k_x, 0, z_s - z_h)|^2 |\tilde{P}(k_x, 0, z_h)|^2] = |G|^2(|P|^2 + \mathcal{E}[|\mathcal{N}|^2], \tag{3.15}$$

where

$$\tilde{P}(k_x, 0, z_h) \equiv P(k_x, 0, z_h) + \mathcal{N}(k_x)$$

and

$$\mathcal{N}(k_x) = \mathcal{F}(\epsilon(x)).$$

If L is the length of the measurement array, then

$$\mathcal{E}[|\mathcal{N}|^2] = \iint e^{ik_x(x'-x)}\mathcal{E}[\epsilon(x)\epsilon^*(x')]dx\,dx' = \gamma \iint e^{ik_x(x'-x)}\delta(x - x')dx\,dx' = L\gamma, \tag{3.16}$$

where we used $\int dx = L$. Finally, Eq. (3.15) becomes

$$\mathcal{E}[|\tilde{\dot{W}}(k_x, 0, z_s)|^2] = |\dot{W}(k_x, 0, z_s)|^2 + L\gamma|G(k_x, 0, z_s - z_h)|^2. \tag{3.17}$$

---

[5] A. Papoulis (1965). *Probability, Random Variables, and Stochastic Processes*. McGraw-Hill, New York.

[6] A. Papoulis (1968). *Systems and Transforms with Application in Optics*. McGraw-Hill, New York.

From Eq. (3.10)

$$|G(k_x, k_y, z_s - z_h)|^2 = \begin{cases} (k_z/\rho_0 ck)^2 & k_x^2 + k_y^2 \leq k^2 \\ (k_z'/\rho_0 ck)^2 e^{2k_z'(z_h - z_s)} & k_x^2 + k_y^2 > k^2. \end{cases}$$

Thus the last term in Eq. (3.17) grows without bound as the wavenumbers increase. Clearly any reconstruction attempt of the actual surface velocity will fail. We now consider how we modify the reconstruction equation to avoid the blow up of $|G|^2$.

3.5   The k-space Filter

To correct this problem NAH imposes a k-space filter on the measured pressure spectrum which limits how far outside the radiation circle one accepts data. The idea is to reduce the high wavenumber content of the sum $|\dot{W}|^2 + L\gamma|G|^2$ by truncating the angular spectrum at a judiciously chosen point outside the radiation circle. Symbolically that operation looks like

$$\tilde{P}(k_x, 0, z_h)\Pi[k_x/(2k_c)]G(k_x, 0, z_s - z_h),$$

where Π is the rectangle window function with a cutoff at $k_x = k_c$ as presented in Section 1.6. The product ΠG remains finite for large k_x and thus is kept in bounds. However, a deleterious effect occurs from the truncation of the actual pressure spectrum represented by $\tilde{P}\Pi$. The proper choice of the break point of the k-space filter is critical to the success of the reconstruction. We will illustrate this with a fairly simple but relevant example.

Consider an example without noise and let $\tilde{w}(x, z_s) \equiv \tilde{w}(x)$ represent the corrupted reconstructed velocity due to the effect of the added filter. Thus the reconstruction equation, Eq. (3.9), is

$$\begin{aligned} \tilde{w}(x) &= \mathcal{F}_x^{-1}\Big[\mathcal{F}_x[p(x, z_h)]G(k_x, 0, z_s - z_h)\Pi[k_x/(2k_c)]\Big] \\ &= \mathcal{F}_x^{-1}\Big[P(k_x, 0, z_h)G(k_x, 0, z_s - z_h)\Pi[k_x/(2k_c)]\Big]. \end{aligned} \quad (3.18)$$

Using the convolution theorem, Eq. (1.12). and the inverse transform of the rectangle function,

$$\frac{1}{2\pi}\int_{-\infty}^{\infty} \Pi[k_x/(2k_c)]e^{ik_x x}dk_x = \frac{k_c}{\pi}\text{sinc}(k_c x), \quad (3.19)$$

yields

$$\tilde{w}(x) = \mathcal{F}_x^{-1}\Big[P(k_x, 0, z_h)G(k_x, 0, z_s - z_h)\Big] * \frac{k_c}{\pi}\text{sinc}(k_c x).$$

The term in the square brackets is recognized as the desired surface velocity, $\dot{w}(x, z_s) \equiv \dot{w}(x)$, so that

$$\tilde{w}(x) = \dot{w}(x) * \frac{k_c}{\pi}\text{sinc}(k_c x). \quad (3.20)$$

We can see that as $k_c \to \infty$ then

$$\frac{k_c}{\pi} \operatorname{sinc}(k_c x) \to \delta(x),$$ (3.21)

and Eq. (3.20) returns $\tilde{w}(x) = \dot{w}(x)$, as it should.

Now we need to make some important observations about the effect of the window on the reconstructed velocity.

- The spatial wavelength corresponding to k_c is $\lambda_c = 2\pi/k_c$. The k-space filter eliminates all wavelengths smaller than this from the reconstruction, and thus no spatial details of the vibrator less than $\lambda_c/2$ can be resolved.

- If the wavenumber content of the surface velocity is limited to components such that $k_x < k_c$, then the effect of the rectangle function in Eq. (3.18) is null, since $P(k_x, 0, z_h)$ is already band limited.

- It is rarely true, however, that a source is wavenumber limited. In this case it is essential that k_c be chosen so that most of the wavenumber content of the source is contained within the window.

We will examine the effect of the window in Eq. (3.20) with some examples.

3.5.1   Examples

Consider a one-dimensional surface velocity field on an infinite baffled plate given by

$$\dot{w}(x) = \sin(3.5\pi x),$$

for $-1 \leq x \leq 1$, and zero outside as shown in Fig. 3.1. The baffle (with zero velocity) extends from $1 < x < \infty$ and from $-\infty < x < -1$. The angular spectrum associated

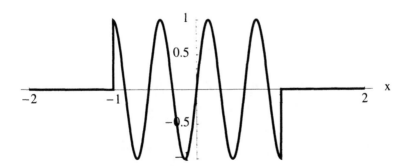

Figure 3.1: One-dimensional velocity distribution $\sin(3.5\pi x)$ with edge discontinuities.

with this velocity field in the source plane,

$$\dot{W}(k_x) = \mathcal{F}_x[\sin(3.5\pi x)],$$

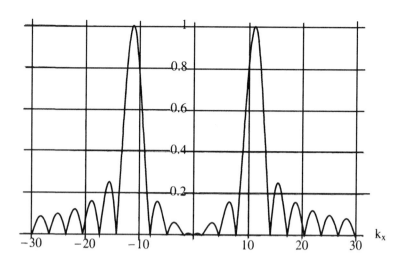

Figure 3.2: Magnitude of angular spectrum of velocity, $|\dot{W}(k_x)|$ for velocity field $\dot{w}(x) = \sin(3.5\pi x)$.

is shown in Fig. 3.2. Note that the spectrum peaks at $k_x = 3.5\pi$ and has the side lobes of a sinc function. These peaks represent the dominant wavenumber on the plate, clearly illustrated in Fig. 3.1. It is logical to conclude, then, that the cutoff for the k-space filter k_c must be chosen to include the two major peaks in the angular spectrum in order to include the important wavenumber information and thus to obtain an accurate reconstruction. For the purposes of illustration we set $k_c = 20$ and use Eq. (3.20) to simulate the reconstruction obtained from this filter. The resulting reconstruction of surface velocity is shown by the dots in Fig. 3.3, superimposed upon the solid curve representing the actual surface velocity. The agreement in the region $-0.9 < x < 0.9$ is excellent. Outside of this region the reconstruction is not as good.

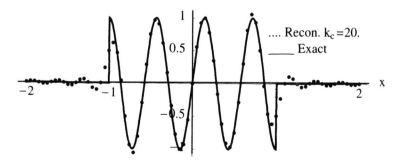

Figure 3.3: Reconstruction versus exact surface velocity for a rectangular k-space filter with cutoff $k_c = 20$.

It is interesting to note that the sharp corners of the velocity field at the ends of the vibrator are smoothed in the reconstruction. Since the edges represent a point of discontinuity (as might occur, for example, at the ends of a free beam) they represent a region of very high spatial wavenumber content. These high values of k_x are lost to the filter, however. The "ringing" of the surface velocity field outside of the vibrator, as indicated by the ripples of low amplitude from $1 < |x| < 2$, is also due to the discontinuity. Fortunately the effect on the reconstruction is localized to the region of the discontinuities.

We present another example, to illustrate the catastrophic effect of choosing a k-space cutoff which excludes the dominant wavenumber on the plate. Using the same vibrator as shown in Fig. 3.1 we set the cutoff. $k_c = 10$. Reference to Fig. 3.2 indicates that this cutoff just misses the peak in the spectrum (at $k_x = 3.5\pi$). The resulting reconstruction of the surface velocity field is shown in Fig. 3.4. It is clear from this figure that the reconstruction has completely failed.

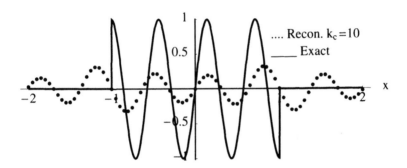

Figure 3.4: Reconstruction with $k_c = 10$ compared with the exact surface velocity field.

3.6 Modification of the Filter Shape

The "ringing" seen in Fig. 3.3 occurs due to the sharp spatial cutoff at the plate edges. In similar fashion one would expect the sharp cutoff of the k-space filter to cause ringing in real space, an effect which may also cause undesirable effects. As a result of this we will find that the reconstruction can be slightly improved by using a tapered window function in k-space, instead of the rectangular one illustrated above. Tapered windows are very popular in signal processing and a fundamental paper on the subject was written by Harris.[7]

The window used in the holographic processing is built upon the exponential in k-space, that is,

$$\bar{\Pi}(k_x/(2k_c)) = \begin{cases} 1 - \frac{1}{2}e^{-(1-|k_x|/k_c)/a} & |k_x| < k_c \\ \frac{1}{2}e^{(1-|k_x|/k_c)/a} & |k_x| > k_c \end{cases} \qquad (3.22)$$

[7]Fredric J. Harris (1978). "On the Use of Windows for Harmonic Analysis with the Discrete Fourier Transform", Proc IEEE, **66**.

where the bar over $\bar{\Pi}$ indicates that it is similar to the rectangular window except for an exponential taper on either side of k_c. Note that the reason for the exponential taper is the rising exponential in Eq. (3.10) for subsonic wavenumbers. Figure 3.5 shows the shape of this window for three typical values of α. α controls the rate of the decay of

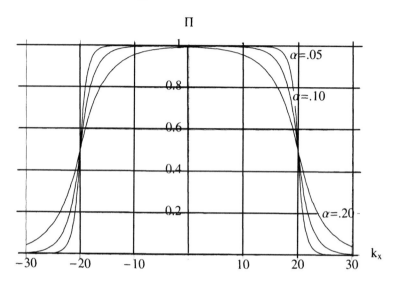

Figure 3.5: Taper for the exponential window used in NAH for $k_c = 20$.

the window and it is clear that

$$\lim_{\alpha \to 0} \bar{\Pi} = \Pi[k_x/(2k_c)],$$

the latter being the rectangular window.

For a value of $\alpha = 0.1$ the velocity reconstruction for a source with the one-dimensional velocity field shown in Fig. 3.1 was repeated for the same case as presented above, $k_c = 20$. The only difference, compared to Fig. 3.3 is the use of the tapered window instead of the rectangular window.

Comparing the result in Fig. 3.6 with Fig. 3.3 indicates a slight improvement near the ends of the vibrator. Note that the ripple outside of the vibrator is slightly diminished when the exponential window is used. Better agreement near the ends can be obtained by increasing k_c which allows higher spatial frequencies in the reconstruction. In practice one is prevented from unlimited increase of k_c due to noise in the measurement.

3.7   Measurement Noise and the Standoff Distance

In the example above there was no restriction on the maximum value chosen for the k-space cutoff for the k-space filter. In practice k_c will be limited by the size of the standoff distance, $z_h - z_s$, and the signal-to-noise ratio of the measured pressure. To derive this relationship we assume that $k_c > k$, which is almost always the case for

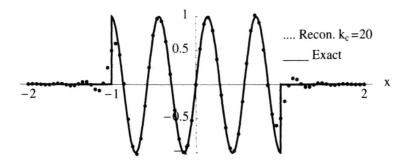

Figure 3.6: Repeat of the case shown in Fig. 3.3 above using the exponential
k-space window with $\alpha = 0.1$ and $k_c = 20$.

practical problems. Then at cutoff, $k_x = k_c$,

$$k'_{zc} = \sqrt{k_c^2 - k^2} \approx k_c.$$

The relationship between the subsonic angular spectrum components in the plane $z = z_h$
and $z = z_s$ was given in Eq. (3.11). If we assume $k'_z >> k$ then this equation becomes

$$P(k_x, 0, z_s) \approx P(k_x, 0, z_h)e^{2\pi(z_h - z_s)/\lambda_x},$$

which is independent of frequency. The greatest exponential increase occurs when $k_x = k_c$ so that

$$P(k_c, 0, z_s) \approx P(k_c, 0, z_h)e^{2\pi(z_h - z_s)/\lambda_c}.$$

If the standoff distance is equal to the cutoff wavelength so that $(z_h - z_s) = \lambda_c$ then
the exponential growth of the evanescent waves is almost 55 dB, $(20\log_{10}(e^{2\pi}) = 54.6)$.
However, the supersonic, non-evanescent waves have

$$P(k_x, 0, z_s) = P(k_x, 0, z_h)e^{-ik_z(z_h - z_s)}$$

so that

$$|P(k_x, 0, z_s)| = |P(k_x, 0, z_h)|,$$

that is, the magnitudes of the supersonic waves in the measurement and reconstruction
planes are the same. Thus, for example, if we assume that the levels of the subsonic and
supersonic waves are about equal in the measurement plane, then in the reconstruction
process the subsonic waves increase almost 1000 times in magnitude compared with the
supersonic waves when $(z_h - z_s) = \lambda_c$. In order for the supersonic part of the spectrum
not to be masked by the subsonic part, then the measurement system must be capable
of resolution of 1 part in 1000. Thus for a measurement system to be able to resolve
the spatial pressure at the surface with a resolution of λ_c, it must have at least 55 dB
of dynamic range, between the pressure in the measurement plane and the noise level
of the measurement system when the standoff distance is $z_h - z_s = \lambda_c$.

In general then, let D be the dynamic range or the SNR of the measurement system measured in decibels. To be able to measure the evanescent component k_c, the dynamic range must be be greater than $e^{k_c(z_h - z_s)}$, that is,

$$10^{D/20} > e^{k_c(z_h - z_s)}.$$

If we let $R_x = \lambda_c/2$ be the spatial resolution of the reconstruction, then

$$R_x = 20\pi(z_h - z_s)\log(e)/D = 27.3(z_h - z_s)/D. \tag{3.23}$$

3.8 Determination of the Cutoff Frequency for the k-space Filter

We have seen in the examples above the deleterious effect of choosing the k-space cutoff k_c to be too small in relation to the important wavenumber content of the source. The value of k_c depends on the experiment: the dynamic range of the measurement system, the signal-to-noise ratio, the measurement offset $z_h - z_s$, the wavenumber content of the radiator, among other things. It is impossible to decide on a value of k_c a priori, and it is best to select it after the measurement has taken place, assuming that $z_h - z_s$ is not too large. We illustrate this process in this section with the one-dimensional example presented above.

Returning to the reconstruction equation, Eq. (3.4), and introducing the effect of the k-space window we can see that the wavenumber content of the source's normal velocity is

$$\tilde{W}(k_x, k_y, z_s) = \mathcal{F}_x\mathcal{F}_y[p(x, y, z_h)]G(k_x, k_y, z_s - z_h)\bar{\Pi}\left(\frac{\sqrt{k_x^2 + k_y^2}}{2k_c}\right), \tag{3.24}$$

where G is given by Eq. (3.10) and the modified filter presented in Section 3.6 extended to two dimensions. $\bar{\Pi}\left(\frac{\sqrt{k_x^2+k_y^2}}{2k_c}\right)$ represents a circular filter of radius $k_r \equiv \sqrt{k_x^2 + k_y^2} = k_c$ defined by

$$\bar{\Pi}\left(\frac{k_r}{2k_c}\right) = \begin{cases} 1 - \frac{1}{2}e^{-(1-|k_r|/k_c)/\alpha} & |k_r| < k_c \\ \frac{1}{2}e^{(1-|k_r|/k_c)/\alpha} & |k_r| > k_c. \end{cases} \tag{3.25}$$

Returning to the example of Section 3.5, we add white spatial noise to the simulated measured pressure in our one-dimensional example.

First we plot the noiseless case, that is, the logarithm of the k-space surface velocity (linear plot shown in Fig. 3.2). The dB plot provides more dynamic range so that the effects of the spatial noise can be better illustrated. The result is shown in Fig. 3.7. The dashed curve in this figure is the noise background which is to be added to the measured pressure $p(x, 0, z_h)$ and is given by the second term in Eq. (3.17), $10\log_{10}(L\gamma|G|^2)$. In this case we set $k = 1$ and $z_h - z_s = 0.1$. With $k = 1$ the acoustic wavelength is 2π so that the radiator covers only a fraction of a wavelength. This source is strongly evanescent.

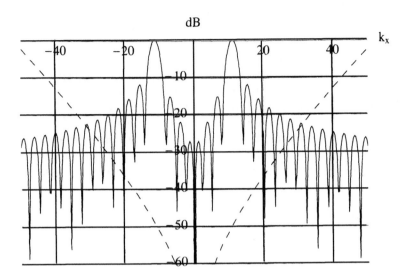

Figure 3.7: Logarithmic plot of the angular spectrum of velocity, $20\log_{10}|\dot{W}(k_x)|$ for velocity field $\dot{w}(x) = \sin(3.5\pi x)$.

Figure 3.8 shows the resulting k-space surface velocity with random noise added to the measured pressure $p(x, 0, z_h)$. The standard deviation of the noise was chosen to be 50 dB below the maximum value of the measured pressure. The dashed line in Fig. 3.8

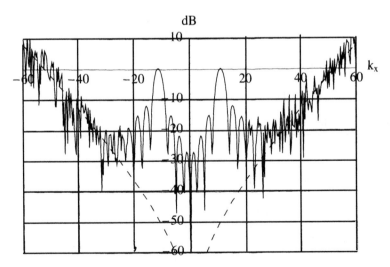

Figure 3.8: k-space surface velocity, $20\log_{10}|\tilde{\dot{W}}(k_x)|$, including the effects of measurement noise. Dashed line shows the magnitude of $10\log_{10}(L\gamma|G|^2)$.

shows the level of noise (without the source) when multiplied by the inverse propagator, $10\log_{10}(L\gamma|G|^2)$. The solid curve is the sum of the noise and source. Comparison with the plot of the source alone, Fig. 3.7, shows that the noise level reaches the level of the source at $|k_x| \approx 30$, and at this point the sum of the source and noise is increased by a factor of two (6 dB). For larger values of $|k_x|$ the noise takes over, growing exponentially, and dominates the k-space velocity.

Before the inverse Fourier transform, $\dot{w}(x) = \mathcal{F}^{-1}[\dot{W}(k_x)]$, can be carried out, it is necessary to apply the k-space filter, $\tilde{\Pi}$ discussed above. In practice one studies the k-space result, as shown in Fig. 3.8, and places the cutoff k_c at the point where the k-space spectrum contains the inflection point. This is the point where the level of noise is equal to the level of the source (without noise), $k_c = 30$, in the example here. It is important to realize that the selection of cutoff is a function of temporal frequency as a result of the dependence of the signal-to-noise on ω. Lower signal-to-noise, for example, will lead to a narrower filter.

For this case we use the exponential filter, shown in Fig. 3.5 with $\alpha = 0.1$ and cutoff $k_c = 25$. Figure 3.9 shows the result of the mathematical operations given in Eq. (3.24) with $k_y = 0$. This should be compared with Fig. 3.7, the k-space surface velocity of the source without noise. This comparison shows that the exponential filter has done an

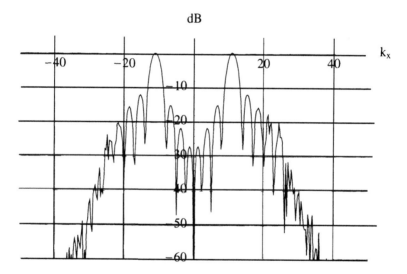

Figure 3.9: k-space surface velocity, $20 \log_{10} |\tilde{\dot{W}}(k_x)|$, after applying the k-space, exponential filter on the source + noise spectrum shown in Fig. 3.8.

excellent job at removing the effects of the noise in the reconstruction.

At this point the final step of the holographic equation is carried out; $\dot{w}(x) = \mathcal{F}^{-1}[\dot{W}(k_x)]$. Figure 3.10 shows this result for the reconstructed surface velocity $\tilde{w}(x)$ which is almost identical to the exact result (without noise), Fig. 3.6.

To illustrate the deleterious effect of noise on the reconstruction, the following reconstruction was carried out with $k_c = 60$, shown in Fig. 3.11. The desired velocity is almost totally obscured by the noise. Reference to Fig. 3.8 shows that at this cutoff value the noise in the reconstruction is 10 dB higher than the main signal peak. For even higher cutoffs, the reconstruction is nonsensical, with noise levels orders of magnitude larger than the desired result.

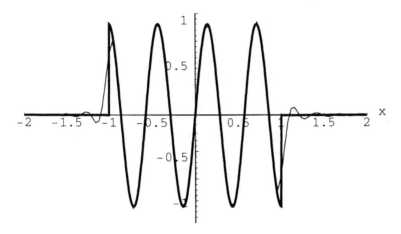

Figure 3.10: Reconstructed surface velocity compared with actual (bold line). The reconstruction is excellent except at the edges where the k-space cutoff restricts the resolution so that the discontinuity is smoothed.

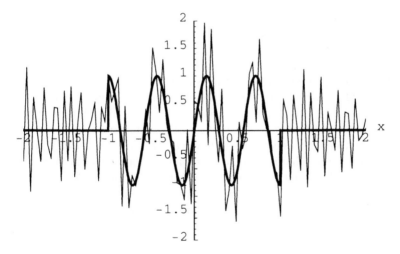

Figure 3.11: Reconstructed surface velocity with $k_c = 60$ versus desired result (bold line) showing the deleterious effect of the noise. The desired reconstruction is almost totally obscured.

3.9   Finite Measurement Aperture Effects

Up to this point we have assumed that the pressure is measured over the complete plane at $z = z_h$. This is impossible in practice. The rectangular area over which the pressure measurement is made is called the measurement aperture. In the reconstruction equation, Eq. (3.4), this aperture can be described by a rectangular window function

$\Pi[x/L_x]\Pi[y/L_y]$ given an aperture of length L_x and width L_y. Thus Eq. (3.4) becomes

$$\tilde{\dot{w}}(x,y,z_s) = \mathcal{F}_x^{-1}\mathcal{F}_y^{-1}\Big[\mathcal{F}_x\mathcal{F}_y\{p(x,y,z_h)\Pi(x/L_x)\Pi(y/L_y)\}G(k_x,k_y,z_s-z_h)\Big], \quad (3.26)$$

where the tilde indicates that $\tilde{\dot{w}}$ is an estimate of the true function $\dot{w}$ but differs due to the finite measurement aperture. This estimate translates into k-space as

$$\mathcal{F}_x\mathcal{F}_y[p(x,y,z_h)\Pi(x/L_x)\Pi(y/L_y)] = P(k_x,k_y,z_h) * *L_x L_y \operatorname{sinc}(k_x L_x/2)\operatorname{sinc}(k_y L_y/2).$$

The convolutions with the sinc functions spread the k-space spectrum of P. For example, if P is a delta function $\delta(k_x)\delta(k_y)$ the result of the aperture is to convert this delta function into a sinc function. Thus wavenumbers appear in the spectrum *which really do not exist*, and since the inverse propagator is a strong function of wavenumber, these wavenumbers are backpropagated to the source surface with different and erroneous amplitudes. Of course, when L_x and L_y tend to infinity the sinc functions become delta functions so that the spread of wavenumbers goes to zero, and the backpropagation is exact.

Unfortunately, it is difficult to derive any closed form expressions for the amount of error introduced into Eq. (3.26), as was possible for the k-space window. However some rules have evolved over the years through experience with the NAH technique. In practice the measurement aperture is always larger than the actual source. In the worst case L_x is chosen to be double the size of the actual source in the x direction and similarly for L_y. This almost always guarantees that the measured pressure field drops off significantly towards the edges of the aperture, as long as $z_h - z_s$ is small. Smallness is also required to capture evanescent waves and to satisfy the resolution requirement given in Eq. (3.23). The measurement then is essentially infinite, since the difference between the small pressure field at the ends and zero is insignificant, compared with the maximum value of pressure in the interior of the aperture.

There is one caveat to this conclusion, however. The inverse velocity propagator G is a deconvolving function. That is, instead of spreading spatial discontinuities it condenses the acoustic field so that what appears to be smooth variations in the measurement plane may turn out to be velocity with sharp edges in the source plane, as we have seen in the examples above. Formally we can write this process, from Eq. (3.4) as

$$\dot{w}(x,y,z_s) = p(x,y,z_h) * *g_v^{-1}(x,y,z_s-z_h), \quad (3.27)$$

where

$$g_v^{-1}(x,y,z_s-z_h) \equiv \mathcal{F}_x^{-1}\mathcal{F}_y^{-1}[G(k_x,k_y,z_s-z_h)]. \quad (3.28)$$

Although this process is a convolution, we must remember that g_v^{-1} is a singular function if $z_s < z_h$. G is the inverse of the forward velocity propagator,

$$G_f = \rho_0 c k e^{ik_z(z_h-z_s)}/k_z.$$

In the forward problem, G_f constantly smoothes the sound field as z_h increases due to the operation of the decaying exponential $e^{-k_z'z}$ which progressively eliminates the higher wavenumbers from the pressure field until only the supersonic wavenumbers are left. When z_h reaches the farfield of the vibrator the acoustic field can vary no less than one half an acoustic wavelength, since only supersonic wavenumbers are present in the spectrum. In the inverse problem, the reverse must be true. Given z_h in the farfield, the convolution in Eq. (3.27) increases the resolution of the reconstructed velocity field dramatically, so that it equals the resolution of the actual source, $\dot{w}(x, y, z_s)$.

Now the ends of the finite measurement aperture appear to the inverse propagator as a region where the slope of the pressure field is infinite, since the pressure field drops to zero outside this aperture. This discontinuity represents a region in space where there is a very high concentration of high wavenumbers. The deconvolution process of the inverse propagator then plays havoc with these high wavenumbers, increasing their levels dramatically. Of course the k-space filter is critical in controlling this process. Even though the measurement aperture itself does not contain any zero values of pressure, the lack of data outside of the aperture appears mathematically to the propagator as if the measurement were surrounded by an infinite pressure release baffle, and thus a rim discontinuity appears.

Thus, even though the aperture is large compared with the source, the pressure discontinuity at the ends must be dealt with. Again, as we did for the k-space filter, we use a tapered window to reduce the size of the discontinuity and to bring the pressure to zero at the ends gracefully. The taper must be confined to as small a region as possible, so as not to alter the measured pressure over the source. The window which has been used in NAH at the Naval Research Laboratory is the 8-point Tukey window.[8] This window is a raised cosine given by

$$
f(x) = \begin{cases} \frac{1}{2} - \frac{1}{2}\cos[\pi(x - L_x/2)/x_w] & \text{if } L_x/2 - x_w < x < L_x/2 \\ 1 & x \le L_x/2 - x_w \\ 0 & x > L_x/2. \end{cases}
\tag{3.29}
$$

Here x_w is the width of the window and $x = L_x/2$ is the right end of the measurement aperture. The taper is applied at each of the four edges of the measurement aperture with x_w (and y_w) chosen so that eight measurement points are contained within the taper. Figure 3.12 shows the Tukey window for $L_x/2 = 2.0$.

Experience has shown that the deconvolution process of the inverse velocity propagator is very localized in space. In other words, the tapered rim of the measurement aperture causes erroneous values of reconstructed surface velocity only in the same region (near $x = \pm L_x/2$ and $y = \pm L_y/2$) below on the surface. Since this region is beyond the actual sources of interest, this rim region can be discarded in the reconstruction. In any case the reconstructed velocity in the rim region is meaningless.

3.10 Discretization and Aliasing

We have assumed up to this point that the pressure and velocity are sampled continuously (with an infinitesimal interval) in the measurement aperture. Of course, any measurement system will sample with a discrete interval. In the use of the Fourier transform this discretization leads to aliasing. We consider only equal sampling of the

---

[8]Harris, "On the use of Windows for Harmonic Analysis with the Discrete Fourier Transform".

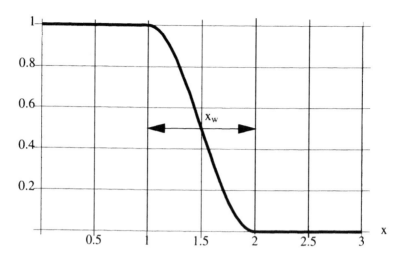

Figure 3.12: Tukey window taper for an aperture ending at $x = L_x/2 = 2.0$.

integrand of the Fourier transform with a lattice constant $\Delta x = a$. Consider the one-dimensional Fourier transform

$$P(k_x) = \int_{-\infty}^{\infty} p(x)e^{-ik_x x}dx,$$

where $P(k_x)$ is the desired exact transform. The sampling process is carried out by using the comb function, so that the approximation to the Fourier transform $\tilde{P}$ is now

$$
\begin{aligned}
\tilde{P}(k_x) &= \int_{-\infty}^{\infty} p(x)III(x/a)e^{-ik_x x}dx \\
&= \frac{a}{2\pi}P(k_x) * III(\frac{k_x}{2\pi/a}),
\end{aligned}
\tag{3.30}
$$

using Eqs (1.15) and (1.45). Using the definition of the comb function (in k-space) the last equation becomes

$$\tilde{P}(k_x) = \sum_{m=-\infty}^{\infty} P(k_x - \frac{2\pi m}{a}).
\tag{3.31}$$

This equation is the mathematical statement of aliasing. The right hand side consists of a sum of the exact transforms, each shifted by $2\pi/a$. Clearly, if $P(k_x)$ is not band limited in k_x then the part of $P(k_x)$ with $|k_x| > 2\pi/a$ will overlap the adjacent transforms ($m = \pm 1$). Similarly, $P(k_x \pm 2\pi/a)$ will overlap $P(k_x)$, corrupting the desired transform. As $a \to 0$, the limit of continuous sampling, the adjacent replications become infinitely far apart and the aliasing disappears.

3.11   Use of the DFT to Solve the Holography Equation

Nearfield acoustical holography is in large part successful due to the speed at which it can forward and backpropagate sound fields. This is due to the use of the FFT (fast Fourier transform) algorithm. The need for the DFT and the FFT arises from the discretization of the holography equation, a prerequisite step for any practical measurement scheme. Williams and Maynard[9] present an extensive analysis of the use of the FFT to solve the forward radiation problem using Rayleigh's first integral formula, Eq. (2.75). We present a numerical solution of the inverse problem using the same approach.

The purpose of this section is to evaluate the errors introduced by replacing the continuous Fourier transforms in the holography equation, Eq. (3.4),

$$\dot{w}(x,y,z_s) = \mathcal{F}_{x,y}^{-1}\Big[\mathcal{F}_{x,y}[p(x,y,z_h)]G(k_x,k_y,z_s-z_h)\Big],$$

with the discrete Fourier transforms (represented by the subscript D),

$$\dot{w}_D(x,y,z_s) = \mathcal{D}_{x,y}^{-1}\Big[\mathcal{D}_{x,y}[p(x,y,z_h)]G(k_x,k_y,z_s-z_h)\Big], \qquad (3.32)$$

where the operators are defined:

$$\mathcal{F}_{x,y} \equiv \mathcal{F}_x\mathcal{F}_y$$
$$\mathcal{F}_{x,y}^{-1} \equiv \mathcal{F}_x^{-1}\mathcal{F}_y^{-1},$$

and similarly for $\mathcal{D}_{x,y}$ and $\mathcal{D}_{x,y}^{-1}$. The discrete Fourier transform is defined by

$$\mathcal{D}_x[p(x)] \equiv P_D(n\Delta k_x) = a \sum_{m=-N/2}^{N/2-1} p(ma)e^{-2\pi inm/N}, \qquad (3.33)$$

where $N = L_x/a$, a is the sample spacing of the spatial function (N samples), and $k_x = n\Delta k_x = n(2\pi/L_x)$. Here Δk_x represents the smallest spatial frequency which can be resolved, represented by a single wavelength (2π radians) across the measurement aperture L_x. The DFT is defined to exclude the point at $m = N/2$. We can almost always assume that the field $p(Na/2)$ at the end point will be negligibly small so that the inclusion or exclusion of this point will make little difference in the end result.

Now we can cast the continuous Fourier transform in terms of the DFT. The continuous transform is

$$P(k_x) = \int_{-\infty}^{\infty} p(x)e^{-ik_x x}dx,$$

which is evaluated at discrete points separated by a over an aperture of length L_x. The DFT, represented by the subscript D, can be constructed from the following operation inside the transform integral:

$$P_D(k_x) \equiv \int_{-\infty}^{\infty} p(x)\amalg(x/a)\Pi(x/L_x)e^{-ik_x x}dx, \qquad (3.34)$$

---

[9]Earl G. Williams and J.D. Maynard (1982). "Numerical evaluation of the Rayleigh integral for planar radiators using the FFT", J. Acoust. Soc. Am. **72**. pp. 2020–2030.

which leads directly to a DFT (ignoring slight discrepancies at the end points $\pm L_x/2$:

$$P_D(n\Delta k_x) = a \sum_{m=-N/2}^{N/2-1} p(ma)e^{-2\pi inm/N}. \tag{3.35}$$

Since the comb function is an infinite series of delta functions spaced a apart, these delta functions act to sample $p(x)$ at equal intervals along the x axis. The rectangular window function Π represents the fact that the measurement aperture extends over L_x. The exponent of the DFT is constructed using

$$\begin{aligned} x &= m\Delta x = ma, \\ k_x &= n\Delta k_x, \\ \Delta k_x &= 2\pi/L_x = 2\pi/Na, \\ k_x x &= 2\pi nm/N. \end{aligned} \tag{3.36}$$

Note that $-N/2 \le n \le N/2 - 1$.

The inverse Fourier transform is related to the DFT in the same way, that is,

$$p(x) = \frac{1}{2\pi} \int_{-\infty}^{\infty} P(k_x)e^{ik_x x}dk_x,$$

is approximated by

$$p_D(x) = \frac{1}{2\pi} \int_{-\infty}^{\infty} P(k_x)\Pi(k_x/\Delta k_x)\Pi(k_x/2k_m)e^{ik_x x}dk_x, \tag{3.37}$$

or

$$p_D(ma) = \mathcal{D}_x^{-1}[P(k_x)] = \frac{1}{Na} \sum_{n=-N/2}^{N/2-1} P(n\Delta k_x)e^{2\pi inm/N}, \tag{3.38}$$

where Equations (3.36) apply, $\Delta k_x/2\pi = 1/Na$ and $k_m \equiv |-\frac{N}{2}\Delta k_x| = \pi/a$ is the maximum spatial frequency (one wavelength across two samples). Δk_x is the smallest spatial frequency (outside of zero) given by one wavelength across the complete aperture L_x. Symbolically we write for the two-dimensional case

$$\mathcal{D}_{x,y}^{-1}[P_D(k_x, k_y, z_h)] = \mathcal{F}_{x,y}^{-1}[P_D(k_x, k_y, z_h)\Pi(\frac{k_x}{\Delta k_x})\Pi(\frac{k_y}{\Delta k_y})\Pi(\frac{k_x}{2k_m})\Pi(\frac{k_y}{2k_m})], \tag{3.39}$$

where $k_m \equiv \pi/a$ (the maximum wavenumber), $\Delta k_x = 2\pi/L_x$, and $\Delta k_y = 2\pi/L_y$.

Now we are in a position to illuminate the errors of the DFT version of the holographic reconstruction equation, Eq. (3.32), using $\mathcal{D}_{x,y}[p] \equiv P_D$ and Eq. (3.39):

$$\begin{aligned} \dot{w}_D(x, y, z_s) &= \mathcal{F}_{x,y}^{-1}\Big[P_D(k_x, k_y, z_h)G(k_x, k_y, z_s - z_h) \\ &\quad \times \Pi(k_x/\Delta k_x, k_y/\Delta k_y)\Pi(k_x/2k_m, k_y/2k_m)\Big]. \end{aligned} \tag{3.40}$$

Also we have made the definitions

$$\text{III}(k_x/\Delta k_x, k_y/\Delta k_y) \equiv \text{III}(k_x/\Delta k_x)\text{III}(k_y/\Delta k_y) \tag{3.41}$$

$$\Pi(k_x/2k_m, k_y/2k_m) \equiv \Pi(k_x/2k_m)\Pi(k_y/2k_m). \tag{3.42}$$

Note that the one-dimensional form of P_D is given by Eq. (3.35). Having applied the DFT (and FFT) in this manner we can now manipulate the continuous Fourier transform operations in Eq. (3.40) to examine the errors which have been introduced by using the DFT to solve the reconstruction equation.

With this in mind we now apply the convolution theorem to Eq. (3.40), inverting the order of G and III:

$$\dot{w}_D(x, y, z_s) = \mathcal{F}_{x,y}^{-1}[P_D] * *\mathcal{F}_{x,y}^{-1}[\text{III}] * *\mathcal{F}_{x,y}^{-1}[G] * *\mathcal{F}_{x,y}^{-1}[\Pi].$$

We need the following relationships, (in Sections 1.6 and 1.7),

$$\mathcal{F}_{x,y}^{-1}[\text{III}(k_x/\Delta k_x, k_y/\Delta k_y)] = \frac{1}{L_x L_y}\text{III}(x/L_x, y/L_y), \tag{3.43}$$

$$\mathcal{F}_{x,y}^{-1}[\Pi(k_x/2k_m, k_y/2k_m)] = \frac{1}{a^2}\text{sinc}(\pi x/a)\,\text{sinc}(\pi y/a), \tag{3.44}$$

so that

$$\dot{w}_D(x, y, z_s) = p_D(x, y, z_h) * *\frac{1}{L_x L_y}\text{III}(\frac{x}{L_x}, \frac{y}{L_y}) * *\mathcal{F}_{x,y}^{-1}[G] * *\frac{1}{a^2}\text{sinc}(\frac{\pi x}{a})\,\text{sinc}(\frac{\pi y}{a}), \tag{3.45}$$

where, from Eq. (3.34), the inverse (continuous) Fourier transform of P_D is

$$p_D(x, y, z_h) = p(x, y, z_h)\text{III}(x/a, y/a)\Pi(x/L_x, y/L_y) \tag{3.46}$$

so that p_D is just a sampled version of the measured pressure p over the measurement aperture, and is zero outside of it. The errors which arise from convolution with the inverse propagator were discussed in Section 3.9 above, without regard to sampling issues. In addition, the spatial sampling in Eq. (3.46) must be dense enough to avoid aliasing, as discussed in the last section. That is, the highest spatial wavenumbers containing significant energy must be sampled at least at the rate of two samples per wavelength to prevent spatial aliasing which causes high wavenumbers to be converted to low wavenumbers. If p is oversampled then we are assured that the angular spectrum computed from the DFT, $P_D(k_x, k_y, z_h)$, is a close representation of the actual wavenumber spectrum.

Note that the effect of the $\text{sinc}(\pi x/a)\,\text{sinc}(\pi y/a)$ term in Eq. (3.45) is small as Fig. 3.13 indicates, since the width of the main lobe spans only two sample points. In fact note that $\lim_{a \to 0} \frac{1}{a}\text{sinc}(\pi x/a) = \delta(x)$. We proceed by neglecting the small effect of the convolution with the sinc function in Eq. (3.45) so that

$$\tilde{w}_D(x, y, z_s) \approx \left[p_D(x, y, z_h) * *\frac{1}{L_x L_y}\text{III}(x/L_x, y/L_y) \right] * *\mathcal{F}_{x,y}^{-1}[G] \tag{3.47}$$

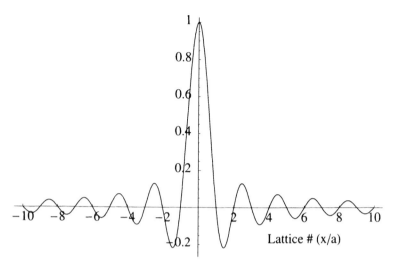

Figure 3.13: Plot of $\mathrm{sinc}(\pi x/a)$ where a is the spacing between adjacent measurement points.

The III function in the double convolution in Eq. (3.47) is a series of impulses separated by a distance L_x and L_y in the x and y directions, respectively:

$$\frac{1}{L_x L_y}\mathrm{III}(x/L_x, y/L_y) = \sum_{p=-\infty}^{\infty} \delta(x - pL_x) \sum_{q=-\infty}^{\infty} \delta(y - qL_y), \qquad (3.48)$$

so that convolution in Eq. (3.47) becomes

$$p_D(x,y,z_h) * * \frac{1}{L_x L_y}\mathrm{III}(x/L_x, y/L_y) = \sum_{p,q=-\infty}^{\infty} p_D(x - pL_x, y - qL_y, z_h). \qquad (3.49)$$

To understand what the right hand side means consider just two terms of Eq. (3.49), for $p = q = 0$ and $p = 1, q = 0$:

$$p_D(x,y,z_h) + p_D(x - L_x, y, z_h).$$

Because of the window in Eq. (3.46) we can see that outside of $-L_x/2 \leq x \leq L_x/2$ and $-L_y/2 \leq y \leq L_y/2$, $p_D(x,y,z_h) = 0$. The second term represents a shifted replica of the first, shifted to $x = L_x$ and is non-zero over the region $L_x/2 \leq x \leq 3L_x/2$, starting exactly where the first source dropped off. Thus we can view the summation in Eq. (3.49) as an infinite chess board with the $p = q = 0$ term illustrated by a king sitting on the central square surrounded by an otherwise empty chess board representing the actual truncated measurement. The result of the summation is then to place an infinite set of kings one on every square of the chess board.

Figure 3.14 shows the source and measurement planes, illustrating this concept. The infinite chess board is represented in part by a 3×3 set of apertures, the central one being the actual measurement aperture. The replicated apertures extend out to infinity.

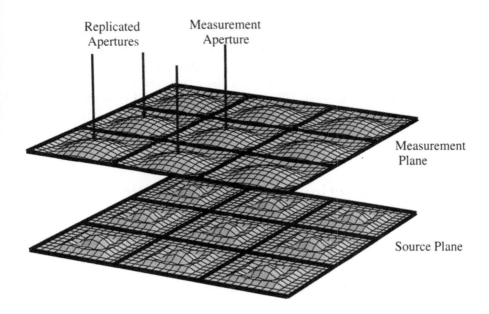

Figure 3.14: Replicated measurement apertures due to the use of the DFT (FFT) to solve the inverse problem. The real aperture, labeled "measurement aperture", is in the center with the actual source below it. The DFT uses the replicated, as well as the real, measurement apertures to reconstruct the source. As shown, the reconstructed sources are also replicated. Although only a 3×3 grid is shown, the replication extends out to infinity in both directions.

Each replicated aperture is identical to the measurement aperture, although shifted in space.

Equation (3.47) now becomes

$$\tilde{w}_D(x, y, z_s) = \sum_{p,q=-\infty}^{\infty} p_D(x - pL_x, y - qL_y, z_h) * *\mathcal{F}_{x,y}^{-1}[G] \qquad (3.50)$$

which is our final result. It shows the inverse propagator operates on an infinite plane of replicated measurements, not just the actual one. The replicated measurements outside the actual one are reconstructed in the same fashion as the actual measurement aperture by convolution with the inverse propagator. However, these replicated measurements contribute less to the velocity in the reconstruction aperture because they are further away from it. (Remember the forward velocity propagator appears in Rayleigh's first formula and depends on $1/R$ where R is the distance from source point to measurement point.) Since the left hand side of Eq. (3.50) is also periodic (verified by replacement of x with $x + L_x$ and y with $y + L_y$), the reconstructed sources also form an infinite lattice of replicated apertures, as shown in Fig. 3.14. These are called replicated sources.

We can see that if the separation between the measurement and source planes is small then the replicated sources should introduce a small error, and as $z_h - z_s$ increases the contributions of these sources will increase the error.

A simple technique to reduce the errors due to replicated measurements is to append zeros outside the actual measurement, enough to double the length and width of the measurement aperture. Since the Tukey window has already taken care of reducing the measured pressure field to zero at the edges then this zero addition only improves the reconstruction result. The improvement is gained through the increased separation between the replicated pressure measurements (due to the $1/R$ effect) and the corresponding reduction in their influence in the reconstruction over the real source.

3.12 Reconstruction of Other Quantities

The reconstruction equation, Eq. (3.4), resulting from Eq. (3.2) provided only the normal velocity in the source plane at $z = z_s$. Assume that the reconstruction plane, given by $z = z_r$ is on a different (but parallel) plane from the source plane. Actually in Eq. (3.2) there is no restriction on the value for z_r, and in fact we can have $z_r > z_h$ as we have discussed above. Of course, we can not go below the source plane (z_r can not be less than z_s); this would violate Eq. (2.1). Thus, if the sources are located at $z = z_s$ then $z_r \geq z_s$ always. Reference to Eq. (2.60) indicates that we can define a *vector* velocity propagator given by

$$\vec{G}(k_x, k_y, z_r - z_h) \equiv (k_x \hat{i} + k_y \hat{j} + k_z \hat{k}) e^{ik_z(z_r - z_h)}/\rho_0 ck, \qquad (3.51)$$

where the $\hat{k}$ component of $\vec{G}$ is the velocity propagator that we have seen before, in Eq. (3.10). With the vector propagator we can rewrite Eq. (3.2), defining

$$\vec{\Upsilon}(k_x, k_y, z_r) \equiv \dot{U}(k_x, k_y, z_r)\hat{i} + \dot{V}(k_x, k_y, z_r)\hat{j} + \dot{W}(k_x, k_y, z_r)\hat{k},$$

as

$$\vec{\Upsilon}(k_x, k_y, z_r) = \vec{G}(k_x, k_y, z_r - z_h)P(k_x, k_y, z_h). \qquad (3.52)$$

The spatial velocity vector is determined in exactly the same way as was presented in the sections above for the normal surface velocity, using the inverse DFT or FFT. One detail of the presentation above changes, however. The cutoff for the k-space filter is a strong function of $z_r - z_h$. When $z_r > z_h$ (the forward problem) the k-space filter need not be applied at all, as a general rule, since there is no amplification of evanescent waves.

The pressure in any plane parallel to the hologram plane is given by Eq. (3.1) with the spatial pressure given by the inverse DFT or FFT as with the velocity. Thus on any plane, $z = z_r$, we are able to determine both the pressure and velocity fields, $p(x, y, z_r)$ and $\vec{v}(x, y, z_r)$. This leads, via Eq. (2.16) page 19, to the acoustic intensity,

$$\vec{I}(x, y, z_r) = \frac{1}{2}p(x, y, z_r)\vec{v}(x, y, z_r)^* \qquad (3.53)$$

where the real part gives the active intensity and the imaginary part the reactive intensity. The total power radiated into the farfield is given by Eq. (2.17) or

$$\Pi(\omega) = \int_{-\infty}^{\infty} \int_{-\infty}^{\infty} Re[I_z(x,y,z_r)] \, dx \, dy. \tag{3.54}$$

The total power can also be computed from just a knowledge of $\dot{W}$ using Eq. (2.111), page 52. Finally, the pressure in the farfield and the farfield directivity pattern are given by Eqs (2.84) and (2.86) on page 39, computed from the k-space velocity spectrum, $W(k_x, k_y, 0)$ at the surface of the source.

3.12.1   Time Domain

When the holographic measurements are made at multiple frequencies and with sufficient frequency resolution, then the time domain responses of the various acoustic quantities can be found through the use of the inverse Fourier transform, Eq. (2.10). Thus the instantaneous pressure and vector velocity can be determined in the reconstruction volume. The instantaneous intensity may then be determined through

$$\vec{I}(x,y,z,t) = p(x,y,z,t)\vec{v}(x,y,z,t).$$

Problems

3.1 To do this problem you will need a computer and a Fourier transform program. Use whatever you have at your disposal. This problem will verify the existence of edge and corner modes and simulates some of the steps taken in the nearfield acoustical holography technique. Using Eq. (2.162) consider the $m = n = 17$ mode of the simply supported plate.

(a) Using a DFT or FFT algorithm compute the two-dimensional Fourier transform of $\Phi_{mn}(x,y)$ that is,

$$W(k_x, k_y) = \mathcal{F}_x \mathcal{F}_y [\Phi_{17.17}(x,y)].$$

Use a square plate for simplicity, so that $L_x = L_y$. You should compare your result with Eq. (2.177) in order to make sure that you have not made any mistakes. You will need to review Section 3.11 in detail so that you apply the transform properly. Note that in most transform algorithms the first point in the output array is the $k_x = 0$ term, and the last point is $k_x = -\Delta k_x$. I recommend that you reshuffle the data so that the center point of the array is $k_x = 0$, the first point $-k_{max}$ and the last point $+k_{max}$. Make a plot of the magnitude of the result so you can understand it.

(b) Apply a k-space, circular filter as given in Eq. (3.25). To make it easier you can specify no taper ($\alpha = 0$) so that Π is either 1 or 0. Choose k_c as large as possible but small enough so that the four peaks in the transform are **outside** the circle. This step is crucial in order to obtain the corner radiation mode. If you make k_c too small you will not have enough resolution for part (c). What you are simulating is the case where the acoustic wavenumber is $k = k_c$, so that the filter cutoff corresponds to the acoustic wavenumber and so that the normal mode is outside the radiation circle, see Fig. 2.34. This choice of filter cutoff is called a supersonic filter since it passes all the wavenumbers which radiate to the farfield and filters those that do not. Plot the magnitude of the filtered data.

(c) Compute the 2-D inverse DFT or FFT of the supersonic-filtered result from above, and plot the magnitude of the results. You should see the four corners are regions that dominate. This is the corner mode. This filtering process identifies, in general, regions of a plate which radiate to the farfield and is extremely powerful since it identifies farfield radiating source regions on the vibrator.

(d) Repeat step (a) for a (5.17) normal mode.

(e) Repeat step (b) choosing a cutoff $5\pi/L_x \ll k_c < 17\pi/L_x$.

(f) Compute the inverse 2-D transform as in (c). You should produce an edge mode with high values on two strips at the $y = 0$ and $y = L_y$ edges of the plate. This is an edge radiation mode.

Chapter 4

Cylindrical Waves

4.1 Introduction

The application of the wave equation in cylindrical coordinates leads to the solution of many problems of interest in acoustics. Of great interest is the vibration and radiation from submarines, a problem which conforms closely to cylindrical symmetry. The aircraft fuselage also is generally cylindrical and the wave equation is of importance in the interior and exterior domains. Noise control in piping systems and circular ducts also finds extensive use of the wave equation in cylindrical coordinates. Nearfield acoustical holography has been used extensively in this geometry, and has piloted some very significant researches in underwater acoustics since the early 1980s. Some of this work will be presented in detail later in Chapter 5.

4.2 The Wave Equation

Of interest are the solutions of the homogeneous, time-dependent wave equation

$$(\nabla^2 - \frac{1}{c^2}\frac{\partial^2}{\partial t^2})p = 0 \tag{4.1}$$

in cylindrical coordinates (r, ϕ, z). The cylindrical coordinate system is shown in Fig. 4.1. The Laplace operator is[1]

$$\nabla^2 = \frac{\partial^2}{\partial r^2} + \frac{1}{r}\frac{\partial}{\partial r} + \frac{1}{r^2}\frac{\partial^2}{\partial \phi^2} + \frac{\partial^2}{\partial z^2}, \tag{4.2}$$

and the gradient operator, in cylindrical coordinates is

$$\vec{\nabla} = \frac{\partial}{\partial r}\hat{e}_r + \frac{1}{r}\frac{\partial}{\partial \phi}\hat{e}_o + \frac{\partial}{\partial z}\hat{e}_z, \tag{4.3}$$

where the unit vectors point in the positive coordinate directions.

---

[1] P. M. Morse and H. Feshbach (1953). *Methods of Theoretical Physics.* McGraw-Hill, New York.

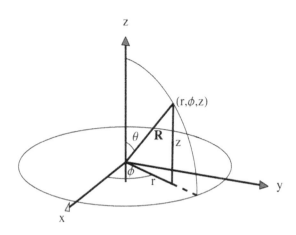

Figure 4.1: Definition of cylindrical coordinates relative to Cartesian coordinates. ϕ is measured in the x, y plane from the x axis.

Given the velocity vector,

$$\vec{v} = \dot{w}\hat{e}_r + \dot{v}\hat{e}_o + \dot{u}\hat{e}_z, \tag{4.4}$$

Euler's equation in the time domain, which has the same form as in rectangular coordinates (Eq. (2.2) on page 15), is

$$-\rho_0 \frac{\partial}{\partial t} \vec{v}(r, \phi, z) = \vec{\nabla} p(r, \phi, z).$$

Similarly in the frequency domain:

$$i\rho_0 ck\vec{v}(r, \phi, z) = \vec{\nabla} p(r, \phi, z). \tag{4.5}$$

To derive the solutions of the wave equation we use a technique called separation of variables. This approach assumes that the solution can be written as a product of solutions of functions of each coordinate and of time. That is,

$$p(r, \phi, z, t) = R(r)\Phi(\phi)Z(z)T(t). \tag{4.6}$$

Introduce this solution into Eq. (4.1) and divide out by $R\Phi ZT$. This leads to

$$\left(\frac{1}{R} \frac{d^2 R}{dr^2} + \frac{1}{rR} \frac{dR}{dr} + \frac{1}{r^2 \Phi} \frac{d^2 \Phi}{d\phi^2} \right) + \left(\frac{1}{Z} \frac{d^2 Z}{dz^2} \right) = \frac{1}{c^2 T} \frac{d^2 T}{dt^2}.$$

Note that the partial derivatives are now total derivatives. The terms in the first set of brackets depend only on the variables r and ϕ, in the second set of brackets upon z only and on the right hand side upon t only. Thus since r, ϕ, z and t are all independent of each other, each of these terms must be equal to a constant. We choose the following arbitrary constants, k and k_z, satisfying the following equations,

$$\frac{1}{c^2 T} \frac{d^2 T}{dt^2} = -k^2 \tag{4.7}$$

$$\frac{1}{Z} \frac{d^2 Z}{dz^2} = -k_z^2 \tag{4.8}$$

$$\frac{1}{R} \left(\frac{d^2 R}{dr^2} + \frac{1}{r} \frac{dR}{dr} \right) + \frac{1}{r^2 \Phi} \frac{d^2 \Phi}{d\phi^2} = -k^2 + k_z^2 \equiv -k_r^2, \tag{4.9}$$

where the constant

$$k_r = \sqrt{k^2 - k_z^2}. \tag{4.10}$$

Equation (4.9) can be written as

$$\frac{r^2}{R} \left(\frac{d^2 R}{dr^2} + \frac{1}{r} \frac{dR}{dr} \right) + k_r^2 r^2 = -\frac{1}{\Phi} \frac{d^2 \Phi}{d\phi^2}, \tag{4.11}$$

which displays the fact that the right and left hand sides must be equal to constants (since the left hand side is a function of r alone, and the right hand side of ϕ only). Choosing n^2 as one of these constants leads to

$$\frac{1}{\Phi} \frac{d^2 \Phi}{d\phi^2} = -n^2, \tag{4.12}$$

resulting in Bessel's equation on the left hand side,

$$\frac{d^2 R}{dr^2} + \frac{1}{r} \frac{dR}{dr} + \left(k_r^2 - \frac{n^2}{r^2} \right) R = 0. \tag{4.13}$$

Since Eq. (4.7), Eq. (4.8) and Eq. (4.12) are second order differential equations, each has a general solution with two arbitrary constants:

$$T(t) = T_1 e^{-i\omega t} + T_2 e^{i\omega t}, \tag{4.14}$$

$$Z(z) = Z_1 e^{ik_z z} + Z_2 e^{-ik_z z}, \tag{4.15}$$

$$\Phi(\phi) = \Phi_1 e^{in\phi} + \Phi_2 e^{-in\phi}, \tag{4.16}$$

with arbitrary constants T_1, T_2, Z_1, Z_2, Φ_1 and Φ_2.

Following our convention for time we set the constant $T_2 = 0$. Since $\Phi(\phi+2\pi) = \Phi(\phi)$ then n must be an integer. From Eq. (4.7) we see that $k = \omega/c$. However, there is no restriction on the independent constant k_z.

4.2.1   Bessel Functions

The solutions of Eq. (4.13) are well known and are given by the Bessel functions of the first and second kinds, $J_n(k_r r)$ and $Y_n(k_r r)$. Y_n is call the Neumann function and is sometimes represented by the symbol N_n. The general solution to Eq. (4.13) uses these two independent functions with arbitrary constants R_1 and R_2:

$$R(r) = R_1 J_n(k_r r) + R_2 Y_n(k_r r). \tag{4.17}$$

Note that the index of the Bessel function corresponds to the separation constant in the Φ equation. J_n and Y_n are called the standing wave solutions of Eq. (4.13) because of their asymptotic behavior (as $x \to \infty$) given by

$$J_n(x) \sim \sqrt{\frac{2}{\pi x}} \cos(x - n\pi/2 - \pi/4) \tag{4.18}$$

and

$$Y_n(x) \sim \sqrt{\frac{2}{\pi x}} \sin(x - n\pi/2 - \pi/4). \tag{4.19}$$

A linear combination of these functions is judicious for traveling wave solutions, given by Hankel functions of the first and second kind:

$$\begin{align} H_n^{(1)}(k_r r) &= J_n(k_r r) + i Y_n(k_r r), \tag{4.20} \\ H_n^{(2)}(k_r r) &= J_n(k_r r) - i Y_n(k_r r). \tag{4.21} \end{align}$$

With time dependence $e^{-i\omega t}$, $H_n^{(1)}(k_r r)$ corresponds to a diverging outgoing wave and $H_n^{(2)}(k_r r)$ to an incoming and converging wave. Again the asymptotic behavior dictates this definition since

$$H_n^{(1)}(x) \sim \sqrt{\frac{2}{\pi x}} e^{i(x - n\pi/2 - \pi/4)} \tag{4.22}$$

and

$$H_n^{(2)}(x) \sim \sqrt{\frac{2}{\pi x}} e^{-i(x - n\pi/2 - \pi/4)}. \tag{4.23}$$

The general traveling wave solution to Eq. (4.13) is then

$$R(r) = R_1 H_n^{(1)}(k_r r) + R_2 H_n^{(2)}(k_r r). \tag{4.24}$$

The series representation of J_n is

$$J_n(z) = \left(\frac{z}{2}\right)^n \sum_{k=0}^{\infty} \frac{(-1)^k}{k!(n+k)!} \left(\frac{z}{2}\right)^{2k} \tag{4.25}$$

where z can be complex and $|\arg z| < \pi$. From this series we can see that for small arguments

$$\begin{align} J_0(x) &= 1 + O(x^2) \\ J_1(x) &= x/2 + O(x^3) \tag{4.26} \\ J_n(x) &= x^n/(2^n n!) + O(x^{n+2}). \end{align}$$

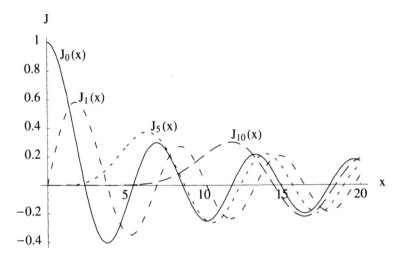

Figure 4.2: Bessel functions of the first kind of orders 0, 1, 5 and 10.

Figure 4.2 shows plots of J_n for $n = 0$, 1, 5 and 10. The plots of J_5 and J_{10} show that the Bessel functions are not only oscillatory, but also have an exponential-like region defined by $x < n$.

The series representation of Y_n is a bit more complicated, and can be found in standard math tables. The small argument expressions are

$$Y_0(x) \sim \frac{2}{\pi}[\ln \frac{x}{2} + \gamma] \tag{4.27}$$

$$Y_n(x) \sim -\frac{(n-1)!}{\pi}(\frac{2}{x})^n. \tag{4.28}$$

In the above equation $\gamma = 0.57721 \cdots$ is Euler's constant. These expressions bear out the singular nature of Y_n at $x = 0$. Figure 4.3 shows plots of Y_n for $n = 0$, 1, 5 and 10. The small argument results for the derivative of H_n will be useful in this chapter:

$$H_n'(x) \approx \frac{in!}{\pi \epsilon_n}(\frac{2}{x})^{n+1}, \tag{4.29}$$

where $\epsilon_0 \equiv 1$ and $\epsilon_n \equiv 2$ for $n \geq 1$.

The following formulas are useful

$$J_{-n}(x) = (-1)^n J_n(x) \tag{4.30}$$

$$Y_{-n}(x) = (-1)^n Y_n(x). \tag{4.31}$$

There is no restriction to the separation constant k_z in Eq. (4.8) and thus when $k_z > k$, Eq. (4.10) leads to an imaginary value for k_r:

$$k_r = i\sqrt{k_z^2 - k^2}.$$

In this case Eq. (4.13) becomes

$$\frac{d^2R}{dr^2} + \frac{1}{r}\frac{dR}{dr} - (k_r^2 + \frac{n^2}{r^2})R = 0. \tag{4.32}$$

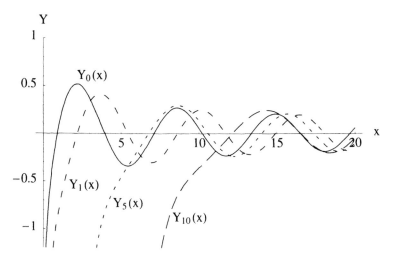

Figure 4.3: Bessel functions of the second kind of orders 0, 1, 5 and 10.

This is the modified Bessel equation leading to solutions called the modified Bessel functions of the first and second kinds, $I_n(k'_r r)$ and $K_n(k'_r r)$ where $k'_r \equiv \sqrt{k_z^2 - k^2}$. These are defined by

$$I_n(x) = i^{-n} J_n(ix) \tag{4.33}$$

$$K_n(x) = \frac{\pi}{2} i^{n+1} H_n^{(1)}(ix). \tag{4.34}$$

Figure 4.4 shows plots of I_n and K_n for $n = 0, 1, 5$.

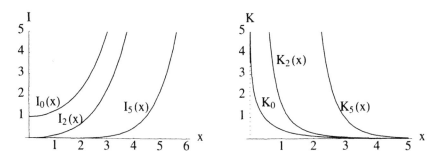

Figure 4.4: Modified Bessel functions of the first and second kind of orders 0, 1 and 5.

The small-argument-limit forms for the modified Bessel functions are

$$I_n(x) \sim \left(\frac{x}{2}\right)^n \frac{1}{n!}, \tag{4.35}$$

$$K_0(x) \sim -\gamma + \ln\left(\frac{2}{x}\right). \tag{4.36}$$

and

$$K_n(x) \sim \frac{1}{2}(n-1)! \left(\frac{2}{x}\right)^n,$$ (4.37)

where γ is Euler's constant.

The asymptotic formulas $(x \to \infty)$ for these functions are

$$I_n(x) \sim \frac{e^x}{\sqrt{2\pi x}}.$$ (4.38)

$$K_n(x) \sim \sqrt{\frac{\pi}{2x}} e^{-x}.$$ (4.39)

The modified Bessel functions become simple exponential functions for large argument. Finally, some recursion relations for the Bessel functions are

$$Z_{n-1}(z) + Z_{n+1}(z) = \frac{2n}{z} Z_n(z)$$ (4.40)

$$Z_{n-1}(z) - Z_{n+1}(z) = 2\frac{dZ_n(z)}{dz},$$ (4.41)

where Z denotes J, Y, $H^{(1)}$, or $H^{(2)}$. Some Wronskian relations are

$$W[J_n(z), Y_n(z)] = 2/(\pi z)$$ (4.42)
$$W[H_n^{(1)}(z), H_n^{(2)}(z)] = -4i/(\pi z)$$ (4.43)
$$W[I_n(z), K_n(z)] = -1/z,$$ (4.44)

where the Wronskian is defined as

$$W[f(z), g(z)] \equiv f(z)g'(z) - f'(z)g(z).$$ (4.45)

4.3 General Solution

Returning to the separation of variables, we combine the solutions back together as dictated by Eq. (4.6). There are six possible combinations with the two independent solutions for each coordinate $(T_2 = 0)$:

$$p(r, \phi, z, t) \propto H_n^{(1),(2)}(k_r r) e^{\pm in o} e^{\pm i k_z z} e^{-i\omega t}.$$

We can include these six combinations in the general solution by summing over all possible positive and negative values of n and k_z with arbitrary coefficient functions (functions of n, k_z and ω) replacing the pairs of constants, $Z_1, Z_2, \Phi_1, \Phi_2, R_1$ and R_2. Thus, $\sum_{n=-\infty}^{\infty}$ takes care of all possible n values and $\int_{-\infty}^{\infty} dk_z$ takes care of all possible values of k_z (a continuum). The most general solution of Eq. (4.1) in the frequency domain is then given by

$$p(r, \phi, z, \omega) = \sum_{n=-\infty}^{\infty} e^{in o} \frac{1}{2\pi} \int_{-\infty}^{\infty} \left[A_n(k_z, \omega) e^{ik_z z} H_n^{(1)}(k_r r) + \right.$$
$$\left. + B_n(k_z, \omega) e^{ik_z z} H_n^{(2)}(k_r r) \right] dk_z.$$ (4.46)

where the arbitrary constants, $A_n(k_z, \omega)/2\pi$ and $B_n(k_z, \omega)/2\pi$ replace Z_1, Z_2, Φ_1, Φ_2, R_1 and R_2.

Equation (4.46) is an inverse Fourier transform in k_z, and a discrete Fourier series in n. Since k_z spans all real values positive and negative, we can see that k_r is either real or imaginary, the latter corresponding to the case $|k_z| > k$, leading to Hankel functions of imaginary argument which we express as MacDonald functions defined in Eq. (4.34).

The time domain solution is obtained from the inverse Fourier transform as usual:

$$p(r, \phi, z, t) = \frac{1}{2\pi} \int_{-\infty}^{\infty} p(r, \phi, z, \omega) e^{-i\omega t} d\omega. \qquad (4.47)$$

In cases where the standing wave solutions are more appropriate Eq. (4.46) is written as

$$p(r, \phi, z, \omega) = \sum_{n=-\infty}^{\infty} e^{in\phi} \frac{1}{2\pi} \int_{-\infty}^{\infty} \Big[C_n(k_z, \omega) e^{ik_z z} J_n(k_r r) +$$
$$+ \quad D_n(k_z, \omega) e^{ik_z z} Y_n(k_r r) \Big] dk_z. \qquad (4.48)$$

Equation (4.46) or (4.48) present completely general and equivalent solutions to the wave equation in a source-free region. In order to determine the arbitrary coefficients, boundary conditions are specified on coordinate surfaces, for example, $r = $ constant. Boundary condition with $z = $ constant lead to discrete solutions in k_z instead of the continuous ones formulated above. We will deal with this problem in a later chapter. The boundary condition on r alone leads to the solution of many practical problems of general interest. We will deal with this case exclusively in this chapter.

Consider the case in which the boundary condition is specified at $r = a$ and $r = b$, for example, $p(a, \phi, z, \omega) = f(\phi, z, \omega)$ and $p(b, \phi, z, \omega) = g(\phi, z, \omega)$, where f and g are the given boundary conditions. Figure 4.5 represents a cross-section of the volume (perpendicular to the z axis) showing a representative boundary value problem (infinite in the z direction). In this case the sources are located in the two regions labeled Σ_1 and Σ_2. These regions may be finite in the z direction (not shown in the figure). The homogeneous Helmholtz equation is not valid in these regions, but is valid in the annular disk region shown. In this region Eq. (4.46) or Eq. (4.48) can be used to solve for the pressure field. The boundary conditions on the surfaces at $r = a$ and $r = b$ yield a unique solution (for all values of z and ϕ). Two boundary conditions are necessary because there are two unknown functions, A_n and B_n in the equations. In direct analogy to the planar problem, one can specify either pressure or one of the components of the velocity on these surfaces, the latter being almost always the normal component (radial component). Note that no part of the source region is allowed to cross the infinite cylinder surfaces defining the annular disk region.

It is important to understand the significance of the two parts to the solution in Eq. (4.46) corresponding to the two Hankel functions. The first term represents an outgoing wave expressed in Eq. (4.22) due to sources which must be on the interior of the volume of validity (Σ_1); these cause waves to diverge outward. The function A_n, once determined, provides the strength of these sources. The second Hankel function in Eq. (4.46) represents incoming waves and is needed to account for sources external

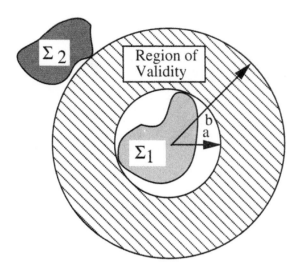

Figure 4.5: An example volume for a boundary value problem in cylindrical coordinates. The regions Σ_1 and Σ_2 represent source regions, and the annular disk region shows the domain of validity of the homogeneous Helmholtz equation.

to the annular region, such as those shown as Σ_2 in Fig. 4.5. Likewise, B_n provides the strength of these sources.

4.3.1   The Interior and Exterior Problems

There are two other kinds of boundary value problems which arise and are of interest to us, each being a subset of the solutions Eq. (4.46) or Eq. (4.48). The first is called the interior problem in which the sources are located completely outside the boundary value surface $r = b$. This situation is shown in Fig. 4.6. What distinguishes this case from the annular region case is that (1) there is only a single boundary surface and (2) the region of validity includes the origin. Noting that the pressure must be finite at the origin (since the homogeneous differential equation is valid there) then we can see that Eq. (4.46) is an ill-suited solution for this case since the Hankel functions are infinite at the origin. Thus we use Eq. (4.48) since the first term J_n is finite at the origin. The second term with Y_n is infinite at the origin and we set $D_n(k_z) = 0$ to arrive at the general solution

$$p(r,\phi,z,\omega) = \sum_{n=-\infty}^{\infty} e^{ino} \frac{1}{2\pi} \int_{-\infty}^{\infty} C_n(k_z,\omega) e^{ik_z z} J_n(k_r r) dk_z. \qquad (4.49)$$

Only one unknown function needs to be determined from the boundary field on the surface at $r = b$.

The second boundary value problem is called the exterior problem because the boundary surface completely encloses all the sources. This situation is shown in Fig. 4.7. Now the requirement for a finite field at the origin is no longer valid and we turn to

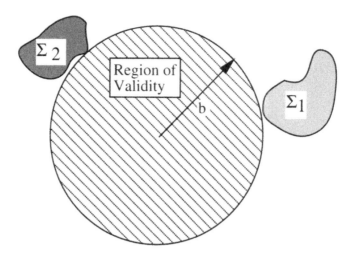

Figure 4.6: Interior domain problem where all sources are located outside of boundary value surface at $r = b$.

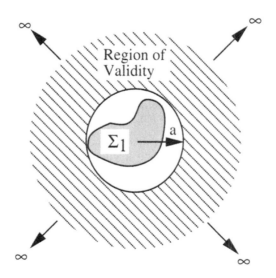

Figure 4.7: Exterior domain problem with all sources inside the boundary value surface at $r = a$.

Eq. (4.46) for the solution. In this equation. however. the second term represents an in-coming wave which can not be excited when all the sources are within the boundary. Thus we set the second coefficient function to zero. $B_n(k_z) = 0$. The general solution now becomes

$$p(r, \phi, z, \omega) = \sum_{n=-\infty}^{\infty} e^{in o} \frac{1}{2\pi} \int_{-\infty}^{\infty} A_n(k_z, \omega) e^{ik_z z} H_n^{(1)}(k_r r) dk_z. \qquad (4.50)$$

Again if the pressure or any component of the velocity is specified on the infinite boundary at $r = a$ then A_n can be determined and Eq. (4.50) can be used to solve for the pressure field in the region from the surface at $r = a$ out to infinity.

Of course, any of these boundary surfaces can be considered also as measurement surfaces, as is done in nearfield acoustical holography as discussed for planar holography in Chapter 3. We need one measurement surface for each unknown coefficient function in Eq. (4.46), Eq. (4.49) and Eq. (4.50).

We proceed to discuss the exterior domain problem in detail. It is especially important since it forms the basis of nearfield acoustical holography in cylindrical coordinates and leads us to a very important concept called the helical wave spectrum, analogous to the angular spectrum of plane waves.

4.4 The Helical Wave Spectrum: Fourier Acoustics

Consider the exterior problem in which the acoustic pressure is specified on the infinite boundary at $r = a$ and proceed to find solutions of Eq. (4.50). We will drop the ω variable from the functions for simplicity of notation. Let the measured pressure at $r = a$ (the boundary value) be $p(a, \phi, z)$ and, since Eq. (4.50) must be satisfied, then

$$p(a, \phi, z) = \sum_{n=-\infty}^{\infty} e^{in\phi} \frac{1}{2\pi} \int_{-\infty}^{\infty} A_n(k_z, \omega) e^{ik_z z} H_n^{(1)}(k_r a) dk_z. \tag{4.51}$$

Let $P_n(r, k_z)$ be the two-dimensional Fourier transform (actually a Fourier series and a transform) in ϕ and z of the acoustic pressure at r:

$$P_n(r, k_z) \equiv \frac{1}{2\pi} \int_0^{2\pi} d\phi \int_{-\infty}^{\infty} p(r, \phi, z) e^{-in\phi} e^{-ik_z z} dz. \tag{4.52}$$

From Section 1.3 on page 4 the relationships for the Fourier series of a function $f(\phi)$ and its inverse F_n are

$$F_n = \frac{1}{2\pi} \int_0^{2\pi} f(\phi) e^{-in\phi} d\phi \tag{4.53}$$

$$f(\phi) = \sum_{n=-\infty}^{\infty} F_n e^{in\phi}. \tag{4.54}$$

With these, and the inverse Fourier transform in k_z, we can write down the inverse relationship for Eq. (4.52):

$$p(r, \phi, z) = \sum_{n=-\infty}^{\infty} e^{in\phi} \frac{1}{2\pi} \int_{-\infty}^{\infty} P_n(r, k_z) e^{ik_z z} dk_z. \tag{4.55}$$

Comparison of Eq. (4.55) at $r = a$ with Eq. (4.51) shows that

$$P_n(a, k_z) = A_n(k_z) H_n^{(1)}(k_r a). \tag{4.56}$$

Using Eq. (4.56) to eliminate A_n in Eq. (4.50) yields the important relationship,

$$p(r, \phi, z) = \sum_{n=-\infty}^{\infty} e^{in\phi} \frac{1}{2\pi} \int_{-\infty}^{\infty} P_n(a, k_z) e^{ik_z z} \frac{H_n^{(1)}(k_r r)}{H_n^{(1)}(k_r a)} dk_z. \qquad (4.57)$$

This expression is quite similar in form to the plane wave spectrum definition with the boundary field at $z = z_0$, Eq. (2.50) on page 32:

$$p(x, y, z) = \frac{1}{4\pi^2} \int_{-\infty}^{\infty} dk_x \int_{-\infty}^{\infty} dk_y P(k_x, k_y, z_0) e^{i(k_x x + k_y y)} e^{ik_z(z - z_0)}.$$

Comparing this to Eq. (4.57) reveals the following correspondences between planar and cylindrical expansions:

$$P(k_x, k_y, z_0) \quad \leftrightarrow \quad P_n(r, k_z)$$

$$e^{ik_z(z - z_0)} \quad \leftrightarrow \quad \frac{H_n^{(1)}(k_r r)}{H_n^{(1)}(k_r a)}$$

$$k_x \quad \leftrightarrow \quad k_z$$

$$k_y \quad \leftrightarrow \quad n/r$$

$$k_z \quad \leftrightarrow \quad k_r.$$

Thus in view of the fact that we called $P(k_x, k_y, z)$ the plane wave (angular) spectrum, we call $P_n(r, k_z)$ the *helical wave spectrum*.

Since the two-dimensional Fourier transform (Eq. (4.52)) of the left hand side of Eq. (4.57) is $P_n(r, k_z)$ then

$$P_n(r, k_z) = \frac{H_n^{(1)}(k_r r)}{H_n^{(1)}(k_r a)} P_n(a, k_z). \qquad (4.58)$$

Equation (4.58) provides the relationship between the helical wave spectrum at different cylindrical surfaces in the same way that $e^{ik_z(z - z_0)}$ provided the relationship between planar surfaces. Note that in Eq. (4.58) r may be less than or greater than a, corresponding to back projection and forward projection, respectively. This equation is the key to projecting pressure fields from one cylindrical contour to another. We still can not, however, cross the sources Σ_1 in the back projection process.

The helical wave is perhaps a bit more difficult to visualize than the plane wave. In the plane wave case we looked at the trace wave on the surface $z = 0$. The complex amplitude of the trace wave was $P(k_x, k_y, 0)$ with a spatial field given by $e^{i(k_x x + k_y y - \omega t)}$. In direct analogy to this we have the trace wave on the cylindrical surface $r = a$ of the helical wave defined by a complex amplitude $P_n(a, k_z)$, with the spatial field $e^{i(k_z z + n\phi - \omega t)}$. To make the analogy even closer we can write the ϕ part in terms of the circumferential arc coordinate s where $s = a\phi$, and define the circumferential wavenumber $k_s \equiv n/a$. Now the form of the trace of the helical wave on the surface $r = a$ becomes

$$P_n(a, k_z) e^{i(k_z z + k_s s - \omega t)}.$$

The phase fronts of the helical wave on the cylinder are defined by

$$k_z z + k_s s - \omega t = \text{constant.}$$

At a particular instant in time the phase fronts appear to spiral around the cylinder as shown in Fig. 4.8. This wave travels very much like the apparent motion in a turning

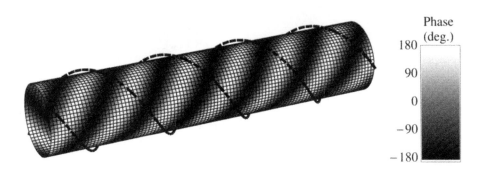

Figure 4.8: Lines of constant phase shown in gray scale at an instant in time on a cylindrical surface rapping around the cylinder like strips on a candy cane or threads on a screw. The direction of propagation is illustrated by the dashed line perpendicular to the phase fronts.

drill bit with the grooves in the bit lining up with the phase fronts of the wave.

We can visualize this better by unwrapping the field on the cylinder and plotting it on a plane as shown in Fig. 4.9. The unwrapping means that the top and bottom horizontal portions of the plot are continuous and represent the same azimuthal position on the cylinder. Note that, as a result, the wavefronts must match from top to bottom (for constant z) as is clear in the figure. The trace angle (angle on the $r = a$ cylinder) of the direction of propagation of the helical wave is indicated by θ.

The circumferential variation of the pressure for this helical wave is shown on a cross cut of the cylinder in Fig. 4.10.

The radial propagation of the helical wave is a bit more complicated than the plane wave case. It travels from one surface $r = a$ to another surface at r by undergoing an amplitude and phase change (the Hankel function is complex) given by the ratio of Hankel functions, Eq. (4.58):

$$\frac{H_n^{(1)}(k_r r)}{H_n^{(1)}(k_r a)} e^{i(k_z z + n\sigma - \omega t)}.$$

This is simplified, however, when $k_r r$ is large enough ($k_r r \gg n$) so that we can use the asymptotic expression for the Hankel function in the numerator. From Eq. (4.22)

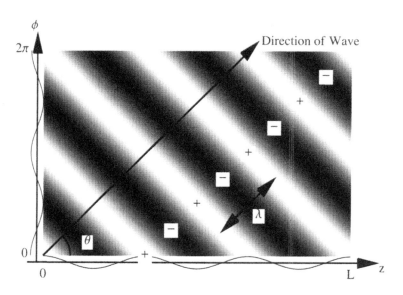

Figure 4.9: Density plot of the pressure in a helical wave, $n = 2$ and $k_z = 6\pi/L$, unwrapped onto a flat plane. The direction of propagation is θ.

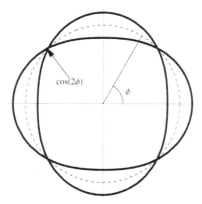

Figure 4.10: The circumferential pressure variation in a cross cut of the helical wave for $n = 2$, where $p(\theta) \times \cos(2\theta)$. The unperturbed, dashed circle is provided for reference. The level of pressure is indicated by the radial excursion from the reference.

the r dependence of the helical wave is $e^{ik_r r}/\sqrt{k_r r}$, where $k_r = \sqrt{k^2 - k_z^2}$. This looks similar to the plane wave case now except that the amplitude is decaying by the square root of the distance, indicating that the wave is diverging as it travels outward. Thus the helical wave has the asymptotic form

$$e^{i(k_r r + k_z z + k_s s - \omega t)}.$$

As would be expected, the angle of launch from the cylinder surface depends on the frequency. Consider the case of $n = 0$: no variation in pressure in the circumferential direction. This is called the breathing mode. The direction θ of the helical wave launched from the surface is shown in Fig. 4.11 which shows the conical wavefronts and the pressure variation (in gray scale) on a cylindrical surface. The figure indicates that

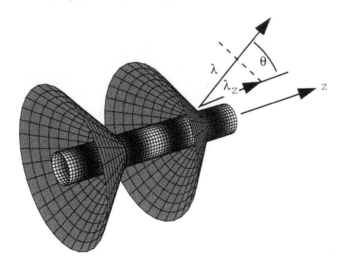

Figure 4.11: Pressure field for an $n = 0$ helical wave radiating away from the cylinder when $k_r r$ is large. The wavefronts form cones around the cylinder.

the angle θ is given by

$$\cos \theta = \lambda/\lambda_z \quad \text{or} \quad k \cos \theta = k_z.$$

When $n \neq 0$ the picture is similar, except that the surfaces of constant phase now spiral around the cylinder like the threads of a screw so that they do not join as in the $n = 0$ case.

The above was true for $k_r r \gg n$. When this is not the case, it is difficult to present such a simple picture since the constant phase term of the Hankel function does not follow a simple exponential. This difficulty arises from the fact that the wave is diverging outward, so that the pressure field spreads in the circumferential direction interacting in a complex fashion with the pressure field in the axial direction.

4.4.1 Evanescent Waves

Up to this point we have considered the case in which $k \geq k_z$, so that the wavelength in the axial direction is greater than the acoustic wavelength (see Fig. 4.11). In view of the results from the study of plane waves we would expect evanescent waves to be generated when the acoustic wavelength is larger than the wavelength in the axial and/or circumferential direction. However, unlike the plane wave case, there is a difference between the axial and circumferential cases, the former giving rise to true exponential decay and the latter to a power law decay. We will look at both cases. The wavelength in the circumferential direction is given by

$$\lambda_o = 2\pi a/n, \tag{4.59}$$

where $2\pi a$ is the circumference and n is the number of complete cycles in the circumference.

First we consider the axial case. When the wavelength in the axial direction is smaller that the acoustic wavelength λ ($\lambda = 2\pi/k$) then one would expect a decay of energy from the surface at $r = a$. These decaying, non-propagating waves are called subsonic or evanescent waves and exhibit an exponential decay away from the surface. That is, when $\lambda_z < \lambda$ then $k_z > k$, and k_r given in Eq. (4.58) becomes a pure imaginary number. In this case Eq. (4.58) can be written as

$$P_n(r, k_z) = \frac{K_n(k_r' r)}{K_n(k_r' a)} P_n(a, k_z), , \tag{4.60}$$

with

$$k_r' \equiv \sqrt{k_z^2 - k^2}, \tag{4.61}$$

and K_n is the modified Bessel function which arises when the argument of the Hankel function is imaginary, Eq. (4.34) on page 120. Figure 4.4 shows that $K_n(k_r' r)$ in the numerator of Eq. (4.60) exhibits a strong decay as r increases. To reveal this mathematically we assume the arguments of the modified Bessel functions are large and use their asymptotic forms, Eq. (4.39), to yield

$$\frac{K_n(k_r' r)}{K_n(k_r' a)} \approx \sqrt{\frac{a}{r}} e^{-k_r'(r-a)}. \tag{4.62}$$

Thus the helical wave amplitude P_n decays exponentially in r indicating an evanescent wave. One can show that the radial velocity for this wave is in phase quadrature with the pressure so that no energy is carried away from the shell by this wave.

Now we consider evanescent conditions in the circumferential direction which occur when the circumferential wavelength λ_o is less than λ. Assume that the axial wave is supersonic, that is, $k_z < k$, and k_r is real. In particular, set $k_z = 0$ (infinite axial wavelength) and note that Eq. (4.58) applies, that is, the Hankel functions of real argument govern the decay. When $r >> n$ the ratio of Hankel functions approaches

$$\sqrt{\frac{a}{r}} e^{ik_r(r-a)},$$

and the field decays as expected for a cylindrical wave, proportional to the square root of the radial distance. There is no evanescent behavior here. However, since $\lambda_o < \lambda$ one anticipates some kind of short circuit of the radiation of this wave from the surface $r = a$, since the medium only supports radiation at the characteristic wavelength λ as implied by the Helmholtz wave equation. Furthermore, this short circuit should become more complete as the index of the Hankel function n becomes larger, since n is the number of wavelengths which fit around the circumference of the cylinder (see Eq. (4.59)).

This short circuit can be demonstrated mathematically by keeping the argument of the Hankel functions fixed and allowing the order to increase so that we can use asymptotic expansions for large orders.[2] In this case the asymptotic expansion $(n \to \infty)$ for the Hankel function is

$$H_n(\zeta) \approx \frac{1}{\sqrt{2\pi n}} \left(\frac{e\zeta}{2n}\right)^n - i\sqrt{\frac{2}{\pi n}} \left(\frac{e\zeta}{2n}\right)^{-n}, \qquad (4.63)$$

where $\zeta \equiv k_r r = kr$, since we have set $k_z = 0$. When $\zeta/n < 1$ then we can ignore the real part of Eq. (4.63) and the second term predicts that the Hankel function will decay as $(1/kr)^n$. Using this result for the two Hankel functions in Eq. (4.58) we find that the nth component of the pressure P_n becomes

$$P_n(r,0) \approx \left(\frac{a}{r}\right)^n P_n(a,0). \qquad (4.64)$$

This equation holds when $kr < n$, which is equivalent to the evanescent wave condition

$$\frac{2\pi r}{\lambda} < n. \qquad (4.65)$$

The ratio on the left hand side is the number of wavelengths which fit around the circumference of the wavefront at the radius r. Thus, whenever the number of wavelengths is less than n, P_n will decay inversely with the nth power of the distance. This is the sought-after evanescent-like condition. However, unlike the evanescent waves generated in the axial case, Eq. (4.62), these waves do not decay exponentially but decay obeying a power law. In addition, one can show that the radial velocity is no longer 90 degrees out of phase with the pressure, so that a small part of the energy radiates away from the cylinder.

Figure 4.12 illustrates the power law decay. Here the exact values of the ratio of Hankel functions are plotted as a function of $20 \log(r/a)$, with k_z zero and $ka = 5$, for three different values of n. The logarithmic abscissa is chosen so that the power law decay of the nearfield would be indicated by lines of constant slope. Note that the maximum abscissa value represents an r value of $10a$, corresponding to 20 dB. The figure shows that each curve can be approximately broken into two straight line segments, the power-law nearfield and the cylindrical-spreading farfield. The asymptotes shown in the figure represent lines of exact power law as labeled. One can see from the figure, for example, that the $n = 20$ component of the pressure has decayed about 110 dB at a distance of twice a (abscissa value of 6 dB). The vertical line segments drawn on each curve represent the abscissa value when the number of wavelengths in a circumference is just equal to n, the equality condition of Eq. (4.65) above. Note that these lines separate the differing slope regions on each curve. To the right of these lines the wave is cylindrically spreading, and to the left it is evanescent.

Another way of explaining the curves in Fig. 4.12 is to note that as the helical wave travels outward, the circumferential wavelength (given by $2\pi r/n$) increases due to the expanding circumference. At some point Eq. (4.65) is no longer valid and the wavelength in the circumferential direction becomes larger than the acoustic wavelength. At this point $\lambda_\phi = \lambda$ and the evanescent propagation turns nonevanescent, spreading cylindrically from that point to the farfield. The helical wave is no longer in a short circuit condition.

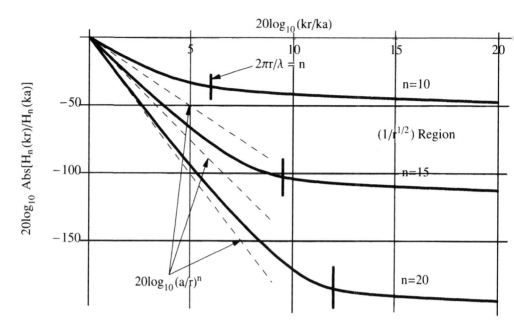

Figure 4.12: dB ratio of Hankel functions when $ka = 5.0$ and $k_z = 0$. The asymptotes drawn in show how the nearfield is dominated by a power-law decay in pressure, proportional to $(r/a)^n$. The vertical tick marks indicate the approximate point at which the propagation changes from power law to cylindrical spreading, indicating a transition from evanescent to nonevanescent propagation.

4.4.2 The Relationship Between Helical Wave Velocity and Pressure

Equation (4.58) provides the basic relationship between the helical wave amplitudes on two different cylindrical surfaces. We use Euler's equation, Eq. (4.5), in cylindrical coordinates to derive the relationship between velocity and pressure.

We define a velocity vector amplitude of a helical wave $\vec{V}_n$ analogous to P_n in Eq. (4.52) as a double Fourier transform of $\vec{v}$:

$$\vec{V}_n(r, k_z) = \frac{1}{2\pi} \int_0^{2\pi} d\phi\, e^{-ino} \int_{-\infty}^{\infty} dz\, e^{-ik_z z} \vec{v}(r, \phi, z). \tag{4.66}$$

Taking double transforms of both sides of Eq. (4.5) and using the fact that $\mathcal{F}_z(\frac{\partial}{\partial z}) = ik_z$ and $\mathcal{F}_\phi(\frac{\partial}{\partial \phi}) = in$ yields

$$\vec{V}_n(r, k_z) = \frac{1}{\rho_0 c} \left(\frac{-i}{k} \frac{\partial}{\partial r} \hat{e}_r + \frac{n}{kr} \hat{e}_o + \frac{k_z}{k} \hat{e}_z \right) P_n(r, k_z). \tag{4.67}$$

Using Eq. (4.58) we can relate the helical wave velocity to pressure on a concentric surface at $r = r_0$. Thus, inserting Eq. (4.58) into Eq. (4.67) yields. for the radial component of the velocity,

$$\dot{W}_n(r,k_z) = \vec{V}_n(r,k_z) \cdot \hat{e}_r = \frac{-ik_r}{\rho_0 ck} \frac{H'_n(k_r r)}{H_n(k_r r_0)} P_n(r_0,k_z), \qquad (4.68)$$

and for evanescent waves (using Eq. (4.60))

$$\dot{W}_n(r,k_z) = \frac{-ik'_r}{\rho_0 ck} \frac{K'_n(k'_r r)}{K_n(k'_r r_0)} P_n(r_0,k_z), \qquad (4.69)$$

where H'_n and K'_n are the derivatives with respect to the argument of the Hankel functions and k'_r is defined by Eq. (4.61).

Equations (4.68) and (4.69) are valid for both $r > r_0$ and $r \leq r_0$. When $r > r_0$ we are solving an inverse problem, and in the second case, $r \leq r_0$, a forward problem. We will use the second case to derive the analog to Rayleigh's first integral in cylindrical coordinates which will provide the relationship between the radial velocity field on the surface of an infinite cylinder and the pressure field outside of the cylinder. To this end we rewrite Eq. (4.68) putting $r = a$ and $r_0 = r$ ($r_0 \geq a$):

$$P_n(r,k_z) = \frac{i\rho_0 ck}{k_r} \frac{H_n(k_r r)}{H'_n(k_r a)} \dot{W}_n(a,k_z). \qquad (4.70)$$

This equation can also be used to formulate the radiation impedance $Z_n(k_z)$ of a helical wave, that is, the ratio of pressure to velocity in k-space. This is called the specific acoustic impedance since it involves a force per unit area to differentiate from the usual impedance definition which involves just a force. Thus,

$$Z_n(k_z) \equiv P_n(a,k_z)/\dot{W}_n(a,k_z) = \frac{i\rho_0 ck}{k_r} \frac{H_n(k_r a)}{H'_n(k_r a)}. \qquad (4.71)$$

Plots and various low and high frequency expressions for the specific radiation impedance are given in Junger and Feit.[3]

4.5 The Rayleigh-like Integrals

Similar integrals to the Rayleigh integrals in Cartesian coordinates, Section 2.10 on page 34, are obtained by inverse transforming Eq. (4.70) using the 2-D Fourier transform defined by Eq. (4.55). The analog of Rayleigh's second integral was already given in Eq. (4.57) above. A relation similar in concept to Rayleigh's first integral formula is obtained by taking the inverse Fourier transforms of Eq. (4.70):

$$p(r,\phi,z) = i\rho_0 ck \sum_{n=-\infty}^{\infty} e^{in\sigma} \frac{1}{2\pi} \int_{-\infty}^{\infty} \dot{W}_n(a,k_z) \frac{H_n(k_r r)}{k_r H'_n(k_r a)} e^{ik_z z} dk_z, \qquad (4.72)$$

[3]M. C. Junger and D. Feit (1986). *Sound, Structures, and their Interaction.* 2nd ed., MIT Press, Cambridge, MA., pp. 168–172.

where

$$\dot{W}_n(a, k_z) = \frac{1}{2\pi} \int_0^{2\pi} d\phi' \int_{-\infty}^{\infty} \dot{w}(a, \phi', z') e^{-in\phi'} e^{-ik_z z'} dz'. \tag{4.73}$$

These equations are notationally simplified by recognizing them as Fourier transforms. We represent Eq. (4.53) and Eq. (4.54) symbolically as

$$\mathcal{F}_\phi[f(\phi)] \equiv \frac{1}{2\pi} \int_0^{2\pi} f(\phi) e^{-in\phi} d\phi, \tag{4.74}$$

and inverse

$$\mathcal{F}_o^{-1}[F_n] \equiv \sum_{n=-\infty}^{\infty} F_n e^{in\phi}. \tag{4.75}$$

Thus we write Eq. (4.72) as

$$p(r, \phi, z) = i\rho_0 c k \mathcal{F}_o^{-1} \mathcal{F}_z^{-1} [\dot{W}_n(a, k_z) \frac{H_n(k_r r)}{k_r H_n'(k_r a)}], \tag{4.76}$$

with

$$\dot{W}_n(a, k_z) = \mathcal{F}_o \mathcal{F}_z [\dot{w}(a, \phi, z)].$$

We can rewrite Eq. (4.72) as a convolution integral in both ϕ and z using the relations Eq. (1.28) on page 5 for ϕ and Eq. (1.14) on page 3. Thus Eq. (4.72) viewed as a convolution looks very much like Rayleigh's first integral formula, Eq. (2.71) on page 36 and Eq. (2.75) on page 36:

$$p(r, \phi, z) = \frac{1}{2\pi} \dot{w}(a, \phi, z) * *g_v(r, a, \phi, z), \tag{4.77}$$

where $**$ stands for a two-dimensional convolution in z and ϕ and

$$\begin{aligned} g_v(r, a, \phi, z) &= \mathcal{F}_z^{-1} \mathcal{F}_o^{-1} \left[\frac{i\rho_0 c k H_n(k_r r)}{k_r H_n'(k_r a)} \right] \\ &= \frac{i\rho_0 c k}{2\pi} \sum_{n=-\infty}^{\infty} e^{in\phi} \int_{-\infty}^{\infty} \frac{H_n(k_r r)}{k_r H_n'(k_r a)} e^{ik_z z} dk_z. \end{aligned} \tag{4.78}$$

Unfortunately, unlike the planar case (Eq. (2.74) on page 36), Eq. (4.78) can not be reduced any further. The integral is unknown. We now present some examples for the application of Eq. (4.72).

4.5.1 Radiation from an infinite length cylinder with an arbitrary surface velocity distribution independent of z

Assume that the radial surface velocity is given as

$$\dot{w}(a, \phi, z) = \dot{w}_0 f(\phi). \tag{4.79}$$

Since $\dot{w}$ does not depend on z we recognize this as the condition $k_z = 0$ or $\lambda_z \to \infty$. Inserting Eq. (4.79) into Eq. (4.73) yields

$$\dot{W}_n(a, k_z) = 2\pi\dot{w}_0 F_n \delta(k_z),$$

where

$$F_n = \frac{1}{2\pi} \int_0^{2\pi} f(\phi)e^{-ino}d\phi.$$

Introducing this result into Eq. (4.72) yields

$$p(r, \phi, z) = i\rho_0 c \dot{w}_0 \sum_{n=-\infty}^{\infty} e^{ino} F_n \frac{H_n(kr)}{H_n'(ka)}. \tag{4.80}$$

In order to obtain the pressure in the farfield we use the asymptotic formula for the Hankel function, Eq. (4.22), in the numerator of Eq. (4.80) to obtain

$$p(r, \phi) \approx \sqrt{\frac{2}{\pi kr}} \rho_0 c \dot{w}_0 e^{i\pi/4} e^{ikr} \sum_{n=-\infty}^{\infty} e^{ino} \frac{(-i)^n F_n}{H_n'(ka)}. \tag{4.81}$$

The radial component of the particle velocity in the farfield (differentiating Eq. (4.81) with respect to r) is

$$\dot{w}(r, \phi) = \frac{1}{i\rho_0 ck} \frac{\partial p}{\partial r} = \frac{1}{i\rho_0 ck}(ik - \frac{1}{2r})p(r, \phi) \approx p(r, \phi)/\rho_0 c. \tag{4.82}$$

Therefore, $p = \rho_0 c\dot{w}$ as it is for a plane wave.

The intensity in the farfield is $I_r = |p|^2/2\rho_0 c$ and the power radiated per unit length of the cylinder is

$$\Pi/L = \frac{1}{L} \int_0^L \int_0^{2\pi} I_r(r, \phi)rd\phi dz = \frac{1}{2\rho_0 c} \int_0^{2\pi} p(r, \phi)p^*(r, \phi)rd\phi. \tag{4.83}$$

Inserting Eq. (4.81) into this equation yields

$$\Pi/L = \frac{\rho_0 c|\dot{w}_0|^2}{\pi k} \int_0^{2\pi} d\phi \sum_{n=-\infty}^{\infty} e^{ino} \frac{(-i)^n F_n}{H_n'(ka)} \sum_{m=-\infty}^{\infty} e^{-imo} \frac{i^m F_m^*}{H_m'(ka)^*}. \tag{4.84}$$

The integration over ϕ is

$$\int_0^{2\pi} e^{i(n-m)o} = 2\pi\delta_{mn}. \tag{4.85}$$

so that the cross terms in the product of the two summations all disappear. This significant fact results from the orthogonality of the circumferential harmonics and it implies that the powers of each of the individual circumferential harmonics are uncoupled. The sum of the powers of each individual mode is the total power radiated. Finally, the power per unit length is

$$\Pi/L = \frac{2\rho_0 c|\dot{w}_0|^2}{k} \sum_{n=-\infty}^{\infty} \frac{|F_n|^2}{|H_n'(ka)|^2} \tag{4.86}$$

Examples

We now present some specific cases. Consider first the case of a monopole breathing mode of the cylinder surface velocity. In this case the surface velocity is a constant,

$$\dot{w}(a, \phi) = \dot{w}_0.$$

and $F_n = \delta_{n0}$. Thus Eq. (4.80) and Eq. (4.86) yield

$$p(r, \phi) = i\rho_0 c \dot{w}_0 \frac{H_0(kr)}{H_0'(ka)}.$$

and

$$\Pi/L = \frac{2\rho_0 c |\dot{w}_0|^2}{k} \frac{1}{|H_0'(ka)|^2},$$

respectively.

As a second example consider a line velocity source at $\phi = \phi_0$ on the surface of the cylinder. In this case

$$\dot{w}(a, \phi) = \dot{w}_0 \delta(\phi_0).$$

and considering Eq. (4.79),

$$F_n = e^{-in\phi_0}/2\pi.$$

All the harmonics are of equal magnitude, a result of the delta function. In this case the pressure radiated to the farfield, Eq. (4.81), is

$$p(r, \phi) \approx \sqrt{\frac{2}{\pi kr}} \frac{\rho_0 c \dot{w}_0}{2\pi} e^{i\pi/4} e^{ikr} \sum_{n=-\infty}^{\infty} e^{in(\phi-\phi_0)} \frac{(-i)^n}{H_n'(ka)}, \qquad (4.87)$$

and the power per unit length

$$\Pi/L = \frac{\rho_0 c |\dot{w}_0|^2}{2\pi^2 k} \sum_{n=-\infty}^{\infty} \frac{1}{|H_n'(ka)|^2}. \qquad (4.88)$$

4.5.2   Radiation from Infinite Cylinder with Standing Wave

Next consider an example of an infinite standing wave on the cylinder surface. We saw in the corresponding case of an infinite planar radiator, that a standing wave pattern had a farfield radiation pattern with two sets of 'rabbit ears' when the standing wave was supersonic in both directions, and no radiation to the farfield when the standing wave was subsonic in either direction. For a standing wave on a cylinder we have,

$$\dot{w}(a, \phi, z) = b \cos(p\phi) \cos(m\pi z/L), \qquad (4.89)$$

where b is a constant. Inserting this equation into Eq. (4.73) yields

$$\dot{W}_n(a, k_z) = \frac{\pi b}{2} (\delta_{np} + \delta_{n(-p)})(\delta(k_z - m\pi/L) + \delta(k_z + m\pi/L)).$$

With Eq. (4.72) we evaluate the pressure radiated from the surface, defining $k_m \equiv m\pi/L$:

$$p(r, \phi, z) = i\rho_0 ck \frac{b}{4}(e^{ip\phi} + e^{-ip\phi})(e^{ik_m z} + e^{-ik_m z})\frac{H_p(\sqrt{k^2 - k_m^2}\, r)}{\sqrt{k^2 - k_m^2}\, H_p'(\sqrt{k^2 - k_m^2}\, a)},$$

where we have used the fact that $H_{-p} = (-1)^p H_p$ and $H'_{-p} = (-1)^p H'_p$. Thus

$$p(r, \phi, z) = i\rho_0 ckb \cos(p\phi) \cos(k_m z)\frac{H_p(\sqrt{k^2 - k_m^2}\, r)}{\sqrt{k^2 - k_m^2}\, H_p'(\sqrt{k^2 - k_m^2}\, a)}. \tag{4.90}$$

The farfield behavior in the ϕ and z directions is the same as the mode on the surface, as we saw in the plate example on page 26. The behavior in r is obscured by the Hankel functions. We use the asymptotic formula, Eq. (4.22), for the Hankel function in the numerator to obtain the farfield:

$$\begin{aligned} p(r, \phi, z) &= -(-i)^{p+1} e^{-i\pi/4}\sqrt{\frac{2}{\pi r}} \rho_0 ckb \cos(p\phi) \cos(k_m z) \\ &\times \frac{e^{i(\sqrt{k^2 - k_m^2}\, r)}}{(k^2 - k_m^2)^{3/4} H_p'(\sqrt{k^2 - k_m^2}\, a)}. \end{aligned} \tag{4.91}$$

When the wavenumber in the axial direction k_m is subsonic then

$$e^{i\sqrt{k^2 - k_m^2}\, r} = e^{-\sqrt{k_m^2 - k^2}\, r}$$

and the pressure decays exponentially into the farfield. When the wavenumber is supersonic, Eq. (4.91) shows that the pressure field decays as $1/\sqrt{r}$ in the farfield. We note that the decay is not $1/r$ because the vibration of the cylinder is infinite in the axial direction and thus the farfield can never get sufficiently far from the cylinder surface to reach the decay one would expect from a confined source.

4.6 Farfield Radiation - Cylindrical Sources

For more realistic vibrators of finite extent, it is not so easy to evaluate Eq. (4.72) since the integration over k_z can not be carried out. We turn to an asymptotic method which can be applied to evaluate this integral when r is very large, and the numerator is replaced with the farfield formula for the Hankel function. This method, which will lead to a $1/r$ decay in the radial direction in the farfield is called the stationary phase method. Like the Ewald sphere construction procedure discussed on page 41 this method will provide the farfield in spherical coordinates.

4.6.1 Stationary Phase Approximation

The k_z integration in Eq. (4.72) can be evaluated for an unknown surface velocity using the mathematical technique called the stationary phase approximation when the farfield is desired, that is, when r is large. Although not necessary for the following analysis it should be realized that the k_z integration in this case can be closely approximated by the finite integral

$$\int_{-k}^{k} \ddot{W}_n(a, k_z) \frac{H_n(k_r r)}{k_r H_n'(k_r a)} e^{ik_z z} dk_z,$$

since, when $k_z > k$, the argument of the Hankel function becomes imaginary, converting the Hankel function to the exponentially decaying (in k_r) $K_n(k_r' r)$.

In any case, as $r \to \infty$ we can replace the Hankel function in the numerator with the first term in its asymptotic expansion given in Eq. (4.22). At the same time we convert from cylindrical coordinates, with the transformations $r = R \sin \theta$ and $z = R \cos \theta$, to spherical coordinates R and θ defined in Fig. 4.1, page 116. Note the ϕ coordinate is the same in both cylindrical and spherical coordinate systems. Thus Eq. (4.72) becomes

$$p(R, \theta, \phi) \approx \frac{i\rho_0 c k}{2\pi} \sqrt{\frac{2}{\pi R \sin \theta}} \sum_n (-i)^n e^{in\phi} e^{-i\pi/4}$$

$$\times \int_{-\infty}^{\infty} \frac{\ddot{W}_n}{k_r^{3/2} H_n'(k_r a)} e^{iR(k_r \sin \theta + k_z \cos \theta)} dk_z. \tag{4.92}$$

This equation is now in an ideal form for the stationary phase evaluation. To understand how it works we need to examine the phase term in the exponential function. Consider the $e^{iRk_z \cos \theta}$ term. The exponential implies behavior of the form $\cos(Rk_z \cos \theta)$ and $\sin(Rk_z \cos \theta)$. Viewed as a function of k_z the argument of the cos or sin has a frequency given by the magnitude of R, so that the larger R is, the more the functions oscillate with respect to k_z. This kind of oscillation is illustrated in Fig. 4.13 for two different values of R. When we consider the effect of $e^{iRk_r \sin \theta} = e^{iR\sqrt{k^2 - k_z^2} \sin \theta}$ in addition, however, we find that there is some value of k_z for which the argument no longer oscillates (or oscillates very slowly). This point is called the stationary phase point.

Figure 4.13 illustrates the stationary phase point for two different values of R and demonstrates the effect as R increases; the real part of

$$e^{iR(\sqrt{k^2 - k_z^2} \sin \theta + k_z \cos \theta)}$$

is plotted given $k = 1$, and $\theta = \pi/3$, and two values of R, $R = 20$ and $R = 80$ in order to demonstrate the trend as $R \to \infty$. The stationary phase point is at $k_z = k \cos \theta = 0.5$.

The success of this technique relies on the fact that the rapidly oscillating regions will tend to cancel the contribution to the rest of the integrand, leaving only the integral around the stationary phase point. Around the stationary phase point there is no cancellation. We determine the stationary phase point by finding where the rate of change of the phase is zero, that is, by taking the derivative of the phase term with respect to k_z.

To derive a general formula, we evaluate the integral of the form

$$I(R) = \int_{-\infty}^{\infty} f(z) e^{iRg(z)} dz \tag{4.93}$$

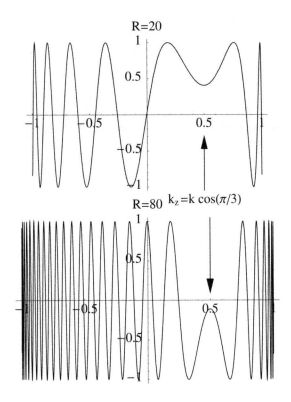

Figure 4.13: Stationary phase illustration.

using the stationary phase approximation. As discussed above, as $R \to \infty$ this formula becomes more accurate. The stationary phase point z_0, where the rate of change of phase is zero, is determined by the solution to the equation

$$\frac{dg(z)}{dz} = 0.$$

We assume that $f(z)$ is smoothly varying around the point z_0, so that it can be replaced by its value at z_0 and removed from the integral. That is,

$$I(R) \approx f(z_0) \int_{z_0 - \epsilon}^{z_0 + \epsilon} e^{iRg(z)} dz.$$

where ϵ is some small interval about the stationary phase point which becomes smaller as R increases. We have assumed that the rest of the integral is negligible due to the rapid oscillations, as already discussed.

To evaluate the integral we expand $g(z)$ in a Taylor series about the stationary phase point,

$$g(z) \approx g(z_0) + \frac{(z - z_0)^2}{2} g''(z_0),$$

where the first term is missing since $g'(z_0) = 0$. Furthermore, define

$$u \equiv (z - z_0)\sqrt{|g''(z_0)|},$$

so that

$$g(z) - g(z_0) \approx \pm u^2/2R, \tag{4.94}$$

where the plus sign is used when $g'' > 0$ and the negative sign when $g'' < 0$. Change variables so that Eq. (4.93) is integrated with respect to u instead of z. Thus, Eq. (4.93) becomes

$$I(R) \approx f(z_0)e^{iRg(z_0)} \int_{-\epsilon\sqrt{R|g''|}}^{\epsilon\sqrt{R|g''|}} e^{\pm iu^2/2} \left(\frac{dz}{du}\right) du. \tag{4.95}$$

From Eq. (4.94) we find that

$$\frac{dz}{du} = \frac{1}{\sqrt{R|g''(z_0)|}},$$

so that Eq. (4.95) becomes

$$I(R) \approx \frac{f(z_0)e^{iRg(z_0)}}{\sqrt{R|g''(z_0)|}} \int_{-\infty}^{\infty} e^{\pm iu^2/2} du,$$

where we have extended the limits of the integral out to infinity since $R \to \infty$. The u integral is known,

$$\int_{-\infty}^{\infty} e^{\pm iu^2/2} du = e^{\pm i\pi/4}\sqrt{2\pi}, \tag{4.96}$$

so that

$$I(R) \approx f(z_0)e^{iRg(z_0)}e^{\pm i\pi/4}\sqrt{\frac{2\pi}{R|g''(z_0)|}}, \tag{4.97}$$

where the positive signs are used when $g'' > 0$ and negative signs otherwise. Equation (4.97) is the general stationary phase formula for an integral in the form of Eq. (4.93).

4.6.2   Farfield of a General Velocity Distribution and k-space

We can now apply the stationary phase approximation to Eq. (4.92) which resulted from the Rayleigh-like integral, Eq. (4.72), when the asymptotic approximation was made for the Hankel function. We rewrite Eq. (4.92) as

$$p(R,\theta,\phi) \approx \frac{i\rho_0 ck}{\pi}\sqrt{\frac{1}{2\pi R \sin\theta}} \sum_n (-i)^n e^{-i\pi/4} e^{in\phi} I_n(R) \tag{4.98}$$

where

$$I_n(R) = \int_{-\infty}^{\infty} \frac{\ddot{W}_n}{k_r^{3/2} H_n'(k_r a)} e^{iR(k_r \sin\theta + k_z \cos\theta)} dk_z.$$

To use the formula for the stationary phase, Eq. (4.97), we replace z with k_z; and, in view of Eq. (4.93),

$$f(k_z) = \frac{\ddot{W}_n(a, k_z)}{k_r^{3/2} H_n'(k_r a)},$$

with

$$g(k_z) = k_r \sin\theta + k_z \cos\theta = \sqrt{k^2 - k_z^2} \sin\theta + k_z \cos\theta.$$

The stationary phase point is

$$\frac{dg}{dk_z} = -\frac{k_z \sin\theta}{\sqrt{k^2 - k_z^2}} + \cos\theta = 0$$

which leads to the important formula for the stationary phase point k_{z0}:

$$k_z = k_{z0} = k\cos\theta. \tag{4.99}$$

Using this result, and since

$$g(k_{z0}) = k.$$

then

$$f(k_{z0}) = \frac{\ddot{W}_n(a, k\cos\theta)}{(k\sin\theta)^{3/2} H_n'(ka\sin\theta)},$$

and

$$g''(k_{z0}) = -\frac{k^2 \sin\theta}{(k^2 - k_{z0}^2)^{3/2}} = \frac{-1}{k\sin^2\theta}.$$

Finally Eq. (4.97) ($g'' < 0$) yields

$$I_n(R) \approx \frac{\ddot{W}_n(a, k\cos\theta)}{k\sqrt{\sin\theta} H_n'(ka\sin\theta)} e^{ikR} \sqrt{\frac{2\pi}{R}} e^{-i\pi/4}.$$

We insert this result into Eq. (4.98) to obtain our final stationary phase result:

$$p(R, \theta, \phi) \approx +\frac{\rho_0 c}{\pi} \frac{e^{ikR}}{R} \sum_{n=-\infty}^{\infty} (-i)^n e^{in\sigma} \frac{\ddot{W}_n(a, k\cos\theta)}{\sin\theta H_n'(ka\sin\theta)}. \tag{4.100}$$

The stationary phase point, $k_z = k\cos\theta$, links k-space with real space and provides the intuitive trace matching condition as was displayed in Fig. 4.11. Thus at a particular frequency k the radiation to the farfield at an angle θ is dictated by a single component of the k-space transform of the surface velocity. Of course, only those components within the radiation circle radiate to the farfield, and in this case we can see the condition $|k_z| \leq k$ must be satisfied for trace matching and radiation to the farfield. Also the pressure in the farfield varies as $1/R$, as it must for radiation from finite bodies.

When $|n/a| > k$ we expect the resulting hydrodynamic short circuit in the circumferential direction to attenuate radiation to the farfield. We discussed this effect in Section 4.4.1, with regard to evanescent waves and found that this short circuit led to a power law decay. This decay is not explicitly indicated in Eq. (4.100), which instead exhibits a complicated dependence upon the circumferential wavenumber n/a.

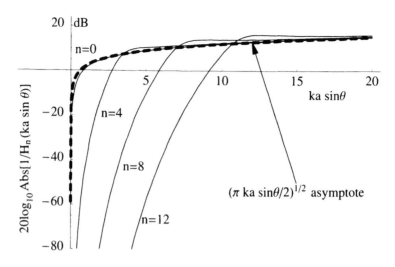

Figure 4.14: $20 \log_{10} |1/H'_n (ka \sin \theta)|$ for $n = 0, 4, 8, 12$. The asymptotic farfield breaks at about $n = ka \sin \theta$. When the argument is less than the order, the magnitude is small.

We can demonstrate the existence of a radiation circle by plotting the dependence of $1/H'_n(ka \sin \theta)$ on n as shown in Fig. 4.14. We see that the Hankel function is a strong filter with respect to n and that when $n > ka \sin \theta$ the radiation to the farfield is greatly diminished since $1/H'_n$ in Eq. (4.100) becomes very small. Given the break point at $ka \sin \theta = n$, this equation forms a circle in k-space for all possible polar angles, $0 \le \theta \le \pi$, and positive and negative n for fixed k. This is shown in Fig. 4.15. The farfield spherical angle θ is the angle between k and the k_z axis and is defined by $\cos \theta = k_z/k$.

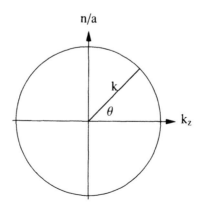

Figure 4.15: Radiation circle for a cylindrical vibrator. If n/a and k_z fall within the circle then there is efficient radiation to the farfield.

Thus we can restrict the infinite limits on the sum and rewrite Eq. (4.100) as

$$p(R,\theta,\phi) \approx +\frac{\rho_0 c}{\pi}\frac{e^{ikR}}{R}\sum_{n=-N}^{N}(-i)^n e^{in\phi}\frac{\dot{W}_n(a,k\cos\theta)}{\sin\theta H_n'(ka\sin\theta)}, \tag{4.101}$$

where $N \approx ka\sin\theta$.

To determine the limit on N in a more quantitative light, we consider the leakage of power from harmonics $n > N$ for the case of an axial line source on an infinite cylinder with no axial variation ($k_z = 0$), given in Eq. (4.88). Outside the radiation circle n/a is subsonic, however, energy still reaches the farfield: the subsonic circumferential waves leak to the farfield. The power per unit length of each harmonic radiated from the axial line source, as given in Eq. (4.88), is proportional to $1/|H_n'(ka)|^2$ which we plot versus n in Fig. 4.16, normalized to the maximum for $ka = 1, 5, 10$. The figure shows that

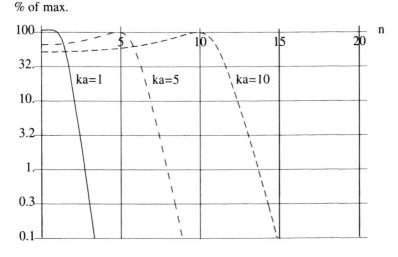

Figure 4.16: Normalized power per unit length as a function of circumferential harmonic for a line source, Eq. (4.88). For $ka = 1, 5, 10$ the percentage of the maximum Π_n/L versus n is plotted to illustrate the power leakage for subsonic circumferential waves.

$\Pi_n(ka)/L$ is maximum near $ka = n$ and then drops sharply. For example, for $ka = 5$, Π_n/L is maximum for $n = 5$, and drops to 10% of this value when $n = 7$ and is less than 0.1% for $n = 9$. In conclusion, because of the sharp drop off in the power outside the radiation circle, defined in Fig. 4.15, we treat the concept of the radiation circle in the same way as we did for plates, realizing, however, that the borderline between radiating and nonradiating helical waves is a bit fuzzy. Or, in other words, there is a taper between the radiating and nonradiating parts of the radiation circle which follows a power law decay when $|k_z| < k$ (for supersonic axial waves). Of course, when the axial wavenumber is subsonic there is no leakage to the farfield: there is no radiation.

By using the stationary phase method one can derive a formula similar to Eq. (4.101), for the farfield given the helical wave amplitude of the pressure $P_n(a, k_z)$ instead of the velocity. That result is

$$p(R, \theta, \phi) \approx \frac{e^{ikR}}{\pi R} \sum_{n=-N}^{N} (-i)^{n+1} e^{in\phi} \frac{P_n(a, k \cos \theta)}{H_n(ka \sin \theta)}. \tag{4.102}$$

There is an important observation to be made yet with respect to Eq. (4.100). What is the behavior of the sum when the farfield pressure is on the axis, $\theta = 0$ or π? In this case the argument of the Hankel function is zero. From the small argument expression for H_n', Eq. (4.29), the denominator of Eq. (4.100) is

$$\sin \theta H_n'(ka \sin \theta) \approx \frac{i}{ka} \frac{2^{n+1} n!}{\pi \epsilon_n} \frac{1}{(ka \sin \theta)^n},$$

where $n \geq 1$. For all values of $n \geq 1$ this expression is infinite on axis, so that the denominator of Eq. (4.100) is infinite. Thus only the $n = 0$ term remains of the sum over n. When $n = 0$, $H_0'(x) \approx 2i/\pi x$ and

$$\sin \theta H_0'(ka \sin \theta) \approx \frac{1}{ka} \frac{2i}{\pi}.$$

The pressure in the farfield is non-zero on the axis. The important conclusion is that only the $n = 0$ mode of the surface velocity contributes to the pressure on the axis of the cylinder. This is generally true, in the nearfield and in the farfield, whether the vibrator is baffled or unbaffled.

Similarly, it is evident that as $ka \to 0$ ($N \to 0$) only the $n = 0$ term will be important, but now at any angle θ. Thus, we conclude that at very low frequencies only $n = 0$ radiation reaches the farfield. This is called breathing mode radiation.

One observation on the finiteness of the vibrator implied in the stationary phase method. Review Eq. (4.91) which provided the farfield pressure for a source which was infinite in extent, carrying a standing wave on its surface. This equation indicated that the farfield pressure decayed only as $1/\sqrt{r}$ not as $1/R$. Clearly this is in contradiction with the stationary phase result, Eq. (4.100). It must be realized that Eq. (4.100) applies only to finite vibrators. that is vibrators which have a vanishing radial velocity as $|z| \to \infty$. This condition was not explicitly stated in the derivation of Eq. (4.100), but is implied in the assumption that $\dot{W}(a, k_z)$ is a slowly varying function in comparison with the phase oscillations. On the contrary. Eq. (4.91) had $\dot{w}(z) = \cos(m\pi z/L)$ which implies that $\dot{W}_n(k_z) \sim \delta(k_z \pm m\pi/L)$, certainly not a slowly varying function. Thus the restriction of $\dot{W}_n$ to slowly varying is equivalent to the assumption of a finite area of vibration on the cylinder.

4.6.3   Piston in a Cylindrical Baffle

As an example of the use of Eq. (4.100) we consider a rigid moving piston in an infinite cylindrical baffle. The baffle is rigid so that the normal velocity is zero everywhere except on the piston. The rigid baffle provides a mathematically tractable solution, as we found for the Rayleigh integral for planar radiators. Otherwise one must use more general techniques to solve for the radiation which are significantly more complicated.

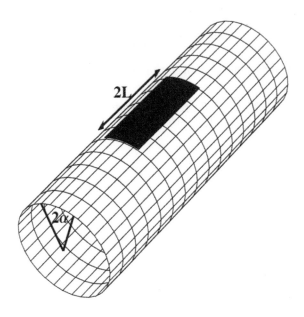

Figure 4.17: Geometry for the piston of length $2L$ and angular width 2α on the surface of an infinite cylinder.

The geometry of the present problem is shown in Fig. 4.17 representing a piston of length $2L$ and width 2α on a cylinder of radius a. The center of the patch is at $\phi = 0$.
If b is the velocity of the piston, the transform, Eq. (4.73) on page 134, is

$$\dot{W}_n(a, k_z) = \frac{b}{2\pi} \int_{-\alpha}^{\alpha} e^{-in\phi} d\phi \int_{-L}^{L} e^{-ik_z z} dz.$$

Integrating yields

$$\dot{W}_n(a, k_z) = \frac{4b\alpha L}{2\pi} \operatorname{sinc}(n\alpha) \operatorname{sinc}(k_z L). \tag{4.103}$$

Substitution of this result back into Eq. (4.100), using the fact that $k_z = k\cos\theta$, yields the farfield radiation from the piston:

$$p(R, \theta, \phi) \approx \frac{\rho_0 c}{2\pi^2} \frac{e^{ikR}}{R} \sum_{n=-N}^{N} (-i)^n e^{in\phi} \frac{4b\alpha L \operatorname{sinc}(n\alpha) \operatorname{sinc}(kL\cos\theta)}{\sin\theta H_n'(ka\sin\theta)}. \tag{4.104}$$

The $\operatorname{sinc}(kL\cos\theta)$ term is maximum when its argument goes to zero, which corresponds to an angle of $\theta = \pi/2$, that is, normal to the axis. This is the same result found for the planar piston radiator, Eq. (2.100) on page 43 with $\phi = 0$. The farfield dependence in the ϕ direction is more evident *in the limit of large ka if we use the* asymptotic expansion of the denominator

$$H'_n(x) \sim (-i)^n \sqrt{\frac{2}{\pi x}} e^{i\pi/4} e^{ix}, \tag{4.105}$$

and assume that α is vanishingly small so that $\mathrm{sinc}(n\alpha) \approx 1$. Then Eq. (4.104) becomes

$$p(R, \theta, \phi) \approx \frac{4b\alpha L \rho_0 c}{\pi} \frac{e^{ikR}}{2\pi R} \sqrt{\frac{\pi ka}{2\sin\theta}} e^{-ika\sin\theta - i\pi/4} \mathrm{sinc}(kL\cos\theta) \sum_{n=-\infty}^{\infty} e^{in\phi}.$$

The summation is just

$$\sum_{n=-\infty}^{\infty} e^{in\phi} = 2\pi\delta(\phi), \tag{4.106}$$

and it is seen that the radiation is directive like a delta function at the angle $\phi = 0$, the circumferential position of the piston vibrator. Thus the radiation from the piston source has its maximum on a beam normal to the axis of the cylinder for large ka.

It is instructive to evaluate Eq. (4.104) in the limit of low frequency when the dimensions of the vibrator are much less than a wavelength in either direction. For this we need to use the small-argument expression for H'_n, given in Eq. (4.29) on page 119. When ka is very small only the $n = 0$ term will dominate in the sum in Eq. (4.104), as discussed in the last section. Using

$$H'_0(ka\sin\theta) \approx \frac{2i}{\pi ka \sin\theta}$$

then

$$p(R, \theta, \phi) \approx \frac{-i\rho_0 ck}{4\pi} Q \frac{e^{ikR}}{R} \tag{4.107}$$

where the volume flow is $Q = 4\alpha a Lb$. Equation 4.107 reveals that in the low frequency limit, the piston in a cylindrical baffle looks like a simple source in free space. Compared to a point source in an infinite planar baffle, Eq. (2.77) on page 38, the pressure is half as much. The lack of an extended baffle in the circumferential direction is the cause of this reduction in pressure.

4.6.4 Radiation from a Confined Helical Wave in a Cylindrical Baffle

Study of the traveling wave radiation from a baffled vibrator is important for the understanding of the waves which travel on real shell vibrators and how they radiate to the farfield. We consider here a helical wave of amplitude b confined only axially to a region $-L/2 \leq z \leq L/2$ on a baffled cylindrical shell of radius a. The wave has a circumferential harmonic $n = p$ and a wavenumber in the axial direction given by k_q. The radial velocity is thus

$$\dot{w}(a, \phi, z) = be^{ip\phi}e^{ik_q z}\Pi(z/L). \tag{4.108}$$

where Π is the window function given in Eq. (1.41). Taking Fourier transforms of Eq. (4.108) yields

$$\dot{W}_n(a, k_z) = 2\pi bL\delta_{pn}\text{sinc}[(k_q - k_z)L/2]. \tag{4.109}$$

The farfield pressure of the confined helical wave, using Eq. (4.100) with the stationary phase point $k_z = k \cos\theta$, is

$$p(R, \theta, \phi) = 2\rho_0 cbL\frac{e^{ikR}}{R}(-i)^p e^{ipo}\frac{\text{sinc}[(k_q - k\cos\theta)L/2]}{\sin\theta H'_p(ka\sin\theta)}. \tag{4.110}$$

Examination of Eq. (4.110) reveals that the maximum in the farfield radiation occurs near the angle $k_q = k\cos\theta$, depending on the variation of the denominator with respect to θ at that point. The sinc function creates a series of side lobes similar to the result obtained for a baffled plate with a traveling wave, Fig. 2.17 on page 47. When k_q is subsonic the main lobe of the sinc function disappears below the horizon and its side lobes determine the radiation pattern in θ.

The circumferential variation of the helical wave is unchanged as it radiates to the farfield, remaining as $e^{ip\phi}$. When it is subsonic $(n/a > k)$ on the surface of the radiator the helical wave amplitude decays, following the power law as illustrated and discussed in Fig. 4.14.

4.7   Radiated Power

The radiated power can be computed from integrals of the radial intensity on the surface of the cylinder or the intensity in the farfield. In the farfield the intensity is integrated over a sphere,

$$\Pi(\omega) = \frac{1}{2\rho_0 c}\int_0^{2\pi}\int_0^\pi |p(R, \theta, \phi)|^2 R^2 \sin\theta \, d\theta d\phi. \tag{4.111}$$

The farfield pressure is determined from the stationary phase result given in Eq. (4.100). The power computed over a cylindrical surface (r constant) is

$$\Pi(\omega) = \frac{1}{2}\text{Re}\int_0^{2\pi}\int_{-\infty}^\infty rd\phi dz \, p(r, \phi, z)\dot{w}^*(r, \phi, z). \tag{4.112}$$

Various formulas for the power in terms of the helical wave amplitudes can be derived. Using the fact that

$$p(a, \phi, z) = \sum_{n=-\infty}^\infty \frac{1}{2\pi}\int_{-\infty}^\infty dk_z P_n(a, k_z)e^{in\phi}e^{ik_z z}, \tag{4.113}$$

and

$$\dot{w}^*(a, \phi, z) = \sum_{n=-\infty}^\infty \frac{1}{2\pi}\int_{-\infty}^\infty dk_z P_n^*(a, k_z)\frac{ik_r H_n^{*'}(k_r a)}{\rho_0 ck H_n^*(k_r a)}e^{-in\phi}e^{-ik_z z}. \tag{4.114}$$

along with the Wronskian formula

$$\mathcal{W}[J_n(z), Y_n(z)] = J_{n+1}(z)Y_n(z) - J_n(z)Y_{n+1}(z) = 2/(\pi z),$$

it can be shown that Eq. (4.112) becomes

$$\Pi = \frac{1}{\pi \rho_0 c k} \int_{-k}^{k} dk_z \sum_{n=-\infty}^{\infty} \frac{|P_n(a, k_z)|^2}{J_n^2(k_r a) + Y_n^2(k_r a)}. \tag{4.115}$$

Given the relationship between $\dot{W}_n$ and P_n in Eq. (4.68) on page 133, we can show that

$$\Pi = \frac{\rho_0 c k}{\pi} \int_{-k}^{k} \frac{dk_z}{k^2 - k_z^2} \sum_{n=-\infty}^{\infty} \frac{|\dot{W}_n(a, k_z)|^2}{J_n'^2(k_r a) + Y_n'^2(k_r a)}. \tag{4.116}$$

Note that the denominator is just

$$J_n'^2(k_r a) + Y_n'^2(k_r a) = |H_n'|^2.$$

(Compare with Eq. (4.88).) This formula can also be derived starting with Eq. (4.100), using Eq. (4.111) and a change of variables.

Interpreted in k-space these last two equations reveal that no power is radiated by helical waves when $k_z > k$ (k_z subsonic). However, all the circumferential orders radiate when k_z is supersonic, although above $n = ka$ the power drops off dramatically, as discussed in Section 4.6.2, and in particular in Fig. 4.16.

Problems

4.1 Consider the general solution, Eq. (4.46), for the case shown in Fig. 4.5. We are given the boundary field at $r = a$,

$$p(r = a, \phi, z) = 0,$$

for all values of ϕ and z. (The surface is pressure release). The outer surface at $r = b$ has an infinite helical wave component traveling on it given by

$$p(r = b, \phi, z) = p_0 e^{ip\phi} e^{ik_q z}.$$

Determine the unknown coefficients. $A_n(k_z)$ and $B_n(k_z)$ in Eq. (4.46) and find the equation for the resulting pressure in the annulus.

4.2 The radial surface velocity on an infinite pipe is known and given by $\dot{w}(a, \phi, z)$. Using Eq. (4.49) derive a Rayleigh-like formula for the pressure in the interior given $\dot{w}(a, \phi, z)$.

4.3 Derive Eq. (4.115)

4.4 Using Eq. (4.100) and Eq. (4.111) derive Eq. (4.116).

4.5 Use the stationary phase technique to find an asymptotic representation of

$$f(x) = \int_0^\infty \cos[x(\omega^3 - \omega)] d\omega,$$

valid as $x \to \infty$.[4]

---

[4]First used by G. G. Stokes (1883). "Mathematical and Physical Papers," vol. 2, p. 329, Cambridge University Press.

Chapter 5

The Inverse Problem: Cylindrical NAH

5.1 Introduction

This chapter is concerned with the implementation of NAH in a cylindrical geometry. In many ways the implementation is identical to the planar case, although there is one significant difference. Finite aperture effects exist only in the axial direction. Again we will discuss the errors associated with cylindrical NAH, similar to the planar developments. Much of this chapter will discuss actual experimental results, which we did not do for the planar case, in order to give the reader a greater appreciation of the power and accuracy of NAH, and to provide examples of actual physical experiments using NAH. All of these examples deal with the external problem–radiation from a vibrating cylindrical structure into the medium outside. The theory for the internal problem will be presented in a later chapter.

5.2 Overview of the Inverse Problem

We consider the exterior problem, as shown in Fig. 4.7, page 124. The inverse problem can be stated in measurement terms: The acoustic pressure is measured on an infinite cylindrical surface at radius $r = r_h$, mathematically backpropagated to an inner surface at $r = a$, $(a < r_h)$, determining (reconstructing) the pressure and vector velocity on the inner surface. These reconstructions are done in the temporal frequency domain. In the time domain the inverse problem is equivalent to back-tracking the acoustic field in time from the outer to the inner surface, with all sources located inside or on the inner surface.

To reconstruct the pressure, we start with Eq. (4.58) (page 126) and consider $r = a$ to be the surface which just encloses the sources in the interior, as shown in Fig. 4.7. The pressure measurement is made on the surface at $r = r_h$. We can invert Eq. (4.58) by solving for $P_n(a, k_z)$,

$$P_n(a, k_z) = \frac{H_n^{(1)}(k_r a)}{H_n^{(1)}(k_r r_h)} P_n(r_h, k_z). \tag{5.1}$$

We simplify the mathematical notation here by using the symbols to represent the Fourier transform and Fourier series with the same definitions as before:

$$\mathcal{F}_z[p(\phi, z)] \equiv \int_{-\infty}^{\infty} p(\phi, z) e^{-ik_z z} dz, \tag{5.2}$$

$$\mathcal{F}_\phi[p(\phi, z)] \equiv \frac{1}{2\pi} \int_0^{2\pi} p(\phi, z) e^{-in\phi} d\phi, \tag{5.3}$$

for the forward transforms. The helical wave spectrum amplitude P_n is thus

$$P_n(r, k_z) \equiv \mathcal{F}_z \mathcal{F}_\phi[p(r, \phi, z)].$$

The inverse transforms are then

$$\mathcal{F}_z^{-1}[P_n(r, k_z)] \equiv \frac{1}{2\pi} \int_{-\infty}^{\infty} P_n(r, k_z) e^{ik_z z} dk_z, \tag{5.4}$$

$$\mathcal{F}_\phi^{-1}[P_n(r, k_z)] \equiv \sum_{n=-\infty}^{\infty} P_n(r, k_z) e^{in\phi}, \tag{5.5}$$

and

$$p(r, \phi, z) = \mathcal{F}_z^{-1} \mathcal{F}_\phi^{-1}[P_n(r, k_z)].$$

Thus Eq. (5.1), applying inverse transforms, becomes

$$p(a, \phi, z) = \mathcal{F}_z^{-1} \mathcal{F}_\phi^{-1} \left\{ \frac{H_n^{(1)}(k_r a)}{H_n^{(1)}(k_r r_h)} \mathcal{F}_z \mathcal{F}_\phi[p(r_h, \phi, z)] \right\}. \tag{5.6}$$

Written in terms of the convolution integrals in ϕ, given by Eq. (1.27) (page 5), and in z, given in Eq. (1.12) on page 3, Eq. (5.6) is (removing the superscripts on the Hankel functions for simplicity of notation)

$$
\begin{aligned}
p(a, \phi, z) &= \frac{1}{2\pi} \mathcal{F}_z^{-1} \mathcal{F}_\phi^{-1} [\frac{H_n(k_r a)}{H_n(k_r r_h)}] * * p(r_h, \phi, z) \\
&= \frac{1}{2\pi} g_p^{-1}(a, r_h, \phi, z) * * p(r_h, \phi, z), \tag{5.7}
\end{aligned}
$$

where g_p^{-1} is called the inverse pressure propagator defined by

$$g_p^{-1}(a, r_h, \phi, z) \equiv \mathcal{F}_z^{-1} \mathcal{F}_\phi^{-1} [\frac{H_n(k_r a)}{H_n(k_r r_h)}]. \tag{5.8}$$

We can also solve the inverse problem to reconstruct the radial velocity on the surface $r = a$ (as well as the axial and circumferential velocity following the same approach) using Eq. (4.68) with $r = a$ and $r_0 = r_h$. Again applying the inverse Fourier transforms Eq. (4.68) yields

$$\dot{w}(a, \phi, z) = \mathcal{F}_z^{-1}\mathcal{F}_\phi^{-1}\left\{\frac{-ik_r H_n'(k_r a)}{\rho_0 ck H_n(k_r r_h)}\mathcal{F}_z\mathcal{F}_\phi[p(r_h, \phi, z)]\right\}, \qquad (5.9)$$

and again using the convolution theorems

$$\dot{w}(a, \phi, z) = \frac{1}{2\pi}g_v^{-1}(a, r_h, \phi, z) * *p(r_h, \phi, z), \qquad (5.10)$$

where the inverse velocity propagator g_v^{-1} is given from Eq. (5.9) as

$$g_v^{-1}(a, r_h, \phi, z) \equiv \mathcal{F}_z^{-1}\mathcal{F}_\phi^{-1}[\frac{-ik_r H_n'(k_r a)}{\rho_0 ck H_n(k_r r_h)}]. \qquad (5.11)$$

Equations (5.6) and (5.9) represent the basic *reconstruction equations of cylindrical NAH*. They summarize the mathematical operations needed to reconstruct the normal velocity and pressure given a measurement of the pressure on a cylindrical surface at $r = r_h$. The corresponding equation for planar holography was given in Eq. (3.4) (page 90). As in the planar case these equations involve four basic operations:

(1) measure the pressure on a cylinder at $z = z_h$, $\rightarrow p(r_h, \phi, z)$,

(2) compute the helical wave spectrum, $\rightarrow P_n(r_h, k_z)$,

(3) multiply by the inverse propagators, given by the terms in square brackets in Eq. (5.8) and Eq. (5.11), $\rightarrow P_n(a, k_z)$ and $\dot{W}_n(a, k_z)$,

(4) compute the inverse transforms, $\rightarrow p(a, \phi, z)$ and $\dot{w}(a, \phi, z)$.

The success in the application of Eq. (5.6) and Eq. (5.9) relies on the inclusion of the evanescent waves generated at the source surface. We have seen that these waves decay exponentially or by a power law as they expand radially from the surface. For the inverse problem, the field increases exponentially or by a power law so that special consideration must be given to the dynamic range of the measurement system in order to capture these waves on the measurement cylinder. The nature of the evanescent waves was already discussed in Section 4.4.1. The resolution in the reconstructed image depends upon the dynamic range of the measurement system as we will now discuss.

The inverse nature of the holographic problem is borne out by the fact that the inverse propagators given in Eq. (5.8) and Eq. (5.11) *do not exist*. They are singular at best and thus any attempt to compute g_p^{-1} or g_v^{-1} will lead to diverging results. This is, of course, no different from the case for the planar geometry, studied in chapter 3. We will see that again we can eliminate the singularities by filtering in k-space so as to eliminate the evanescent waves which are beyond the dynamic range of our measurement system.

5.2.1   Resolution of the Reconstructed Image

It is important to realize that the evanescent waves contain the fine detail, high resolution information about the source. We must be able to measure these components, if we want to reconstruct fields at low temporal frequencies. Since these waves decay rapidly from the surface, we must measure the fields close to the surface with a measurement system with sufficient dynamic range. To develop an approximate relationship for the resolution consider a source with a constant Fourier spectrum in k_z, that is, $P_n(a, k_z) = P_0$ where $r = a$ is the surface of the cylinder. The ratio between the evanescent wave components (represented by k'_z) and the non-evanescent components on the measurement surface $r = r_h$ is

$$\left| \frac{P_n(r_h, k'_z)}{P_n(r_h, k_z)} \right| \approx \left| \frac{\sqrt{a/r_h}\, e^{-k'_r d} P_n(a, k'_z)}{\sqrt{a/r_h}\, e^{ik_r d} P_n(a, k_z)} \right| = e^{-k'_r d}, \tag{5.12}$$

where $d = r_h - a$. To measure the evanescent component k'_r the dynamic range of the measurement system must be better than this ratio, that is,

$$10^{-D/20} < e^{-k'_r d}, \tag{5.13}$$

where D is the dynamic range in decibels. Keeping d and D fixed, clearly there is a maximum value of k'_r for which this inequality will still hold. It is this value which determines the axial resolution. In other words, define the axial resolution Z_r as one half the axial wavelength,

$$Z_r = \lambda_{z0}/2,$$

where $\lambda_{z0} = 2\pi/k_{z0}$ is the smallest axial wavelength corresponding to this maximum value of k'_r. Now we can write $k'_r = \sqrt{k_{z0}^2 - k^2} \approx 2\pi/\lambda_{z0} = \pi/Z_r$, so that Eq. (5.13) becomes, solving for Z_r in terms of k'_r,

$$Z_r = \frac{20\pi d \log(e)}{D} = 27.3(d/D). \tag{5.14}$$

Equation (5.14) displays the important relation between the location of the measurement surface and the resolution Z_r desired in the backward propagation process. Inserting some representative values into this equation one is easily convinced that the measurement surface must be located close to the acoustic source to obtain super resolution; resolution better than the wavelength in the medium. Any hopes of making super resolution measurements far away from the source are proven impossible by Eq. (5.14). It is interesting to note that this result is identical to that obtained for planar holography, Eq. (3.23).

In the above derivation we determined the axial resolution for the case where $\lambda_z \ll \lambda$. Below we carry out a similar development for the corresponding case of circumferential resolution, $\lambda_\phi \ll \lambda$, which, as we have seen, results in a power law decay of the pressure. We turn to Eq. (4.64) (page 131) instead of Eq. (5.12) resulting in

$$D \approx -20\log(a/r_h)^n, \tag{5.15}$$

where D is the dynamic range as before and n is the largest order of interest. Noting that $r_h = a + d$, and assuming that $d < a$. we have

$$(a/r_h)^n \approx 1 - nd/a. \tag{5.16}$$

By analogy to Z_r we define the circumferential resolution C_r on the surface of the cylinder by ($\lambda_c = 2\pi a/n$)

$$C_r = \lambda_c/2 = \pi a/n. \tag{5.17}$$

Inserting Eq. (5.16) into Eq. (5.15), solving for n, and keeping the first term of the Taylor expansion of the logarithm (assuming $nd/a \ll 1$), Eq. (5.17) yields

$$C_r \approx 20\pi d \log(e)/D = 27.3(d/D). \tag{5.18}$$

This result is identical to Eq. (5.14) and, as one might expect, the criteria for axial and circumferential resolution are the same.

5.2.2   The k-space Filter

It should be quite evident now, having considered the resolution limitation question, that when we backward propagate from the measurement surface r_h to the surface $r = a$ as prescribed by Eq. (5.6) and Eq. (5.9), we must limit the values of n and k_z used in the inverse transforms, Eq. (5.4) and Eq. (5.5). This limit is determined by the dynamic range of the measurement system and $d = r_h - a$. Thus, in view of Eq. (5.17) and Eq. (5.18), we must limit n to

$$|n| < \frac{\pi a}{27.3d/D}, \tag{5.19}$$

and, in view of Eq. (5.14), we limit k_z to

$$|k_z| = |2\pi/\lambda_z| < \frac{\pi}{27.3d/D}. \tag{5.20}$$

Limiting the values of n and k_z corresponds to a filter in transform space. Due to the equal roles of these two wavenumbers for the decaying pressure field, the k-space filter is chosen to be circular in k-space, as it was for rectangular coordinates, Eq. (3.25) on page 100. Again the taper of this window is chosen to be an exponential and we use Eq. (3.25) for $\bar{\Pi}$ directly by replacing k_x with k_z. and k_y with $k_o = n/a$. If we define k_t as

$$k_t \equiv \sqrt{k_z^2 + (n/a)^2}, \tag{5.21}$$

then a good estimate of the filter cutoff is

$$k_c = k_{t0} = \sqrt{\frac{k_{z0}^2 + (n_0/a)^2}{2}} \tag{5.22}$$

where

$$k_{z0} = n_0/a = \frac{\pi}{27.3d/D}$$

with k_{z0} and n_0 given by Eq. (5.20) and Eq. (5.19). Thus the reconstruction equations, Eq. (5.6) and Eq. (5.9), are approximated by

$$\tilde{p}(a,\phi,z) = \mathcal{F}_z^{-1}\mathcal{F}_\phi^{-1}\left\{\frac{H_n^{(1)}(k_r a)}{H_n^{(1)}(k_r r_h)}\bar{\Pi}(k_t/2k_c)\mathcal{F}_z\mathcal{F}_\phi[p(r_h,\phi,z)]\right\}, \qquad (5.23)$$

and

$$\tilde{w}(a,\phi,z) = \mathcal{F}_z^{-1}\mathcal{F}_\phi^{-1}\left\{\frac{-ik_r H_n'(k_r a)}{\rho_0 ck H_n(k_r r_h)}\bar{\Pi}(k_t/2k_c)\mathcal{F}_z\mathcal{F}_\phi[p(r_h,\phi,z)]\right\}. \qquad (5.24)$$

Real sources will exhibit a spectrum which is essentially band limited, that is, only components below some maximum value of n and k_z will exist to any significant degree. One must be sure that these dominant wavenumbers of the source are in the passband of this filter specified by Eq. (5.21). We can conclude from Eq. (5.14) and Eq. (5.18) that the source should satisfy the condition

$$\lambda_{\text{structure}} > 54.6(d/D). \qquad (5.25)$$

Fortunately, in most cases the elastic (or stiff) nature of the vibrator limits it from exhibiting extremely small variations in space (high frequency evanescent waves), thus band-limiting its wavenumber spectrum. However, the limitation given in Eq. (5.25) only applies when the frequency of excitation is below the coincidence frequency of the source. Above coincidence these small variations no longer generate evanescent waves, since the wavelengths on the structure become larger than λ. There is generally no need for windowing in this case.

Equation (5.25) provides a useful guide for nearfield acoustical holography. For example, if our measurement system has a dynamic range of 55 dB, then Eq. (5.25) indicates that we must keep the distance from the source to the hologram surface d less than the smallest wavelength of interest of the source, that is,

$$d < \lambda_{\text{structure}}. \qquad (5.26)$$

5.3   Implementation of Cylindrical Nearfield Acoustical Holography

Almost all the details of implementation are identical to the planar case, and thus we discuss in this chapter only a few details and encourage the reader to refer back to Chapter 3 to fill in the gaps.

5.3.1   Use of the Fast Fourier Transform (FFT)

The basic equations Eq. (4.52) (page 125). Eq. (5.6), and Eq. (5.9) which are the foundation of cylindrical nearfield acoustical holography are implemented on a computer using the fast Fourier transform (FFT). The differentials dz and $d\phi$ of Eq. (4.52) are given by $\Delta z = L/N$ and $\Delta \phi = 2\pi/N$, representing the spatial sampling intervals for the pressure field on the cylindrical contour, where N is the number of intervals in both the axial and circumferential directions (which we make equal for sake of simplicity) and L is the total length of the measurement aperture in the axial direction. The translations of these equations into FFT form was presented in Section 1.8. Using the results shown in Eqs (1.57) and (1.58), Eq. (4.52) is then approximated by forward FFTs in ϕ and z as

$$P_n \approx \tilde{P}_{n'-N/2}(r_h, k_{zm'}) = \frac{L}{N^2}(-1)^{m'+n'} \sum_{q'=0}^{N-1} e^{-i\frac{2\pi m' q'}{N}}$$

$$\times \sum_{p'=0}^{N-1} e^{-i\frac{2\pi n' p'}{N}} (-1)^{p'+q'} p(r_h, \phi_{p'}, z_{q'}), \qquad (5.27)$$

where

$$
\begin{aligned}
z_{q'} &= q'\Delta z - L/2, \\
\phi_{p'} &= p'\Delta \phi - \pi, \\
k_{zm'} &= m'\Delta k_z - N\Delta k_z/2, \\
\Delta k_z &= 2\pi/L, \quad \text{and} \\
L/\Delta z &= N.
\end{aligned}
\qquad (5.28)
$$

$p(r_h, \phi_{p'}, z_{q'})$ represents the sampled acoustic pressure on a given cylindrical contour at $r = r_h$. We use the tilde over a variable to represent the FFT result to distinguish it from the exact values (with no tilde). Note that the relation for Δk_z is basic to the FFT, and indicates that the largest wavelength provided by the FFT is equal to L. The discretized spatial frequencies k_z are just integer multiples of 2π divided by this wavelength.

The inverse transforms given in Eq. (5.6) are approximated by inverse FFTs as specified in Eqs (1.63) and (1.64) using n and k_{zm} ($k_{rm} = \sqrt{k^2 - k_{zm}^2}$) yielding an estimate of the pressure on the source surface:

$$p \approx \tilde{p}(a, \phi_q, z_l) = \frac{N^2}{L}(-1)^{q'+p'} \sum_{m'=0}^{N-1} e^{i\frac{2\pi m' q'}{N}}$$

$$\times \sum_{n'=0}^{N-1} e^{i\frac{2\pi n' p'}{N}} (-1)^{n'+m'} \frac{H_{n'-\frac{N}{2}}(k_{rm'}a)}{H_{n'-\frac{N}{2}}(k_{rm'}r_h)} \tilde{P}_{n'-\frac{N}{2}}(r_h, k_{zm'}). \; (5.29)$$

At this point we are interested in determining how close the estimated pressure field $\tilde{p}$ is to the actual pressure field p at $r = a$. (If we set $a = r_h$, then the FFT guarantees that the right hand side of Eq. (5.29) reduces to the sampled pressure $p(r_h, \phi_p', z_q')$ of Eq. (5.27).) Now we investigate the errors produced by the discretization of the reconstruction equation, Eq. (5.6). We will draw upon some experimental results to provide illumination along the way.

5.3.2   Errors Due to Discretization and Finite Scan Length

In experiments described in Section 5.4 a pressure measurement grid of 64×64 points in ϕ and z ($N = 64$) was used. This collected surface of data is called a hologram. It is usually obvious at the time of the experiment whether or not enough points in ϕ and z were taken. A preliminary check near the vibrating cylinder surface indicates the rapidity of variation of the pressure field. From this preliminary data the appropriate discretization, Δz and $\Delta \phi$, can be chosen. In this particular set of experiments both ϕ and z were oversampled for ease of analysis.

Let us first address the issue of the finite axial scan length. For a given scan length L, if we let Δz and $\Delta \phi$ be small (with N correspondingly large), then Eq. (5.27) is an accurate representation of the following approximation (with finite limits in z) to Eq. (4.52):

$$P_n(r_h, k_z) \approx \tilde{P}_n(r_h, k_z) = \frac{1}{2\pi} \int_0^{2\pi} d\phi \, e^{-in\phi} \int_{-L/2}^{L/2} dz \, e^{-ik_z z} p(r_h, \phi, z). \qquad (5.30)$$

Equation (5.30) is accurate for P_n only if the finite limits on the second integration are sufficiently large so that the significant part of the pressure field is included within the limits. Minimally, L must be larger than the extent of the source, with the pressure field decaying towards the ends of the aperture. If it decays sufficiently then we would expect the FFT to provide an accurate wavefunction amplitude P_n. Fortunately, since the pressure field is measured very close to the cylinder, one needs to overscan the cylinder only a relatively small amount to arrive at accurate results. Generally it is difficult to quantify the errors due to the finite aperture, but we hope that the following discussion will show the soundness of our arguments, at least for the case at hand.

We draw upon an experiment example, taken from Section 5.4. The measured hologram shown in Fig. 5.1 provided excellent reconstruction results as documented in Section 5.4. Figure 5.1 shows the measured acoustic pressure at $r = r_h$ in the axial (part a) and circumferential (part b) directions from an actual experiment. In this case a point-driven, finite cylinder (Fig. 5.2) was excited at a resonance frequency of 1761.8 Hz. Note that the logarithm of the magnitude is plotted in Fig. 5.1(a) and the real part in Fig. 5.1(b). First of all, it is evident from the figure that the discretization (lattice points are represented by squares) was fine enough. Secondly, one can also see in Fig. 5.1(a) that the pressure field at the edges of the aperture (positions 0 and 64) is down by at least 35 dB from the peak pressure. The ends of the vibrating shell are indicated by the arrow heads in the figure. The overscan distance is about 37 percent of the cylinder length on either side. The 35 dB drop off in pressure was sufficient to provide an accurate estimation of $P_n(r_h, k_z)$ using Eq. (5.30).

At this point we study carefully the implementation of the inverse transforms using the FFT in Eq. (5.29) which approximate Eq. (5.6). Due to the FFT the limits of the integration in k_z are $k_z = \pm\pi/\Delta z$. With the experimental values $\Delta z = 1.75$ cm and $k = \omega/c = 0.075$ cm^{-1}, these limiting values of k_z correspond to strongly decaying evanescent waves. The rigidity of the cylinder surface does not permit vibrations to exist with the small wavelengths corresponding to these evanescent wavenumbers. The same holds true in the circumferential direction. With $N = 64$ we are limited to $n = \pm 32$.

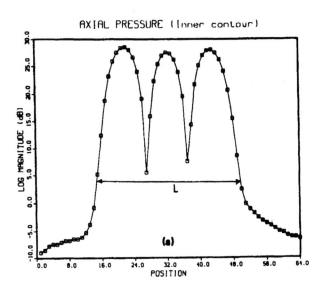

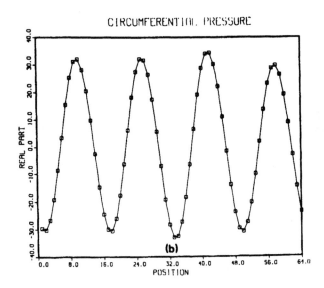

Figure 5.1: (a) Plot of the log magnitude of the pressure from a single scan 1.0 cm from the surface of a vibrating cylinder. The square symbols represent the measurement points (64 across) and the arrowheads indicate the ends of the cylinder. The pressure field outside of the cylinder drops by 35 dB. (b) A scan around the circumference of the cylinder (64 points). The real part of the instantaneous pressure is plotted at the moment of maximum oscillation amplitude. The $n = 4$ mode is very strong.

However, vibrations (at this wavenumber) do not exist. In fact, Fig. 5.1 shows a mode which has its dominant contribution at $n = 4$ and $k_z = 0.174\ \mathrm{cm}^{-1}$, whereas the limiting (maximum) value of k_z was $\pi/1.75 = 1.8$ cm^{-1}.

Next we consider sampling in k_z and the inverse Fourier transform. One must be careful to avoid aliasing due to undersampling in k_z in Eq. (5.6) and Eq. (5.9). If the prescribed Δk_z set by the length of the aperture L is too large, aliasing will occur. Thus we consider the effects now of aliasing in the k_z domain, especially as regards reconstruction of the normal velocity on the surface of the cylinder.

We can model the effect of aliasing in the integral of Eq. (5.9) by using the comb generalized function, Section 1.7 (page 8), as a sampling function. It was defined as

$$\text{III}(\frac{k_z}{\Delta k_z}) = |\Delta k_z| \sum_{n=-\infty}^{\infty} \delta(k_z - n\Delta k_z), \tag{5.31}$$

representing an infinite set of equally spaced delta functions with the required spacing Δk_z. The one-dimensional inverse Fourier transform of $\text{III}(k_z/\Delta k_z)$ (the inverse of Eq. (1.45)) is

$$\mathcal{F}_z^{-1}[\text{III}(k_z/\Delta k_z)] = \frac{1}{L}\text{III}(\frac{z}{L}), \tag{5.32}$$

where $\Delta k_z = 2\pi/L$, as defined in Eq. (5.28). Note that the sampling interval Δk_z is fixed by the length of the aperture, L, in real space. Using the comb function we sample Eq. (5.9), using the tilde over a variable to indicate that it is an approximation to the corresponding variable without the tilde:

$$\tilde{w}(a, \phi, z) = \mathcal{F}_z^{-1}\mathcal{F}_o^{-1}[P_n(r, k_z)\text{III}(\frac{k_z}{\Delta k_z})\frac{-ik_r H_n'(k_r a)}{\rho_0 ck H_n(k_r r_h)}]. \tag{5.33}$$

Whereas Eq. (5.33) provides the effect of sampling in the inverse Fourier transform, it does not model the finite limits of integration imposed by the FFT. The latter is unimportant, however, since the Fourier components outside this range are highly evanescent, contribute little to the vibration of the structure and, besides, are filtered out by the k-space filter.

By using the convolution theorem, Eq. (1.14), and Eq. (1.28) as we did to derive Eq. (5.10), the effect of aliasing can be extracted from Eq. (5.33):

$$\tilde{w}(a, \phi, z) = \frac{1}{2\pi}\mathcal{F}_z^{-1}\mathcal{F}_o^{-1}\left[P_n(r, k_z)\text{III}(\frac{k_z}{\Delta k_z})\right] * *g_v^{-1}(a, r_h, \phi, z), \tag{5.34}$$

where g_v was defined in Eq. (5.11).

Using the one-dimensional convolution theorem again, Eq. (1.14), to compute the product $\mathcal{F}_\phi^{-1}[P_n]\text{III}$, and the result Eq. (5.32) for the inverse Fourier transform of the comb function, we obtain

$$\mathcal{F}_z^{-1}[\mathcal{F}_\phi^{-1}[P_n]\text{III}(\frac{k_z}{2\pi/L})] = p(r_h, \phi_0, z) * \frac{1}{L}\text{III}(\frac{z}{L})$$

$$= \sum_{q=-\infty}^{\infty} p(r_h, \phi_0, z - qL). \tag{5.35}$$

'inally, Eq. (5.34) becomes

$$\tilde{w}(a,\phi,z) = \frac{1}{2\pi} \sum_{q=-\infty}^{\infty} p(r_h,\phi_0, z-qL) * *g_v^{-1}(a,r_h,\phi,z),$$

or

$$\tilde{w}(a,\phi,z) = \frac{1}{2\pi} \sum_{q=-\infty}^{\infty} \int_0^{2\pi} \int_{-\infty}^{\infty} g_v^{-1}(a,r_h,\phi-\phi_0, z-z_0)p(r_h,\phi_0, z_0-qL)\, d\phi_0\, dz_0. \quad (5.36)$$

Now we can see quantitatively the relationship between $\tilde{w}$ and the desired $\dot{w}$. The $q = 0$ term in Eq. (5.36) provides the desired exact velocity field $\dot{w}$ as can be seen by comparison with the exact equation, Eq. (5.10). (Note that $z_0 = 0$ represents the center of the measurement aperture.) The $q \neq 0$ terms are the effect of k-space aliasing. For example, the $q = 1$ term can be pictured as the velocity field created by the measured pressure field shifted to the right a distance L so that it is contained within the limits $L/2 < z_0 < 3L/2$. This term is then backpropagated by the inverse Green's function to the surface to create there a false velocity image field in the reconstruction aperture, $-L/2 < z < +L/2$. As long as the distance from the center to this replication of the measured pressure field is large, one would expect this false image term to be small. This is the one-dimensional analog of the exact same effect for planar holography shown in Fig. 3.14 on page 111 and Eq. (3.50). Similarly, for $q = 2$, a replication of the measured pressure field is moved a distance of $2L$ to the right of the measurement aperture, and backpropagated to create an additional false image, and so on. Thus we see why it is necessary that the measurement aperture extend beyond the ends of the vibrating cylinder. The larger this aperture, the further away the replicated sources are located from the center of the cylinder, reducing the effect of aliasing.

Note that there is no k-space aliasing problem in the circumferential direction, since $\mathcal{F}_\phi^{-1}$ is a sum and is already discretized. In other words, the discrete inverse Fourier transform can represent the sum over n exactly. The only issue here is the cutoff value of n for the k-space filter. (There can be, of course, aliasing in the sampling in ϕ.)

It is possible to reduce aliasing in k_z by artificially extending the measurement aperture with synthetic data so that the measurement plus added data now extend over a distance of $2L$, or $4L$, depending on the amount of added synthetic data. This pushes the replicated fields even further away, at the sacrifice of some accuracy in the real image $\dot{w}$ due to the synthetic data. The simplest synthetic data consists of zeros; although adding zeros represents a rather severe distortion of the physics of the sound field, since radiated pressure fields must propagate according to Huygen's principle which precludes the radiated fields from becoming zero over a finite area. It is best to extend the aperture using some of the known physics of the propagation process, with a close matching to the measured data at the ends of the aperture. It is interesting to note that in the experiments described below adding zeros to extend the aperture had no noticeable effect on the reconstructions of the velocity and the surface intensity. This was due to overscanning the source, as shown in Fig. 5.1.

5.4   Experimental Results

Much of this section is taken from the reference.[1]

5.4.1   Scanning Control and Data Acquisition

Nearfield acoustical holography is often implemented by scanning the sound pressure field over a defined cylindrical contour with a hydrophone/microphone receiver. This task is accomplished experimentally by means of a robotic scanning facility which allows movement of a measuring hydrophone using several degrees of freedom. The heart of this facility is a computer which provides system control, monitoring, data acquisition, mass storage, and intermediate display of data.

A robotic manipulator moves a single, piezoelectric hydrophone along the three coordinate axes, described in detail in the reference. In order to achieve the necessary cylindrical contour, a rotating table is used driven by a stepper motor. The cylinder to be tested is suspended vertically from the rotating table and is rotated about its principle axis. One such arrangement is shown in Fig. 5.2.

All of the experiments in this section were performed with a single frequency signal driving an electromagnetic shaker which was bolted inside a thin cylindrical shell. The computer recorded pressure data during the experiment.

With computer control of the robotic manipulator, virtually any scan contour can be prescribed. In these experiments two cylindrical contours concentric with the shell were used. Two contours were chosen so that the particle velocity midway between them could be calculated and compared with projections made using Eq. (5.9) (with $a > r_h$). The cylindrical contours are achieved as follows. The cylindrical shell of interest is hung vertically at about a 3 m depth in an underwater tank facility. The hydrophone is positioned in the horizontal plane at the desired radial distance from the cylinder, and the hydrophone's acoustic center is located on a ray normal to the cylinder axis (see Fig. 5.2). The hydrophone is moved vertically in successive steps, taking data at each of 64 points along a line parallel to the Z-axis. When the hydrophone has completed a vertical line, it is moved radially outward and a second vertical scan is performed along the same circumferential angle, but at the new radius. The cylinder is then rotated one step, the hydrophone is moved radially inward to the original radius, and the vertical scans are repeated along the new circumferential angle. This process is repeated for a total of 64 different circumferential angles. Thus the cylinder is turned through one revolution and two complete contours are generated, a primary (inner) contour from which nearfield acoustical holography computations are made, and a secondary (outer) contour, which, in conjunction with the inner contour, is used to compute a standard acoustic intensity map for comparative purposes.

5.4.2   Experimental Parameters

The test body used in the experiments was a stainless steel cylindrical shell. The shell dimensions were 55.9 cm length, 16.83 cm O.D., 16.15 cm I.D. (a thickness of 0.68

[1] E. G. Williams, H. D. Dardy, and R. G. Fink (1985). "Nearfield acoustical holography using an underwater, automated scanner," J. Acoust. Soc. Am. **78**, pp. 789–798.

cm). Figure 5.2 shows the shell and the mechanical setup for the experiment. The

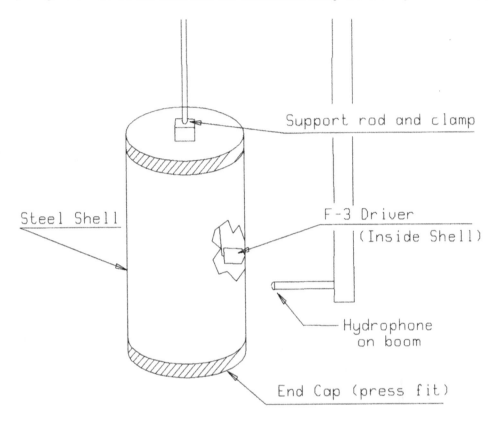

Figure 5.2: Illustration of the shell used in the experiment for the results presented in Figs 5.1 to 5.16. The shell is attached to a vertical boom which allows it to be rotated under water. An internal shaker (F-3) excites the shell at a given frequency. The hydrophone is at the end of a boom which traverses up and down.

shell was sealed with flat, stainless steel endcaps whose diameters match the outside diameter of the shell. Each cap had a total thickness of 3.18 cm and was machined to accommodate an O-ring which formed a press fit into the shell, approximating a simply-supported boundary condition. Tension was applied between the endcaps by means of three rods inside the shell bolted to the endcaps. A 4.5 m long, 1.91 cm O.D. aluminum support tube was clamped into the center of one endcap, providing the means of vertical suspension from the rotating table, as well as acting as a conduit for the cables carrying signals to and from the internal shaker. The shaker was mounted normal to the cylinder on a 10–32 stud located 33.0 cm from the top end of the cylinder.

The measurement contour parameters were identical for all the experiments. The vertical step distance was 1.75 cm, and the incremental rotation angle was 5.625 degrees. Each vertical scan contained 64 measurement points with a total contour length of 111.8 cm. Since these scans were centered with respect to the shell length, the contour extended 25.4 cm (10.0 in.) beyond each end of the cylinder. The primary contour radius was 0.51 cm (0.2 in.) from the outside of the shell. The secondary contour was 1.78 cm from the outside of the shell, giving a separation distance of 1.27 cm. There were 64 vertical scans for each contour, creating two 4096-point arrays. Structural resonance frequencies of the shell were identified using a swept-frequency, drive-point admittance measurement. Scans were then made while the shell was driven at one of the several resonance frequencies.

5.4.3   Comparison to Other Techniques: Two-hydrophone Versus Cylindrical NAH

At this point we will provide experimental results of the cylindrical NAH technique. The actual experimental conditions were described above. To verify the accuracy of cylindrical NAH we compare it with the two-hydrophone technique.

The two-hydrophone technique provides a nearly direct measure of velocity and intensity by using two hydrophone probes spaced a fraction of a wavelength apart. We use this measurement as a check on the reconstruction of the same quantities using cylindrical NAH. The two-hydrophone technique provides the velocity by measurement of the pressure field at two closely spaced points and computing the gradient. The intensity is computed from the in-phase (with the velocity) product of the average of the two pressures (used to compute the pressure at the midpoint) and the velocity. The main source of error is the finite difference approximation of the gradient of the pressure field, computed at the midpoint between the probes, by a subtraction of the two pressure signals. Another severe error can occur when two mismatched probes are used to measure the pressure at the two points. In this case very small phase calibration errors can overwhelm the accuracy of the gradient and intensity computations. To avoid this error we used the same probe to measure at each of the two points, by mechanically moving the hydrophone radially outward. Also, diffraction errors were eliminated by this technique.

Specifically, holograms were taken on two concentric cylindrical contours $r = r_0$ and $r = r_2$, located a fraction of a wavelength apart (generally a 12 mm separation was used). The inner hologram ($r = r_0$), located about 0.5 cm from the surface of the vibrating cylinder, was processed using cylindrical NAH to reconstruct the radial velocity and intensity at the midpoint ($r = r_1$ where $r_1 = (r_0 + r_2)/2$) between the inner and outer holograms. Both the inner and outer holograms were used to compute the radial velocity and intensity at the midpoints using the two-hydrophone technique for comparison with cylindrical NAH. In the former this radial velocity is given by

$$v_r(r_1, z, \phi) \approx \frac{1}{i\omega\rho} \frac{p(r_2, z, \phi) - p(r_0, z, \phi)}{r_2 - r_0}.$$

In the results which follow it should be realized that comparison of radial intensity (power flow per unit area) between the two techniques is a very sensitive test of accuracy since it depends on both the amplitude and phase of the velocity and pressure. In fact, in highly reverberant fields where the phase of the pressure and velocity may be close to 90 degrees apart, it becomes nearly impossible to measure the intensity by any means!

The radial intensity, as usual, is given by

$$I_r = \frac{1}{2}\text{Re}[p(r_1, \phi, z)\dot{w}^*(r_1, \phi, z)].$$

In the first set of experimental results to be examined, the shaker driven shell was excited at 1761.8 Hz corresponding to a resonance mode of the shell. This mode had eight nodal lines in the circumferential direction, with a mode shape proportional to $\cos(4\phi)$, and four nodal lines (including the shell ends) in the axial direction. Because of the low frequency the distance between the two hologram contours was set at 50 mm (standard for low frequency work) instead of 12 mm. Figure 5.3 shows a comparison between the two-hydrophone results and cylindrical NAH.

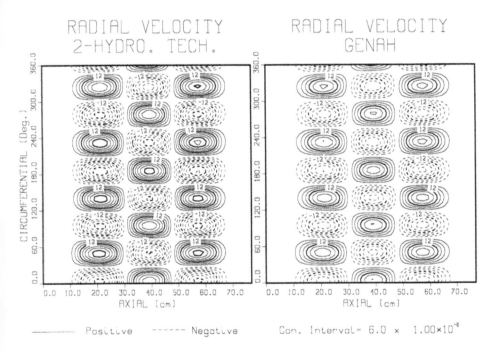

Figure 5.3: Two-hydrophone technique results compared to cylindrical NAH for the radial velocity at a distance of 3 cm from the shell surface. Plotted are contours of constant velocity, with negative velocity indicated by dashed contours. Finite difference errors cause the two-hydrophone technique results to be in error, resulting in generally higher peaks than cylindrical NAH. The hydrophone separation was 50 mm, with a driving frequency of 1761.8 Hz.

Displayed are contours of constant radial velocity, chosen at an instant in time when the field is a maximum (necessary because of oscillation at the driving frequency). Note that each contour represents a value of 6.0×10^{-6} m/s, the nodal lines being suppressed. The negative values of velocity are represented by dotted lines. Study of Fig. 5.3 indicates excellent agreement between the two techniques. The two-hydrophone results, however, have peaks (anti-nodes) which are about 14 percent higher than cylindrical

NAH. It is not hard to show that this error is due to the two point gradient approxima-tion used in the two-hydrophone technique. Cylindrical NAH appears to provide more accurate results. Errors occur because of the strong evanescent wave field which exists near the shell. In fact, if one assumes a pure evanescent pressure field in the form of $p_0 e^{-k_r r}$, and use the value of $k_z = 0.174$ cm^{-1} obtained from Fig. 5.3, one finds after a couple of differentiations that the error term in the gradient approximation could be as large as 38 percent.

In Fig. 5.4 we present the comparison of the radial intensities. The agreement is generally excellent except in a few areas where again the two-hydrophone levels are a bit high compared to cylindrical NAH, a result which is expected due to the velocity gradient errors noted above.

Measuring the acoustic intensity at this very low frequency is by no means a trivial task, a fact which will be more evident to readers who are familiar with the work in the two-microphone area in air acoustics. (The same wavelength occurs below 500 Hz in air.)

The hologram at $r = r_2$ was backprojected, using Eq. (5.23), to the inner mea-surement surface at $r = r_0$, a distance of 5 cm. Comparison to the measured data at $r = r_0$ is shown in Fig. 5.5 for an axial scan across the plane including the shell driver. The ends of the shell are indicated by the arrow heads. Note that the logarithm of the pressure magnitude is plotted against lattice position. Over the shell the agreement is excellent. However, at the ends of the aperture the projected field begins to build up (curve with circles), as a result of the replicated measurement field as discussed above in reference to Eq. (5.34).

Returning to comparisons with the two-hydrophone technique, we present data taken at a different resonance frequency, 8191 Hz, apparently a mode with half a wavelength axially and 9 wavelengths circumferentially. In this case the distance between the inner and outer contours was 12 mm. Again the inner contour hologram was propagated outward 6 mm and compared with the two-hydrophone results in the same area. Figure 5.6 shows a comparison of the radial velocities and Fig. 5.7 the radial intensities. The variation of the mode shape in the axial direction indicates a second mode present, al-though quite a bit weaker than the first one. The agreement between the two techniques is excellent for both velocity and intensity. The finite difference errors in the velocity calculation for the two-hydrophone technique are now quite small. It is significant that the intensity has one large peak. This peak occurs over the shaker location indicating that the driving point is the predominant point of radiation into the water, unlike the low frequency case above in which the whole shell showed intensity activity.

With the accuracy of cylindrical NAH demonstrated from the above analysis we now provide some examples of the breadth of the results obtained from cylindrical NAH.

5.4.4   Pressure, Velocity and Vector Intensity Reconstructions

In the first set of results, the cylinder was driven at a 1.7 kHz resonance. This resonance of the shell was chosen since it was isolated from any adjacent ones. The pressure was measured on a single cylindrical contour located 1.0 cm from the surface. The results of reconstructions using cylindrical NAH of the surface pressure and surface velocity from this measurement are shown in Figs 5.8 and 5.9.

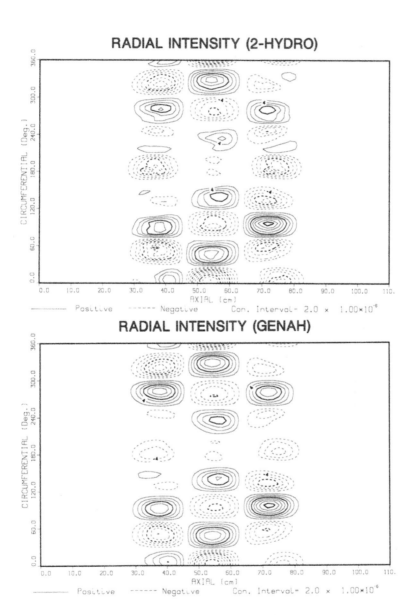

Figure 5.4: Same as Fig 5.3 except that the contours of constant radial active intensity are plotted. The negative contours (dashed lines) indicate energy flowing towards the cylinder. Agreement between the two-hydrophone technique (upper) and cylindrical NAH (lower) is good.

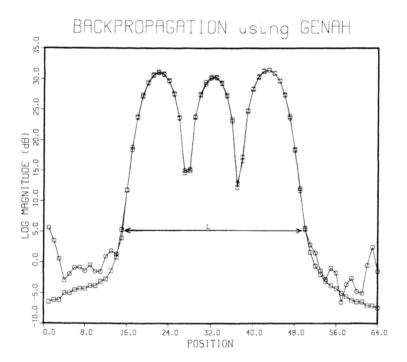

Figure 5.5: Comparison of the pressure in an axial scan 0.5 cm from the cylinder surface (squares) with the backward propagated pressure (circles) from the hologram measured at 5.6 cm from the surface. The ends of the shell are indicated by the arrow heads. The agreement is excellent over the shell and diverges outside due to the replicated sources. The driving frequency was 1761.8 Hz.

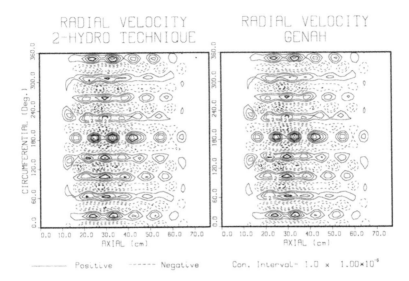

Figure 5.6: Comparison of the two-hydrophone technique to cylindrical NAH for a higher excitation frequency, 8191 Hz. Shown are contours of constant radial velocity for a (1,9) mode of the cylinder. The agreement is excellent between the two techniques.

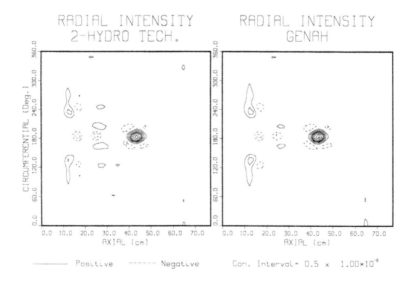

Figure 5.7: Comparison of two-hydrophone and cylindrical NAH results for the case in Fig. 5.6 but for the radial intensity. Agreement is excellent. The location of the shaker is now evident from the dense contours, at 41 cm axially and 180 degrees indicating that the driving point is the predominant region of radiation.

SURFACE PRESSURE

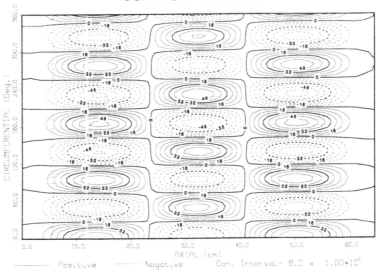

Figure 5.8: Result of cylindrical NAH reconstruction of the surface pressure plotted as contours of constant pressure (contour interval is 8.0 Pa) at an instant in time of maximum field. The length of the vibrating cylinder plus endcaps was 61 cm, f =1761.8 Hz, mode shape = (3,4). Note that 0 and 360 degrees on the vertical axis represent the same points on the cylinder.

SURFACE VELOCITY

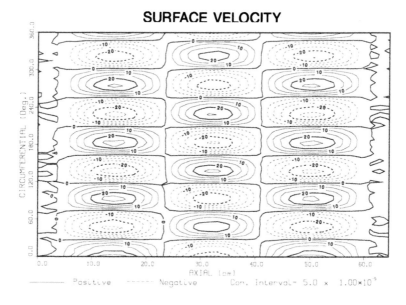

Figure 5.9: Same as Fig. 5.8 except the surface radial velocity is plotted. The contour interval is 5.0×10^{-5} m/s. The surface velocity and pressure have identical shapes. The jagged contours to the left and right represent low level data outside of the cylinder.

The field is plotted when it is at a maximum in time. Note, in consideration of the horizontal scale, that the length of the cylinder was 61 cm. It is very clear from these figures that the shell was vibrating in a (3,4) mode with three half wavelengths in the axial direction and four full wavelengths in the circumferential direction. Note that the surface pressure follows the radial velocity.

Using cylindrical NAH to generate reconstructions in cylindrical contours of increasing radius (using Eq. (5.23) with a replaced with the increased radius), one can obtain a picture of the radial distribution of radiated pressure in the nearfield of the shell. Figure 5.10 shows the instantaneous pressure in a plane perpendicular to the axis, through the center of the cylinder. Plotted are contours of equal pressure scaled by multiplication with $\sqrt{r/a}$.

RADIAL PRESSURE

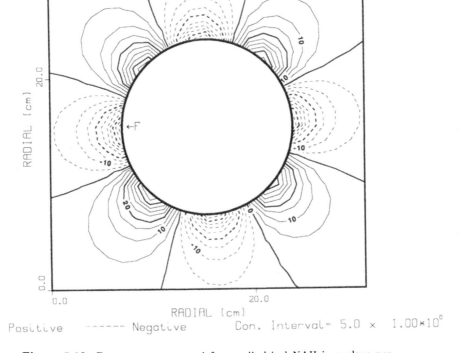

Figure 5.10: Pressure reconstructed from cylindrical NAH in a plane perpendicular to the midpoint of the axis of the cylinder. The pressure has been multiplied by the square root of r/a to remove cylindrical spreading. The $n = 4$ mode is very evident and $f = 1761.8$ Hz. Notice the strong decay of the pressure from the cylinder surface, indicating that this mode is evanescent.

One can see very clearly the $n = 4$ nature of this mode, and the evanescent nature of this mode. (The wavelength in water is 87 cm at this frequency.) One can see better the evanescent character in Fig. 5.11 in which the logarithm of the magnitude of the pressure is plotted instead of the linear pressure. Noting that the contour interval is 5

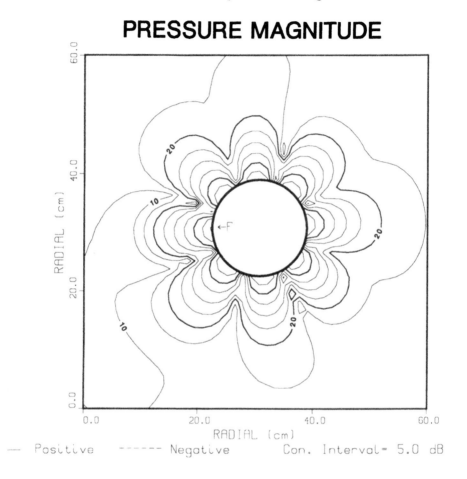

Figure 5.11: Same as Fig. 5.10 except the contours are $20 \log |p| + c$, c is an arbitrary constant. Contour interval is 5 dB. This kind of plot displays better the evanescent waves, and one sees that the pressure field decays over 20 dB at $r = 2a$. It appears that this rapid decay ends outside of the 20 dB contour.

dB, it is evident that the pressure field decays over 20 dB at a distance $r = 2a$. In fact, the spacing between adjacent contours is very close to a constant. This indicates that the pressure field decays, obeying a power law relationship, as we expect for evanescent waves. Note that outside of the 20 dB contour, the decay of the sound field is arrested, and it appears that the region of cylindrical spreading is reached, as was discussed previously in reference to Fig. 4.12 on page 132.

Figure 5.12 shows the same type of plot for a plane through the axis of the cylinder and including the shaker, shown in the figure. Again the distance between the contours is 5 dB, and the pressure field decays very rapidly from the cylinder surface. Note that this kind of plot gives no detail about the phase of the pressure.

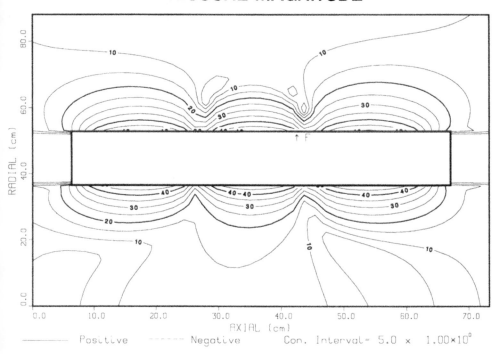

Figure 5.12: Same case as Fig. 5.11, except that a broadside view of the cylinder is shown, a plane coincident with the axis and passing through the point driver labeled F in the figure. dB contours are plotted and again the evanescent nature of the pressure field is evident. At a distance of one diameter from the cylinder surface, the evanescent decay ends.

With reconstructions of the velocity vector from Eq. (4.67) and the pressure at different radii, one can construct an energy flow map using the acoustic, time averaged intensity vector, computed from

$$\vec{I} = \frac{1}{2}\text{Re}[p(\dot{u}^*\hat{e}_z + \dot{v}^*\hat{e}_o + \dot{w}^*\hat{e}_r)].$$

Figure 5.13 is the result. Each vector indicates the level and direction of the energy flow in a plane perpendicular to the axis at the axial coordinate, 50 cm, in Fig. 5.12. The visible length of the intensity vector is 40 dB, and only the components of the vector in the plane are plotted. The mesh formed by the tails of the vectors has a lattice spacing of $a/4$. One can see that energy flows into and out of the shell, although most of the energy flows around the shell surface. A region of circulating intensity can be

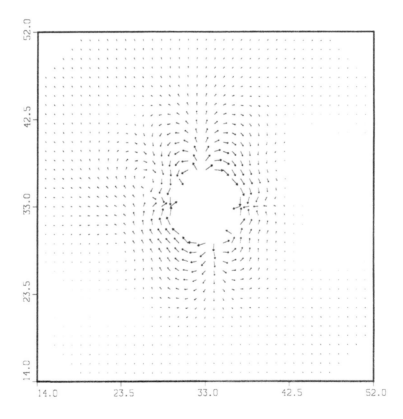

Figure 5.13: Active acoustic (time averaged) intensity plotted (cylindrical spreading removed) in a plane perpendicular to the cylinder axis, through an axial antinode. This data is reconstructed using cylindrical NAH from the measurement of the pressure on the cylindrical contour 0.5 cm from the surface. Each vector indicates the level and direction of the energy flow near the shell surface. The visible length of the vector is 40 dB and only the components in the plane are plotted. The innermost vectors are at $r = a$, and indicate energy flowing in and out of the cylinder. Note the region of circular energy flow at about 10 o'clock. The lattice spacing between each vector is $a/4$.

seen at about 10 o'clock. (The intensity was scaled by multiplication by r/a so that the evanescent fields would be more visible.)

Let us consider now the shell driven at a resonance at 8158 Hz, a mode with one half wavelength axially and $n = 9$ circumferentially. The acoustic intensity in a plane through the center of the shell, perpendicular to the axis, is quite interesting for this case as shown in Fig. 5.14. The energy appears to circulate around the shell, counter clockwise with 9 'spokes' radiating outward.

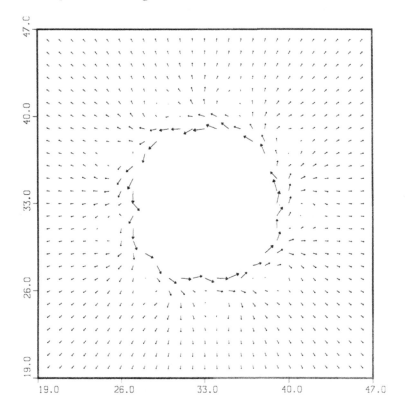

Figure 5.14: Active intensity in a plane perpendicular to the center of the cylinder axis, f =8158 Hz. 25 dB visible vector length. In this case the separation between vectors is $a/6 = 1.4$ cm, and the innermost vectors correspond to the intensity on the surface. Energy appears to flow around the cylinder in a counter clockwise direction.

Above the ring resonance of the shell (which occurred around 10 kHz) the resonant nature of the shell was quite damped, and clearly defined resonances did not occur. The 12 kHz results to be presented now are a good example of the results of this damped nature. Figure 5.15 shows the radiated pressure, on a linear scale, at one instant in time, in the planes containing the driver. On the left is a plane perpendicular to the axis, and on the right a plane coincident with the axis, both planes containing the driver. Now the sound field has a wavelength of 12 cm, and the traveling nature of the pressure field is very clear. One can see that the shaker is now the point of radiation,

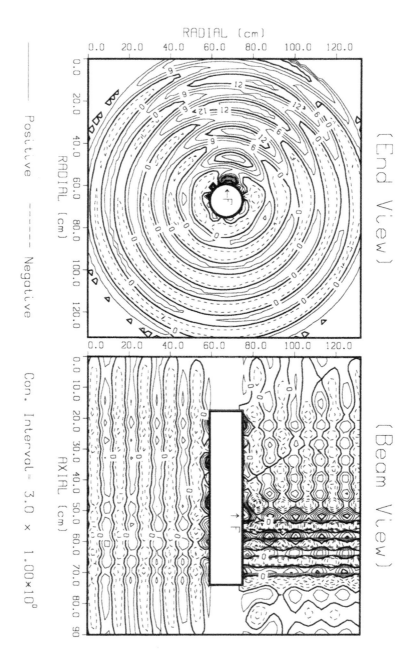

Figure 5.15: High frequency case, $f = 12$ kHz, damped mode, contours of constant pressure. The driver position is indicated with an F and the data is plotted in the plane which contains the driver. The cylindrical NAH reconstructions clearly show the radiation from the driving point, a beam pattern from about 8 to 10 o'clock on the left plot. Axially the beam tends to travel straight outward (right plot). Evanescent waves very close to the shell surface are evident.

with the pressure field beaming to the left (at 9 o'clock). The side view on the right indicates that the field beams upward without much spreading in the nearfield. The evanescent field now is very close to the surface of the cylinder, much closer than at the low frequencies, consistent with Eq. (4.62) in which k_z has a dominant value given by the very small structural wavelength in the axial direction.

Figure 5.16 shows the vector intensity in the same plane as the left figure of Fig. 5.15 above. In this case one can see that the point driver radiates energy directly into the

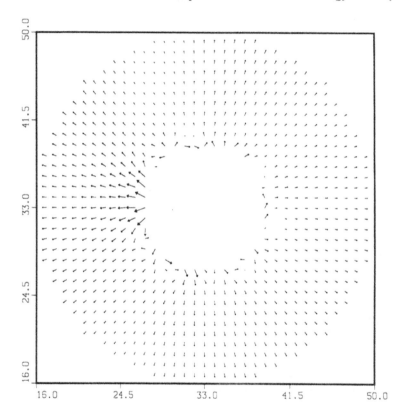

Figure 5.16: Intensity vector plot for the case shown on the left of Fig. 5.15. Radiation from the driving point is shown by the large arrows at 9 o'clock. The visible vector length is 20 dB.

medium. The intensity indicates that the mechanical source is also a dominant acoustic source.

The sometimes overwhelming amount of data generated from an NAH experiment necessitates sophisticated means of graphic display. The use of color in the displays is crucial to displaying the physics, in many cases. Unfortunately, color figures are too costly to reproduce in this text, and we hope that the black and white figures presented have provided a clear and comprehensive look at some representative results from NAH.

5.4.5 Comparisons with a Surface Accelerometer[2]

The following figures were generated in a different experiment on a different shell from the one described above. In this case we used a thin nickel shell with hemispherical-like endcaps, of the same thickness as the shell. Figure 5.17 is a schematic of the shell. The length of the straight section was 8.48 inches, and the shell radius was 2.18 inches.

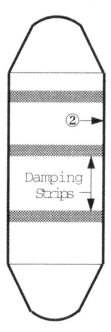

Figure 5.17: Shell used for the accelerometer comparisons. The shell was driven at the point labeled with the circled 2. The accelerometer was located near the center of the shell.

The thickness of the shell was 0.021 in. A holographic scan was made 0.6 in. from the cylindrical surface. The shell was outfitted with a surface accelerometer so that the velocity could be measured at a point on the cylindrical surface and compared with the holographic reconstruction of the velocity at that point. The shell was driven by a small shaker with a broad-band chirp signal so that the frequency range of 2–8 kHz was covered with a single pulse.

The pressure was measured in the hologram cylinder as a function of time. Each time series was Fourier transformed to the frequency domain so that the holographic equation, Eq. (5.9), could be used to reconstruct the surface velocity at the location of the accelerometer. The following figures show the results of these comparisons.

The figures included in this section are intended to provide an illustration of the accuracy of cylindrical NAH in studying radiators. The thin shell used in this study is rich in evanescent, subsonic waves. One point of the holographic reconstruction (resulting from Eq. (5.33)) on the surface of the shell corresponded to the accelerometer

---

[2]Earl G. Williams, Brian H. Houston and Joseph A. Bucaro (1989). "Broadband nearfield acoustical holography for vibrating cylinders", J. Acoust. Soc. Am. **86**, pp. 674–679.

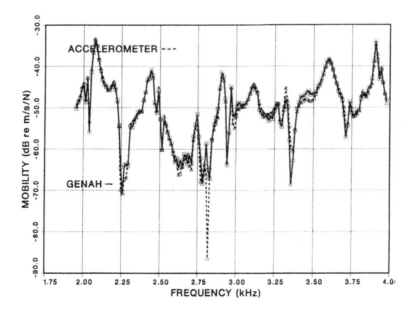

Figure 5.18: Comparison between the reconstructed velocity from cylindrical NAH (squares and solid lines) and the direct measurement using an accelerometer (triangles and dashed lines). The frequency range is 2.0–4.0 kHz. The agreement between the two is excellent.

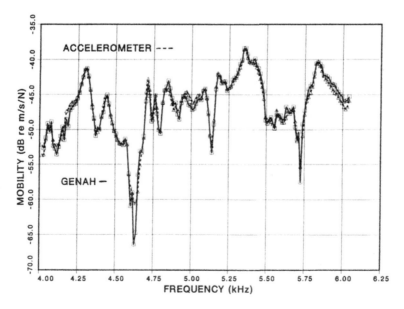

Figure 5.19: Comparison between the reconstructed velocity from cylindrical NAH (squares and solid lines) and the direct measurement using an accelerometer (triangles and dashed lines). The frequency range is 4.0–6.0 kHz. The agreement between the two is excellent.

location. Figure 5.18 shows the comparison in the magnitudes of the NAH reconstruction (labeled GENAH) at this point and the exact measurement from the accelerometer (dashed line), the latter divided by $-j\omega$ to convert the acceleration to velocity. The vertical scale is in decibels and has been normalized by the driving force providing the mobility, that is, velocity divided by force. The agreement between the two curves is excellent.

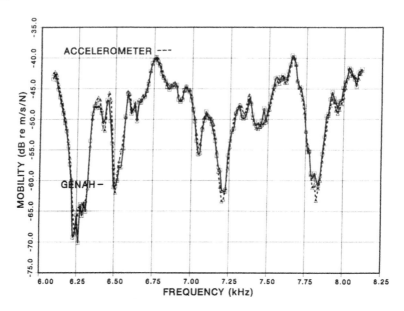

Figure 5.20: Comparison between the reconstructed velocity from cylindrical NAH (squares and solid lines) and the direct measurement using an accelerometer (triangles and dashed lines). The frequency range is 6.0–8.0 kHz. The agreement between the two is excellent.

Figures 5.19 and 5.20 show the comparisons for different frequency ranges, 4–6 kHz and 6–8 kHz, respectively. Again the agreement between the accelerometer and NAH reconstruction is excellent. Note that the shell contained three damping strips which reduced the Q values of the modes in the shell, creating a smoother mobility curve.

5.4.6 Helical Wave Spectrum Examples

We have seen that the helical wave spectrum (k-space) provides a representation of the helical waves which make up radiation from a source. The pressure spectrum was given in Eq. (4.52) and the spectrum of the radial velocity in Eq. (4.73). Presentation of the helical wave spectrum of the radial velocity in the fluid but at the surface of a vibrating shell provides a key to the helical waves which are traveling on the shell, due to the continuity of normal velocity at the boundary of the shell. The wave equation which describes the motion of an infinite shell resulting from an internal excitation admits solutions in terms of helical waves. That is, the radial velocity on an infinite shell can be expressed as a helical wave expansion just as was done for the fluid in Eq. (4.73). When the experimental shell is no longer infinite, its boundaries reflect these helical waves back and forth as they travel on the shell. The standing waves which result can still be described in terms of combinations of infinite helical waves. This fact is the basis of Fourier acoustics.

In Fig. 5.21 we present the helical wave radial velocity spectra at the surface of a shell similar to the one shown in Fig. 5.17, although the former was about three and a half times longer. The eight panels in the figure represent different excitation frequencies (the shell was internally driven) with lowest frequency in the lower left and highest in the upper right. The gray scale represents levels of $20 \log_{10}(|W_n(a, k_z, \omega)|)$, with black representing 50 dB and white 0 dB. The vertical axis is n and the horizontal axis is m where m is defined by $k_z = m\pi/L$, that is, it provides the number of half-wavelengths in the axial direction over the length of the shell. The figure-eights which appear reveal the helical waves which are free to travel on the shell. Clearly the vibration of the shell is composed of a rich mixture of helical waves of different wavenumbers, and many modes are excited at a single frequency. As the frequency increases the figure-eights open up and begin to form a circle. The reader should recall the simply-supported plate, the vibration of which is made up of a sum of orthonormal modes as modeled by Eq. (2.168) on page 66. The resonant modes excited at the driving frequency were illustrated in Fig. 2.34 and were found to form a circle in k-space. What we see in Fig. 5.21 is the experimental analog of the plate case. As in the plate these figure-eights represent waves which are called bending (flexural) waves and are dominated by a motion which is out-of-plane bending. Infinite shell theory predicts this domination, and also predicts the figure-eight shape in excellent agreement with the measurements.[3]

At the center of most of the panels in the figure we see more modes appearing. These are the supersonic helical waves called shear and longitudinal waves. They are within the radiation circle (not shown).

---

[3]E. G. Williams, B. H. Houston, and J. A. Bucaro (1990). "Experimental investigation of the wave propagation on a point-driven, submerged capped cylinder using k-space analysis", J. Acoust. Soc. Am. **87**, pp. 513–522.

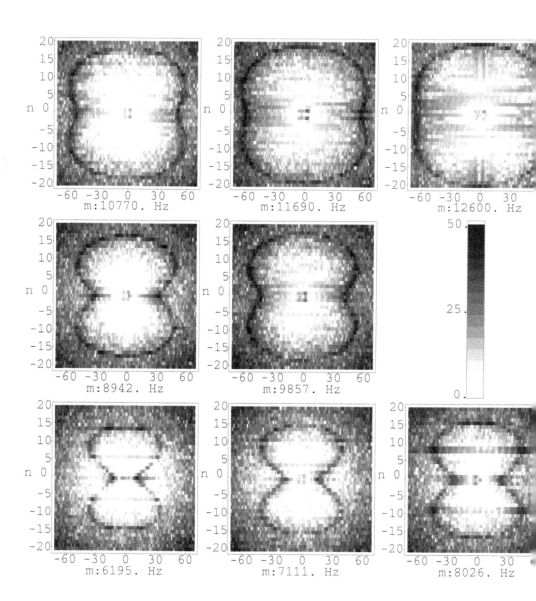

Figure 5.21: Helical wave spectrum for the normal velocity on a measurement cylinder coincident with the actual cylinder surface. The logarithmic gray scale represents $20 \log_{10}[|W_n(a, k_z, \omega)|]$. Black represents 50 dB and white 0 dB (arbitrary reference level). Each panel represents a different frequency.

Problems

5.1 Simplify the expression (final result should contain no Bessel functions in it)

$$J_n(z)\frac{dH_n^{(1)}(z)}{dz} - H_n^{(1)}(z)\frac{dJ_n(z)}{dz}.$$

5.2 A baffled square plate of length and width L is vibrating with normal velocity

$$\dot{w}(x,y) = A\sin(5\pi x/L)\sin(8\pi y/L).$$

Describe the radiation classification of this mode

(a) When $k \gg \sqrt{(5\pi/L)^2 + (8\pi/L)^2}$,

(b) when $k \ll \sqrt{(5\pi/L)^2 + (8\pi/L)^2}$. What is the dependence on k in this region?

(c) When k is slightly larger than $(5\pi/L)$.

5.3 Obtain the asymptotic farfield expression for the time domain acoustic pressure, $p(R,\theta,\phi,t)$, radiated by an infinite length cylinder whose normal velocity (independent of ϕ) is

$$\dot{w}(a,\phi,z,t) = \dot{w}_0\delta(z - z_0)e^{-i\omega_0 t}.$$

5.4 Apply the stationary phase technique to obtain the asymptotic limit as $z \to \infty$ for the following integral representation of the Bessel function

$$J_n(z) = \frac{1}{2\pi}\int_0^{2\pi} e^{i(z\sin\phi - n\phi)}d\phi.$$

Note that there are two stationary phase points. Evaluate each one separately and add the results. The answer is already in the notes and your result should be put in a form identical with it.

5.5 You have discovered a measurement technique which can measure the frequency domain particle velocity in the <u>axial</u> direction, $\dot{u}(a,\phi,z)\hat{e}_z$ on a cylindrical contour of radius a and of infinite extent. You would like to come up with a mathematical formula which will allow you to compute the <u>radial</u> velocity, $\dot{w}(z,\phi,z)$, on the measurement contour from the measurement results. Derive an integral formula for $\dot{w}(a,\phi,z)$ using the two-dimensional inverse Fourier transform of $\dot{U}_n(a,k_z)$ where

$$\dot{U}_n(a,k_z) = \frac{1}{2\pi}\int_0^{2\pi}\int_{-\infty}^{\infty} \dot{u}(a,\phi,z)e^{-in\phi}e^{-ik_z z}dz.$$

5.6 A cylindrical source, infinite in the axial direction, becomes a line source in the limit as the radius $a \to 0$. Also in this limit the radial velocity becomes independent of ϕ so that only the $n = 0$ of $\dot{W}_n(a,k_z)$ is non-zero. Using the small argument approximations for the Hankel functions along with Eq. (4.72), determine the pressure, $p(a,z)$, on the line source when $ka \ll 1$.

5.7 Reconstruction of the pressure on a surface at $r = a$ from a knowledge of the pressure on a surface $r = r_h$ where $r_h > a$ represents the NAH inverse problem. Equation (5.7) provides the reconstruction equation viewed as a convolution in ϕ and z. We stated that the inverse pressure propagator given by Eq. (5.8) was at best singular and thus does not exist. This is a result of the very small wavelength evanescent waves which are included in the propagator. The k-space window, used in Eq. (5.23) serves to make the inverse propagator finite so that it can be computed. We can rewrite Eq. (5.23) as a convolution in z and ϕ,

$$\tilde{p}(a, \phi, z) = \tilde{g}_p^{-1}(a, r_h, \phi, z) * *p(r_h, \phi, z).$$

where

$$\tilde{g}_p^{-1}(a, r_h, \phi, z) = \mathcal{F}_z^{-1}\mathcal{F}_\phi^{-1}\left\{\frac{H_n^{(1)}(k_r a)}{H_n^{(1)}(k_r r_h)}\bar{\Pi}[k_t/2k_c]\right\}.$$

Write a computer program to compute $\tilde{g}_p^{-1}(a, r_h, \phi, z)$ for the given parameters (all in cm): $a = 5.54$, $r_h = 7.11$, $f = 6000.$ Hz, $D = 60$ dB, $\Delta z = 1.27$, $N = 64$ (for both z and ϕ). Use an FFT to discretize the problem as described in Section 1.8. Use a filter cutoff as given after Eq. (5.21) and $\alpha = 0.02$ (in Eq. (3.25)). It is probably a good idea to use the propagator in Eq. (4.60) when $k_z > k$ so that the argument of the Bessel function is kept real instead of trying to compute $H_n(i|x|)$ which may not work, depending on your math package.

Plot the real part and the imaginary part of $\tilde{g}_p^{-1}(a, r_h, \phi, z)$. Is the inverse propagator localized in space? How is the reconstruction resolution related to this?

Repeat the problem with a dynamic range of only $D = 30$ dB keeping all other parameters fixed. How does the resolution change?

Chapter 6

Spherical Waves

6.1 Introduction

Wave expansions in cylindrical and planar geometries consist of wavefronts which extend to infinity in at least one dimension. The spherical geometry provides a finite and compact expansion of wavefronts which allows us to gain an understanding of expanding waves. Furthermore, compact, realistic vibrators are more closely modeled with spherical wave expansions than with planar or cylindrical expansions. We present in this chapter the solution of the wave equation in spherical coordinates, and detailed discussions of the radial functions, spherical harmonics and multipole expansions, the latter being very useful for low frequency modeling of radiation from compact vibrators.

The spherical wave spectrum, analog of the helical wave spectrum, is introduced to follow the formulations for plane and cylindrical geometries with regard to the forward and backward propagation of spherical waves for both internal and external problems. Various radiation problems are solved.

6.2 The Wave Equation

The coordinate system is given in Fig. 6.1 and is related to rectangular coordinates through

$$
\begin{aligned}
x &= r \sin\theta \cos\phi, \\
y &= r \sin\theta \sin\phi, \\
z &= r \cos\theta,
\end{aligned}
\tag{6.1}
$$

so that $r = \sqrt{x^2 + y^2 + z^2}$, $\theta = \tan^{-1}[\sqrt{x^2 + y^2}/z]$, and $\phi = \tan^{-1}[y/x]$.

The time dependent wave equation given in spherical coordinates is

$$
\frac{1}{r^2}\frac{\partial}{\partial r}\left(r^2\frac{\partial p}{\partial r}\right) + \frac{1}{r^2 \sin\theta}\frac{\partial}{\partial \theta}\left(\sin\theta\frac{\partial p}{\partial \theta}\right) + \frac{1}{r^2 \sin^2\theta}\frac{\partial^2 p}{\partial \phi^2} - \frac{1}{c^2}\frac{\partial^2 p}{\partial t^2} = 0.
\tag{6.2}
$$

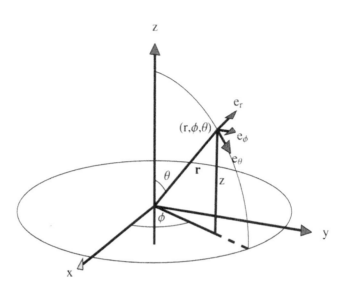

Figure 6.1: Definition of spherical coordinates relative to Cartesian coordinates. ϕ is measured in the x, y plane from the x axis (azimuthal direction). θ is measured from the polar axis, z.

It is useful to note that

$$\frac{1}{r^2}\frac{\partial}{\partial r}(r^2\frac{\partial p}{\partial r}) = \frac{1}{r}\frac{\partial^2}{\partial r^2}(rp). \tag{6.3}$$

The gradient of the pressure given in spherical coordinates is

$$\vec{\nabla}p = \hat{e}_r\frac{\partial p}{\partial r} + \hat{e}_\theta\frac{1}{r}\frac{\partial p}{\partial \theta} + \hat{e}_o\frac{1}{r\sin\theta}\frac{\partial p}{\partial \phi}, \tag{6.4}$$

where $(\hat{e}_r, \hat{e}_\theta, \hat{e}_\phi)$ represent unit vectors in each of the coordinate directions as shown in Fig. 6.1. Euler's equation is given by (Eq. (2.14), page 19)

$$i\rho_0 ck\vec{v}(r, \theta, \phi) = \vec{\nabla}p(r, \theta, \phi), \tag{6.5}$$

where

$$\vec{v} \equiv \dot{u}\hat{e}_\theta + \dot{v}\hat{e}_o + \dot{w}\hat{e}_r. \tag{6.6}$$

The solution of Eq. (6.2) is given by separation of variables:

$$p(r, \theta, \phi, t) = R(r)\Theta(\theta)\Phi(\phi)T(t), \tag{6.7}$$

which leads to four ordinary differential equations:[1]

$$\frac{d^2\Phi}{d\phi^2} + m^2\Phi = 0, \tag{6.8}$$

$$\frac{1}{\sin\theta}\frac{d}{d\theta}(\sin\theta\frac{d\Theta}{d\theta}) + [n(n+1) - \frac{m^2}{\sin^2\theta}]\Theta = 0, \tag{6.9}$$

$$\frac{1}{r^2}\frac{d}{dr}(r^2\frac{dR}{dr}) + k^2 R - \frac{n(n+1)}{r^2}R = 0, \tag{6.10}$$

$$\frac{1}{c^2}\frac{d^2T}{dt^2} + k^2 T = 0. \tag{6.11}$$

The solution to Eq. (6.11) is, with $k = \omega/c$,

$$T(\omega) = T_1 e^{i\omega t} + T_2 e^{-i\omega t}, \tag{6.12}$$

and again we choose the second solution for our time dependence, so that $T_1 = 0$.

The solution to Eq. (6.8) is

$$\Phi(\phi) = \Phi_1 e^{imo} + \Phi_2 e^{-imo}, \tag{6.13}$$

or, alternatively,

$$\Phi(\phi) = \Phi_3 \cos(m\phi) + \Phi_4 \sin(m\phi). \tag{6.14}$$

We keep both solutions in these cases. In both these cases m must be an integer so that there is continuity and periodicity of $\Phi(\phi)$.

The orthogonality relation for the azimuthal functions is

$$\int_0^{2\pi} e^{-im'o}e^{imo}d\phi = 2\pi\delta_{mm'}. \tag{6.15}$$

The solution to Eq. (6.9) is found using a transformation of variables. Let $\eta = \cos\theta$ $(-1 \leq \eta \leq 1)$ so that the differential equation for Θ becomes

$$\frac{d}{d\eta}[(1-\eta^2)\frac{d\Theta}{d\eta}] + [n(n+1) - \frac{m^2}{1-\eta^2}]\Theta = 0. \tag{6.16}$$

The solutions are given by Legendre functions of the first and second kinds,

$$\Theta(\theta) = \Theta_1 P_n^m(\cos\theta) + \Theta_2 Q_n^m(\cos\theta). \tag{6.17}$$

The functions of the second kind, Q_n^m are not finite at the poles where $\eta = \pm 1$ so this solution is discarded ($\Theta_2 = 0$). The functions of the first kind diverge at $\cos\theta = 1$ ($\theta = 0$) unless we restrict n to be an integer. Furthermore, when n is an integer then $P_n^m(\eta) = 0$ when $m > n$. We will discuss the functions of the first kind in more detail later.

Finally for the radial differential equation, Eq. (6.10), the solutions are

$$R(r) = R_1 j_n(kr) + R_2 y_n(kr), \tag{6.18}$$

---

[1] E. Skudrzyk (1971). *Foundations of Acoustics.* Springer-Verlag, New York, pp. 379–380.

where j_n and y_n are spherical Bessel functions of the first and second kind, respectively. Alternatively, the solutions can be written as

$$R(r) = R_3 h_n^{(1)}(kr) + R_4 h_n^{(2)}(kr), \tag{6.19}$$

where $h_n^{(1)}$ and $h_n^{(2)}$ are spherical Hankel functions. Just as with the cylindrical Bessel functions it is true that

$$h_n^{(1)}(kr) \propto e^{ikr},$$

representing an outgoing wave and

$$h_n^{(2)}(kr) \propto e^{-ikr},$$

representing an incoming wave. We may keep one or both of these solutions depending upon the locations of the sources in the problem.

The angle functions are conveniently combined into a single function called a spherical harmonic Y_n^m defined by

$$Y_n^m(\theta, \phi) \equiv \sqrt{\frac{(2n+1)}{4\pi} \frac{(n-m)!}{(n+m)!}} P_n^m(\cos\theta) e^{im\phi}. \tag{6.20}$$

We will study spherical harmonics in detail in Section 6.3.3.

We can write any solution to Eq. (6.2), with $e^{-i\omega t}$ implicit, as

$$p(r, \theta, \phi, \omega) = \sum_{n=0}^{\infty} \sum_{m=-n}^{n} \left(A_{mn} j_n(kr) + B_{mn} y_n(kr)\right) Y_n^m(\theta, \phi) \tag{6.21}$$

for standing wave type solutions and

$$p(r, \theta, \phi, \omega) = \sum_{n=0}^{\infty} \sum_{m=-n}^{n} \left(C_{mn} h_n^{(1)}(kr) + D_{mn} h_n^{(2)}(kr)\right) Y_n^m(\theta, \phi) \tag{6.22}$$

for traveling wave solutions. We now study the properties of these function in more detail.

6.3   The Angle Functions

6.3.1   Legendre Polynomials

Solutions of Eq. (6.16) in which $m = 0$ are very important. These represent fields which have no variation in the azimuthal direction ϕ. The convention is to define

$$P_n^0(x) \equiv P_n(x).$$

These functions are polynomials of degree n. The general form is

$$\begin{aligned}
P_n(x) =\ & \frac{(2n-1)!!}{n!}\left[x^n - \frac{n(n-1)}{2\cdot(2n-1)} x^{n-2} \right. \\
& + \frac{n(n-1)(n-2)(n-3)}{2\cdot 4\cdot(2n-1)(2n-3)} x^{n-4} \\
& \left. - \frac{n(n-1)(n-2)(n-3)(n-4)(n-5)}{2\cdot 4\cdot 6\cdot(2n-1)(2n-3)(2n-5)} x^{n-6} + \cdots \right], \quad n = 0, 1, 2, \cdots.
\end{aligned} \tag{6.23}$$

(The last term in square brackets occurs when the exponent of x is zero or one.) Also, $(2n-1)!! \equiv (2n-1)(2n-3)\cdots 1$. It is clear from Eq. (6.23) that Legendre polynomials obey

$$P_n(-x) = (-1)^n P_n(x). \tag{6.24}$$

The Legendre polynomials take on the following special values:

$$
\begin{aligned}
P_n(1) &= 1 \\
P_n(-1) &= (-1)^n \\
P_n(0) &= (-1)^{n/2}\frac{1\cdot 3\cdot 5\cdots(n-1)}{2\cdot 4\cdot 6\cdots n}, \quad n \text{ even} \\
&= 0 \qquad\qquad n \text{ odd.}
\end{aligned}
\tag{6.25}
$$

The first six Legendre polynomials are

$$
\begin{array}{|ll|}
\hline
P_0(x) = 1 & P_3(x) = \frac{1}{2}(5x^3 - 3x) \\
P_1(x) = x & P_4(x) = \frac{1}{8}(35x^4 - 30x^2 + 3) \\
P_2(x) = \frac{1}{2}(3x^2 - 1) & P_5(x) = \frac{1}{8}(63x^5 - 70x^3 + 15x). \\
\hline
\end{array}
\tag{6.26}
$$

These are plotted in Fig. 6.2 as a function of θ, where $x = \cos\theta$.

Equation (6.23) can be written in more concise form called Rodrigues' Formula:

$$P_n(x) = \frac{1}{2^n n!}\frac{d^n}{dx^n}(x^2 - 1)^n. \tag{6.27}$$

The Legendre polynomials are orthogonal and

$$\int_{-1}^{1} P_n(x)P_m(x)dx = \frac{2}{2n+1}\delta_{nm}. \tag{6.28}$$

6.3.2 Associated Legendre Functions

The associated Legendre functions are given by two integer indices $P_n^m(x)$. For positive m these are related to the Legendre polynomials by the formula,[2]

$$P_n^m(x) = (-1)^m (1 - x^2)^{m/2}\frac{d^m}{dx^m}P_n(x). \tag{6.29}$$

The series representation is[3]

$$
\begin{aligned}
P_n^m(x) &= \frac{(-1)^m (2n-1)!!}{(n-m)!}(1-x^2)^{m/2}\Big[x^{n-m} - \frac{(n-m)(n-m-1)}{2(2n-1)}x^{n-m-2} \\
&\quad + \frac{(n-m)(n-m-1)(n-m-2)(n-m-3)}{2\cdot 4(2n-1)(2n-3)}x^{n-m-4} - \cdots\Big],
\end{aligned}
\tag{6.30}
$$

---

[2] I. S. Gradshteyn and I. M. Ryzhik (1965). *Table of Integrals, Series and Products.* 4th ed., Academic Press, N.Y.

[3] Gradshteyn and Ryzhik, *Table of Integrals, Series and Products.*

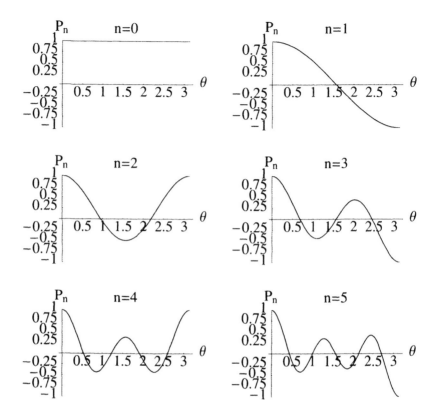

Figure 6.2: Plots of the Legendre polynomials, $P_n(\cos\theta)$, as a function of θ for $n = 0, 1, 2, 3, 4, 5$.

where the series truncates when the numerator goes to zero.

Note that sometimes in the literature the associated Legendre functions are defined differently, without the $(-1)^m$ after the equal sign in Eq. (6.29). Equation (6.29) can not be used to generate the functions for $m < 0$. In this case the associated Legendre function is defined as

$$P_n^{-m}(x) \equiv (-1)^m \frac{(n-m)!}{(n+m)!} P_n^m(x) \tag{6.31}$$

where m is positive.

For each m the functions $P_n^m(x)$ form a complete set of orthogonal functions which obey the relation:

$$\int_{-1}^{1} P_{n'}^m(x) P_n^m(x)\, dx = \frac{2}{2n+1} \frac{(n+m)!}{(n-m)!} \delta_{n'n}. \tag{6.32}$$

The orthogonality integral for the $m = 0$ case was already given in Eq. (6.28).

Following is a partial list of associated Legendre functions, grouped by orthogonal sets:

$m = 0$	$m = 1$
$P_0^0 = 1$	$P_1^1 = -\sin\theta$
$P_1^0 = \cos\theta$	$P_2^1 = -3\cos\theta\sin\theta$
$P_2^0 = \dfrac{1 + 3\cos 2\theta}{4}$	$P_3^1 = \dfrac{3\left(1 - 5\cos^2\theta\right)\sin\theta}{2}$
$P_3^0 = \dfrac{-3\cos\theta + 5\cos^3\theta}{2}$	$P_4^1 = \dfrac{5\left(3\cos\theta - 7\cos^3\theta\right)\sin\theta}{2}$

$$(6.33)$$

$m = 2$	$m = 3\ \&\ 4$
$P_2^2 = 3\sin^2\theta$	$P_3^3 = -15\sin^3\theta$
$P_3^2 = 15\cos\theta\sin^2\theta$	$P_4^3 = -105\cos\theta\sin^3\theta$
$P_4^2 = \dfrac{15\left(5 + 7\cos 2\theta\right)\sin^2\theta}{4}$	$P_4^4 = 105\sin^4\theta$

$$(6.34)$$

Due to the definition, Eq. (6.31), the functions for negative m differ only by a constant from the corresponding functions for positive m:

$m = -1$	$m = -2$
$P_1^{-1} = \dfrac{\sin\theta}{2}$	$P_2^{-2} = \dfrac{\sin^2\theta}{8}$
$P_2^{-1} = \dfrac{\cos\theta\sin\theta}{2}$	$P_3^{-2} = \dfrac{\cos\theta\sin^2\theta}{8}$
$P_3^{-1} = \dfrac{(3 + 5\cos 2\theta)\sin\theta}{16}$	$P_4^{-2} = \dfrac{(5 + 7\cos 2\theta)\sin^2\theta}{96}$

$$(6.35)$$

Figure 6.3 shows an example of associated Legendre functions. The first six of the functions for $m = 1$ are plotted. Note that they are orthogonal, satisfying Eq. (6.32).

The associated Legendre functions for $m = 3$ are plotted in Fig. 6.4 for comparison with Fig. 6.3. The first six orthogonal functions ($n = 3 - 8$) are shown. Note that as m increases the functions are more tapered at the two poles. The increase in the number of oscillations is linearly proportional to the increase in m.

The recurrence relations for the Legendre functions can be written in numerous ways.[4] Four derivative relations are

$$(1 - x^2)\frac{dP_n^m(x)}{dx} = (n + 1)x P_n^m(x) - (n - m + 1)P_{n+1}^m(x)r, \qquad (6.36)$$

$$= -nx P_n^m(x) + (n + m)P_{n-1}^m(x), \qquad (6.37)$$

$$= -\sqrt{1 - x^2}\,P_n^{m+1}(x) - mx P_n^m(x), \qquad (6.38)$$

$$= (n - m + 1)(n + m)\sqrt{1 - x^2}\,P_n^{m-1}(x) + mx P_n^m(x). \qquad (6.39)$$

---

[4]Gradshteyn and Ryzhik, *Table of Integrals, Series and Products*, p.1005.

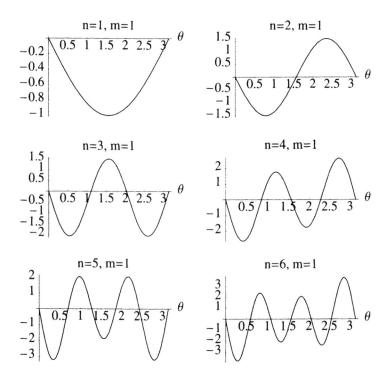

Figure 6.3: Plots of the associated Legendre functions, $P_n^m(\cos\theta)$, as a function of θ for $n = 1, 2, 3, 4, 5, 6$ and $m = 1$.

Four recurrence relations not involving derivatives are

$$(2n+1)xP_n^m(x) = (n-m+1)P_{n+1}^m(x) + (n+m)P_{n-1}^m(x), \qquad (6.40)$$

$$P_n^{m+2}(x) = -2(m+1)\frac{x}{\sqrt{1-x^2}}P_n^{m+1}(x)$$
$$-(n-m)(n+m+1)P_n^m(x), \qquad (6.41)$$

$$P_{n-1}^m(x) - P_{n+1}^m(x) = (2n+1)\sqrt{1-x^2}P_n^{m-1}(x), \qquad (6.42)$$

$$P_{-n-1}^m(x) = P_n^m(x). \qquad (6.43)$$

Many other recurrence relations can be developed from these. In the appropriate relations above, putting $m = 0$ generates the recurrence relationships for the Legendre polynomials.

6.3.3 Spherical Harmonics

Equation 6.20 above defined the spherical harmonics as

$$Y_n^m(\theta,\phi) \equiv \sqrt{\frac{(2n+1)}{4\pi}\frac{(n-m)!}{(n+m)!}}P_n^m(\cos\theta)e^{im\phi}.$$

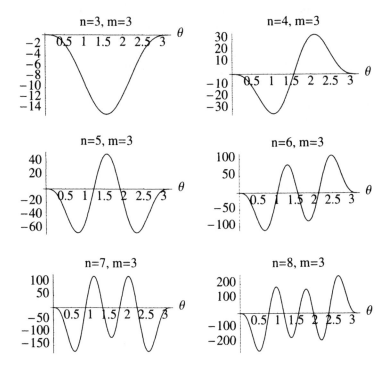

Figure 6.4: Plots of the associated Legendre functions $P_n^m(\cos\theta)$ as a function of θ for $n = 3, 4, 5, 6, 7, 8$ and $m = 3$.

Due to Eq. (6.31) we have

$$Y_n^{-m}(\theta, \phi) = (-1)^m Y_n^m(\theta, \phi)^*. \tag{6.44}$$

The spherical harmonics Y_n^m are orthonormal:

$$\int_0^{2\pi} d\phi \int_0^\pi Y_n^m(\theta, \phi) Y_{n'}^{m'}(\theta, \phi)^* \sin\theta d\theta = \delta_{nn'}\delta_{mm'}. \tag{6.45}$$

Earlier we learned that for any complete set of orthonormal functions $U_n(\zeta)$ there exists a completeness or closure relation given by

$$\sum_{n=1}^\infty U_n^*(\zeta')U_n(\zeta) = \delta(\zeta' - \zeta). \tag{6.46}$$

Applying this equation to the spherical harmonics the completeness relation becomes

$$\sum_{n=0}^\infty \sum_{m=-n}^n Y_n^m(\theta, \phi) Y_n^m(\theta', \phi')^* = \delta(\phi - \phi')\delta(\cos\theta - \cos\theta'). \tag{6.47}$$

The importance of the spherical harmonics rests in the fact that *any arbitrary function on a sphere* $g(\theta, \phi)$ can be expanded in terms of them,

$$g(\theta, \phi) = \sum_{n=0}^{\infty} \sum_{m=-n}^{n} A_{nm} Y_n^m(\theta, \phi), \tag{6.48}$$

where A_{nm} are complex constants. Because of the orthonormality of these functions the arbitrary constants can be found from

$$A_{nm} = \int d\Omega \, Y_n^m(\theta, \phi)^* g(\theta, \phi), \tag{6.49}$$

where Ω is the solid angle defined by

$$\int d\Omega \equiv \int_0^{2\pi} d\phi \int_0^{\pi} \sin\theta d\theta.$$

The delta function in spherical coordinates is given by

$$\delta(\vec{r} - \vec{r'}) = \frac{1}{r^2} \delta(r - r') \delta(\phi - \phi') \delta(\cos\theta - \cos\theta'), \tag{6.50}$$

where $\vec{r} = (r, \theta, \phi)$ and $\vec{r'} = (r', \theta', \phi')$. It is easily verified that this delta function satisfies the relation

$$\int_0^{2\pi} d\phi \int_0^{\pi} \sin\theta d\theta \int_0^{\infty} r^2 \delta(\vec{r} - \vec{r'}) \, dr = 1.$$

The two-dimensional delta function on a sphere is simply

$$\delta(\phi - \phi') \delta(\cos\theta - \cos\theta')$$

which integrates to unity over the solid angle.

The following is a table of some of the spherical harmonics.

$n = 0$ & 1	$n = 2$
$Y_0^0(\theta, \phi) = \dfrac{1}{\sqrt{4\pi}}$	$Y_2^{-2}(\theta, \phi) = 3e^{-2i\phi}\sqrt{\dfrac{5}{96\pi}}\sin^2\theta$
$Y_1^{-1}(\theta, \phi) = e^{-i\phi}\sqrt{\dfrac{3}{8\pi}}\sin\theta$	$Y_2^{-1}(\theta, \phi) = \dfrac{3}{2}e^{-i\phi}\sqrt{\dfrac{5}{24\pi}}\sin 2\theta$
$Y_1^0(\theta, \phi) = \sqrt{\dfrac{3}{4\pi}}\cos\theta$	$Y_2^0(\theta, \phi) = \sqrt{\dfrac{5}{16\pi}}\left(-1 + 3\cos^2\theta\right)$
$Y_1^1(\theta, \phi) = -e^{i\phi}\sqrt{\dfrac{3}{8\pi}}\sin\theta$	$Y_2^1(\theta, \phi) = -\dfrac{3}{2}e^{i\phi}\sqrt{\dfrac{5}{24\pi}}\sin 2\theta$
	$Y_2^2(\theta, \phi) = 3e^{2i\phi}\sqrt{\dfrac{5}{96\pi}}\sin^2\theta$

$$(6.51)$$

$n = 3$	
$Y_3^{-3}(\theta, \phi) = e^{-3i\phi}\sqrt{\dfrac{35}{64\pi}}\sin^3\theta$	$Y_3^1(\theta, \phi) = e^{i\phi}\sqrt{\dfrac{21}{64\pi}}\left(1 - 5\cos^2\theta\right)\sin\theta$
$Y_3^{-2}(\theta, \phi) = 15e^{-2i\phi}\sqrt{\dfrac{7}{480\pi}}\cos\theta\sin^2\theta$	$Y_3^2(\theta, \phi) = 15e^{2i\phi}\sqrt{\dfrac{7}{480\pi}}\cos\theta\sin^2\theta$
$Y_3^{-1}(\theta, \phi) = e^{-i\phi}\sqrt{\dfrac{21}{256\pi}}\left(3 + 5\cos 2\theta\right)\sin\theta$	$Y_3^3(\theta, \phi) = -\dfrac{5}{8}e^{3i\phi}\sqrt{\dfrac{7}{5\pi}}\sin^3\theta$
$Y_3^0(\theta, \phi) = \sqrt{\dfrac{7}{16\pi}}\left(-3\cos\theta + 5\cos^3\theta\right)$	

$$(6.52)$$

Note that for $m = 0$,

$$Y_n^0(\theta, \phi) = \sqrt{\frac{2n + 1}{4\pi}}P_n(\cos\theta). \tag{6.53}$$

The simplicity of the spherical harmonics is borne out in the gray-scale plot of the $n = 8$ terms, shown in Fig. 6.5. In this plot the values of $\mathrm{Re}[Y_n^m(\theta, \phi)]$ ($m = 0, 1, \cdots, 8$) on a unit sphere are projected onto the (y, z) plane, looking down the positive x axis, using a gray scale with white the most positive and black the most negative in the mapping of the value of the function. The nodal lines are drawn in along with the outline of the sphere. The gray background outside the sphere is the color mapped to zero. The beauty and simplicity of the spherical harmonics is illustrated very well in this kind of plot. Note that Y_8^0 has no longitudinal nodal lines, whereas Y_8^1 has its longitudinal nodal lines on the great circle corresponding with the circle outline.

6.4 Radial Functions

6.4.1 Spherical Bessel Functions

We can rewrite Eq. (6.10) as

$$\left[\frac{d^2}{dr^2} + \frac{2}{r}\frac{d}{dr} + k^2 - \frac{n(n + 1)}{r^2}\right]R_n(r) = 0. \tag{6.54}$$

This would be Bessel's equation, Eq. (4.32), except for the coefficient of $2/r$ instead of $1/r$. However, we transform Eq. (6.54) to Bessel's equation with the substitution,

$$R_n(r) = \frac{1}{r^{1/2}}u_n(r) \tag{6.55}$$

yielding

$$\left[\frac{d^2}{dr^2} + \frac{1}{r}\frac{d}{dr} + k^2 - \frac{(n + 1/2)^2}{r^2}\right]u_n(r) = 0. \tag{6.56}$$

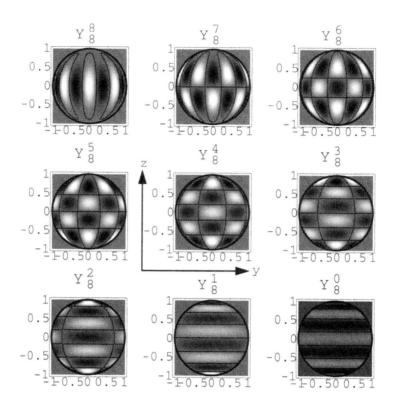

Figure 6.5: $n=8$ spherical harmonics viewed looking down the x-axis. The real part is plotted in gray scale and the nodal lines are indicated by the thin solid lines.

The solutions of Eq. (6.56) are Bessel functions $J_{n+1/2}(kr)$ and $Y_{n+1/2}(kr)$ or the corresponding Hankel functions. Note that the wavenumber k appears alone in the argument now (whereas in cylindrical coordinates we had $\sqrt{k^2 - k_z^2}$). Thus Eq. (6.55) leads us to the following solution of Eq. (6.54).

$$R_n(r) = \frac{A_n}{r^{1/2}} J_{n+1/2}(kr) + \frac{B_n}{r^{1/2}} Y_{n+1/2}(kr).$$

The spherical Bessel and Hankel functions are defined in terms of these solutions:

$$
\begin{aligned}
j_n(x) &\equiv \left(\frac{\pi}{2x}\right)^{1/2} J_{n+1/2}(x) \\
y_n(x) &\equiv \left(\frac{\pi}{2x}\right)^{1/2} Y_{n+1/2}(x) \\
h_n^{(1)}(x) &\equiv j_n(x) + iy_n(x) = \left(\frac{\pi}{2x}\right)^{1/2}[J_{n+1/2}(x) + iY_{n+1/2}(x)] \\
h_n^{(2)}(x) &\equiv j_n(x) - iy_n(x) = \left(\frac{\pi}{2x}\right)^{1/2}[J_{n+1/2}(x) - iY_{n+1/2}(x)].
\end{aligned}
\tag{6.57}
$$

Note that when x is real

$$h_n^{(2)}(x) = (h_n^{(1)}(x))^*. \tag{6.58}$$

Figures 6.6 and 6.7 show plots of $j_n(x)$ and $y_n(x)$, respectively, for $n = 0, 2, 4, 6$.

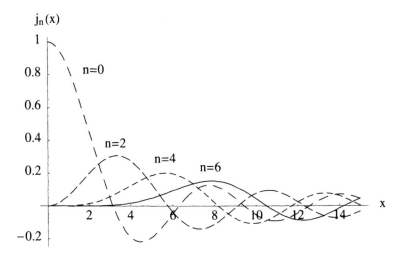

Figure 6.6: Spherical Bessel functions, $j_n(x)$ for $n = 0, 2, 4, 6$.

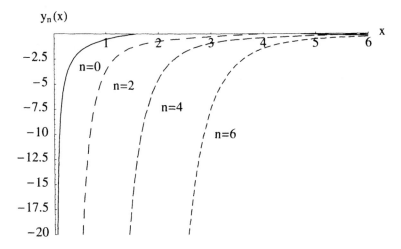

Figure 6.7: Spherical Bessel functions $y_n(x)$ for $n = 0, 2, 4, 6$.

It is important to realize that, unlike the Bessel functions of integer argument, simple expressions exist for the spherical Bessel functions in terms of trigonometric functions.

These are written concisely as

$$\begin{aligned}
j_n(x) &= (-x)^n \left(\frac{1}{x}\frac{d}{dx}\right)^n \left(\frac{\sin x}{x}\right) \\
y_n(x) &= -(-x)^n \left(\frac{1}{x}\frac{d}{dx}\right)^n \left(\frac{\cos x}{x}\right) \\
h_n^{(1)}(x) &= (-x)^n \left(\frac{1}{x}\frac{d}{dx}\right)^n \left(\frac{e^{ix}}{ix}\right) \\
&= (-x)^n \left(\frac{1}{x}\frac{d}{dx}\right)^n [h_0^{(1)}(x)].
\end{aligned}$$
(6.59)

The first few relationships are

$$\begin{aligned}
j_0(x) &= \frac{\sin x}{x} \\
j_1(x) &= -\frac{\cos x}{x} + \frac{\sin x}{x^2} \\
j_2(x) &= \frac{-\sin x}{x} - \frac{3\cos x}{x^2} + \frac{3\sin x}{x^3} \\
j_3(x) &= \frac{\cos x}{x} - \frac{6\sin x}{x^2} - \frac{15\cos x}{x^3} + \frac{15\sin x}{x^4},
\end{aligned}$$
(6.60)

$$\begin{aligned}
y_0(x) &= -\frac{\cos x}{x} \\
y_1(x) &= -\frac{\sin x}{x} - \frac{\cos x}{x^2} \\
y_2(x) &= +\frac{\cos x}{x} - \frac{3\sin x}{x^2} - \frac{3\cos x}{x^3} \\
y_3(x) &= \frac{\sin x}{x} + \frac{6\cos x}{x^2} - \frac{15\sin x}{x^3} - \frac{15\cos x}{x^4},
\end{aligned}$$
(6.61)

and

$$\begin{aligned}
h_0^{(1)}(x) &= \frac{e^{ix}}{ix} \\
h_1^{(1)}(x) &= -\frac{e^{ix}(i+x)}{x^2} \\
h_2^{(1)}(x) &= \frac{ie^{ix}(-3+3ix+x^2)}{x^3} \\
h_3^{(1)}(x) &= \frac{e^{ix}(-15i-15x+6ix^2+x^3)}{x^4}.
\end{aligned}$$
(6.62)

The behavior near the origin of the coordinate system is given by the small argument expressions for the spherical Bessel functions ($x \ll n$):

$$j_n(x) \approx \frac{x^n}{(2n+1)!!}\left(1 - \frac{x^2}{2(2n+3)} + \cdots\right),$$
(6.63)

$$y_n(x) \approx -\frac{(2n-1)!!}{x^{n+1}}\left(1 - \frac{x^2}{2(1-2n)} + \cdots\right),$$

$$h_n(x) \approx -i\frac{(2n-1)!!}{x^{n+1}}$$

$$h'_n(x) \approx i\frac{(n+1)(2n-1)!!}{x^{n+2}}$$

(6.64)

where $(2n+1)!! = (2n+1)(2n-1)(2n-3)\cdots 3\cdot 1$. A useful relation is

$$(2n+1)!! = \frac{(2n+1)!}{2^n n!}.$$

(6.65)

It is clear from these expansions that, as with the cylindrical Bessel functions, only the j_n functions are finite at the origin, and only j_0 is non-zero there.

Two useful Wronskian relations are

$$j_n(x)y'_n(x) - j'_n(x)y_n(x) = \frac{1}{x^2},$$

(6.66)

and

$$j_n(x)h_n^{'(1)}(x) - j'_n(x)h_n^{(1)}(x) = \frac{i}{x^2}.$$

(6.67)

The large argument limits of the spherical Hankel functions are given by the first term (the $1/x$ term) in Eqs (6.60–6.62). In particular,

$$h_n^{(1)}(x) \approx (-i)^{n+1}\frac{e^{ix}}{x},$$

(6.68)

$$h'_n(x) \approx (-i)^n\frac{e^{ix}}{x}.$$

A set of recurrence relations exist for the spherical Bessel and Hankel functions. In particular,

$$\frac{2n+1}{x}h_n(x) = h_{n-1}(x) + h_{n+1}(x)$$

$$h'_n(x) = h_{n-1}(x) - \frac{n+1}{x}h_n(x).$$

(6.69)

In Eq. (6.69) one can replace h_n with either j_n, y_n or $h_n^{(2)}$ due to the fact that $h_n^{(1)} = j_n + iy_n$ and equating real and imaginary parts.

6.5 Multipoles

Radiation from bodies which are located at the origin and which are of finite extent can be characterized by sums of multipoles. Multipole expansions are similar to (but not identical to) the spherical harmonic expansions with the corresponding outgoing radial functions:

$$p(r,\theta,\phi,\omega) = \sum_{n=0}^{\infty}\sum_{m=-n}^{n} C_{mn}h_n^{(1)}(kr)Y_n^m(\theta,\phi),$$

(6.70)

where r is greater than or equal to the largest radial extent of the body. There is not, however, a one-to-one correspondence between the multipoles and the terms in the series in Eq. (6.70). Multipoles are constructed from distributions of point sources, infinitesimally close to the origin, of equal amplitudes but opposite phases. Whereas the spherical harmonics are orthogonal to one another, the multipoles are not. In any case, we will show that the $n = m = 0$ term in Eq. (6.70) does correspond to the simplest multipole, the monopole, and the $n = 1$ terms to the dipoles.

6.5.1   Monopoles

Consider first the simplest multipole, a point source located at the origin called a monopole. The radiated pressure associated with it is omnidirectional (independent of angle) and, if Q_s is its source strength (volume flow), then its pressure is given by

$$p = \frac{-i\rho_0 ck}{4\pi} Q_s \frac{e^{ikr}}{r}. \tag{6.71}$$

Note that due to Eq. (6.62)

$$\frac{e^{ikr}}{r} = ikh_0(kr)$$

so that the monopole is the $n = 0$, $m = 0$ term of Eq. (6.70):

$$p(r, \theta, \phi) = C_{00} \frac{1}{\sqrt{4\pi}} h_0(kr).$$

Comparing this to Eq. (6.71) reveals that

$$C_{00} = \frac{\rho_0 ck^2}{\sqrt{4\pi}} Q_s.$$

To develop expressions for the dipoles and quadrupoles we need to represent combinations of point sources located near the origin. To this end consider a point source located at $r = r'$.[5] The Helmholtz equation is

$$\nabla^2 G(\mathbf{r}|\mathbf{r}') + k^2 G(\mathbf{r}|\mathbf{r}') = -\delta(\mathbf{r} - \mathbf{r}'), \tag{6.72}$$

where both $\mathbf{r}$ and $\mathbf{r}'$ are measured from the origin and

$$G(\mathbf{r}|\mathbf{r}') = \frac{e^{ikR}}{4\pi R} = \frac{ik}{4\pi} h_0(kR). \tag{6.73}$$

The bold face quantities represent vector quantities (so that $\vec{r} \equiv \mathbf{r}$, $\vec{r}' \equiv \mathbf{r}'$) and

$$R^2 \equiv |\mathbf{r} - \mathbf{r}'|^2 = (x - x')^2 + (y - y')^2 + (z - z')^2. \tag{6.74}$$

Thus the radiated pressure at the field point $\mathbf{r}$, in view of Eq. (6.71), is

$$p(\mathbf{r}|\mathbf{r}') = -i\rho_0 ckQ_s G(\mathbf{r}|\mathbf{r}'). \tag{6.75}$$

[5]P. M. Morse and K. U. Ingard (1968). *Theoretical Acoustics.* McGraw-Hill, New York, pp. 310–314.

This is a monopole located at $\mathbf{r}'$.

We can compute the average energy density ($\mathbf{v}$ is a velocity vector) at a field point $\mathbf{r}$ for any of the multipoles using the general relation[6]

$$e_{av} = \frac{1}{4}\rho_0\mathbf{v}\cdot\mathbf{v}^* + \frac{1}{4}\frac{|p|^2}{\rho_0c^2}. \tag{6.76}$$

It is instructive to write the expressions for the radial velocity, average energy density, intensity, and power radiated by a monopole source from the pressure field given by Eq. (6.71).[7] The radial velocity from Euler's equation is

$$\begin{aligned}
\dot{w} &= \frac{1}{i\rho_0ck}\frac{\partial p}{\partial r} = -\frac{ikQ_s}{4\pi}\frac{e^{ikr}}{r}(1-\frac{1}{ikr}) \tag{6.77}\\
&= \frac{p}{\rho_0c}(1+\frac{i\lambda}{2\pi r}).
\end{aligned}$$

Unlike the pressure the radial velocity has a nearfield associated with it, defined by the additional $1/ikr$ term. As kr increases this nearfield becomes negligible. From Eq. (6.76) the average energy density is

$$e_{av} = \frac{\rho_0k^2|Q_s|^2}{(4\pi r)^2}(\frac{1}{2} + \frac{1}{4(kr)^2}). \tag{6.78}$$

The three components of the average acoustic intensity are provided by Eq. (2.16), page 19:

$$\begin{aligned}
I_r &= \frac{|Q_s|^2}{2}\frac{\rho_0ck^2}{(4\pi r)^2} = \frac{|p|^2}{2\rho_0c}, \\
I_\phi &= 0 \tag{6.79}\\
I_\theta &= 0.
\end{aligned}$$

Finally, the total power defined by $\Pi = \int I_r r^2 d\Omega$ is

$$\Pi = (4\pi r^2)I_r = \frac{\rho_0ck^2}{8\pi}|Q_s|^2. \tag{6.80}$$

These relations contrast the nearfield and farfield regions of the monopole.

6.5.2 Dipoles

Next we consider the representation of a dipole. A dipole consists of two point sources opposite in phase and separated by an infinitesimal distance d oriented along one of the rectangular coordinate axes. Consider the axial dipole as shown in Fig. 6.8. We define

---

[6]A. Pierce (1981). *Acoustics: An Introduction to Its Physical Principles and Applications*. McGraw-Hill, New York, p. 40 Eq. (1-11.11a).

[7]P. M. Morse and K. U. Ingard (1968). *Theoretical Acoustics*. McGraw-Hill, New York, p. 311.

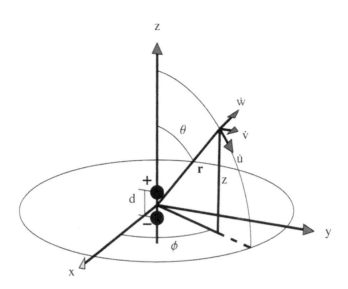

Figure 6.8: Dipole along the z axis. The positive source is at $\mathbf{r}' = \mathbf{d}/2$ and the negative point source is at $\mathbf{r}' = -\mathbf{d}/2$. The field point, $\mathbf{r}$, shows the orientations of the three components of the velocity vector.

the dipole strength, $D_s = Q_s d$, and assume that d is vanishingly small. Two point sources of opposite phase, as shown in Fig. 6.8, create a pressure

$$
\begin{aligned}
p(r,\theta,\phi) &= -i\rho_0 ck Q_s[G(\mathbf{r}|\tfrac{1}{2}\mathbf{d}) - G(\mathbf{r}| - \tfrac{1}{2}\mathbf{d})] \\
&= -i\rho_0 ck Q_s d \frac{\partial G}{\partial z'}\Big|_{\mathbf{r}'=0} \\
&= -\rho ck^2 D_s \cos\theta (1 + \frac{i}{kr})\frac{e^{ikr}}{4\pi r}.
\end{aligned}
\tag{6.81}
$$

Because the dipole was oriented along the z axis the pressure field is independent of ϕ. Unlike the monopole there is a nearfield pressure associated with the dipole. Also the pressure field depends on $\cos\theta$. The two non-zero components of velocity are

$$
\begin{aligned}
\dot{w} &= -k^2 D_s \cos\theta \left(1 + \frac{2i}{kr} - \frac{2}{(kr)^2}\right)\frac{e^{ikr}}{4\pi r}, \\
\dot{u} &= -ik D_s \sin\theta \left(1 + \frac{i}{kr}\right)\frac{e^{ikr}}{4\pi r^2}.
\end{aligned}
\tag{6.82}
$$

The average energy density is

$$
e_{av} = \frac{\rho_0}{2}\left(\frac{k^2 D_s}{4\pi r}\right)^2 \left[\frac{1}{2}\left(\frac{1}{(kr)^2} + \frac{1}{(kr)^4}\right) + \left(1 + \frac{3}{2(kr)^4}\right)\cos^2\theta\right].
\tag{6.83}
$$

The only non-zero active intensity component is the radial intensity given by

$$I_r = \frac{\rho_0 c k^4 |D_s|^2 \cos^2 \theta}{2(4\pi r)^2}.$$ (6.84)

Whereas the monopole power is proportional to the square of the frequency, the total power for the dipole is proportional to the fourth power:

$$\Pi = \frac{\rho_0 c k^4 |D_s|^2}{24\pi}.$$ (6.85)

If the dipole is along the x axis then the positive and negative sources create a pressure field given by

$$p = -i\rho_0 c k Q_s d \frac{\partial G}{\partial x'}\Big|_{r'=0}.$$ (6.86)

Evaluating the partial derivative and using Eq. (6.1) leads to the angle factor $\sin\theta \cos\phi$ which is proportional to $\text{Re}[Y_1^1]$. The dipole oriented along the y axis turns out to have an angle factor of $\sin\theta \sin\phi$ which is proportional to $\text{Im}[Y_1^1]$. The directivity patterns of the three dipoles are shown in Fig. 6.9.

To relate the dipole pressure field to the spherical harmonic expansion, Eq. (6.70), we need to recast Eq. (6.81) in terms of Hankel functions. From Eq. (6.59)

$$h_n(kR) = -(\frac{R}{k})^n (\frac{1}{R}\frac{d}{dR})^n [h_0(kR)].$$ (6.87)

We can see that a dipole is generated if $n = 1$ in Eq. (6.87) since this leads to the first derivative of $h_0(kR)$, the latter being a monopole source. That is, since

$$G = \frac{ik}{4\pi} h_0(kR)$$

and $R = |\mathbf{r} - \mathbf{r}'|$, then $\frac{\partial G}{\partial z'}$ in Eq. (6.81) is

$$\frac{\partial G}{\partial z'}\Big|_{r'=0} = \frac{\partial h_0(kR)}{\partial z'}\Big|_{r'=0} = \frac{d}{dR}h_0(kR)\frac{\partial R}{\partial z'}\Big|_{r'=0} = -kh_1(kr)\cos\theta.$$

Note that $\cos\theta = \sqrt{4\pi/3}Y_1^0$ (see Eq. (6.52)) so that, returning to Eq. (6.70), we see that the axial dipole corresponds to the $m = 0$, $n = 1$ term of the general spherical harmonic expansion and thus can be represented by a single spherical harmonic. Note that the x and y oriented dipoles are both constructed from the $m = 1$, $n = 1$ term using the real and imaginary parts of Y_1^1, respectively.

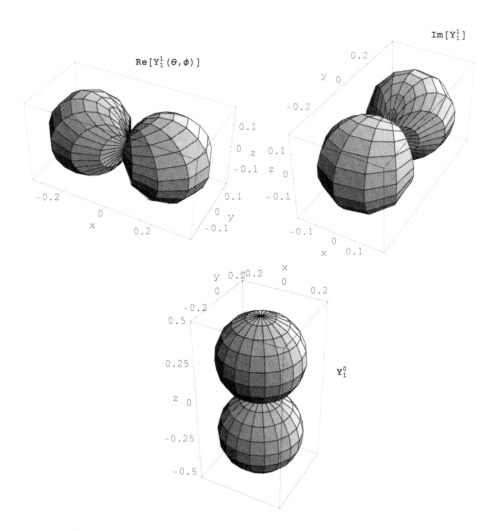

Figure 6.9: Angular dependence of the pressure field generated by the three dipoles.

6.5.3   Quadrupoles

Opposing orientations of dipoles lead to quadrupoles with strengths $Q_s d^2$. The pressure due to the configuration shown in Fig. 6.10 is

$$p = -i\rho_0 ck Q_{xy} \left[\frac{\partial^2}{\partial x' \partial y'} G(\mathbf{r}|\mathbf{r}') \right]\Big|_{r'=0}, \tag{6.88}$$

with the two opposing dipoles placed in the (x, y) plane. Using analysis similar to the dipole case it turns out that this pressure is proportional to $h_2(kr)$ and the imaginary part of $Y_2^2(\theta, \phi)$. The angular dependence of this quadrupole is shown in the left-top plot of Fig. 6.11, labeled $\text{Im}[Y_2^2]$. Note the pressure field vanishes on the coordinate

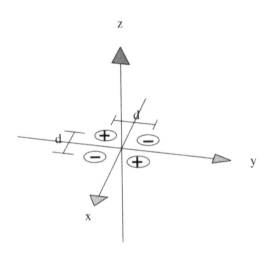

Figure 6.10: Two opposing dipoles in the (x, y) plane which form a quadrupole.

axes. Similarly, two opposing dipoles in the (x, z) plane lead to the real part of Y_2^1, and in the (y, z) plane to the imaginary part of Y_2^1 as shown in the right plot in the figure.

It is interesting to compute the so called longitudinal quadrupole, given by

$$[\frac{\partial^2}{\partial z'^2} G(\mathbf{r}|\mathbf{r}')]\Big|_{r'=0}.$$

We find that this is nearly equal to $Y_2^0(\theta)$ (see Fig. 6.11 for Y_2^0) except that we need to subtract a monopole term from it in order to arrive at the correct pressure field of a longitudinal quadrupole. This points out the important fact that the terms of Eq. (6.70) do not have a one-to-one correspondence with multipoles. In fact, multipoles are not even orthogonal, but are built out of terms in the spherical harmonic series of Eq. (6.70).

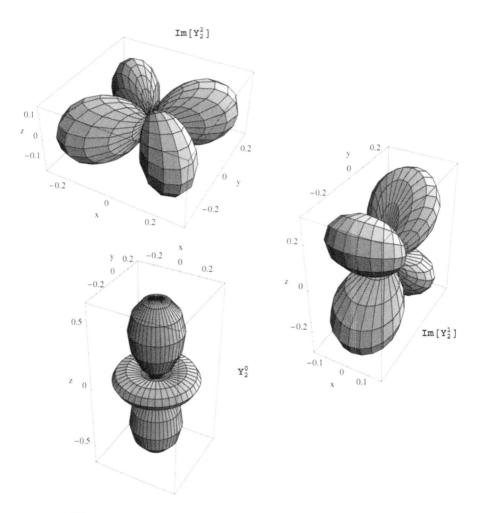

Figure 6.11: Farfield directivity patterns for the $n = 2$ spherical harmonics.

6.6   Spherical Harmonic Directivity Patterns

As we have seen with plates and cylinders the farfield directivity pattern is defined by removing the e^{ikr}/r factor so that the directivity pattern $D(\theta, \phi)$ is defined by

$$\lim_{r \to \infty} [p(r, \theta, \phi)] = \frac{e^{ikr}}{r} D(\theta, \phi). \tag{6.89}$$

Note that the expansion in Eq. (6.70) differs from the functions in cylindrical and planar coordinates in that the directivity pattern is explicitly part of the wave functions. That is, from Eq. (6.70)

$$D(\theta, \phi) = \lim_{r \to \infty} [re^{-ikr} h_n(kr)] Y_n^{-m}(\theta, \phi). \tag{6.90}$$

Using Eq. (6.68) the directivity pattern of the (n, m)th spherical harmonic is

$$D(\theta, \phi) = \frac{(-i)^{n+1}}{k} Y_n^{*m}(\theta, \phi). \tag{6.91}$$

The spherical harmonics provide the farfield directivity pattern of each term in the series, Eq. (6.70). This exposes the remarkable fact that for each spherical harmonic the nodes of pressure which arise from $P_n^m(\cos\theta)$ extend uninterrupted from the origin to the farfield.

The directivity patterns for some of the higher order spherical harmonics are shown in Figs 6.12 and 6.13. The first of these figures represents the patterns for the Y_3^m harmonics, and the second figure the Y_4^m harmonics. We have plotted the absolute value of the imaginary parts (ϕ dependence is $\sin(m\phi)$). The vertical axis is the z axis. Note that Fig. 6.11 provided, in similar fashion, the Y_2^m directivity patterns.

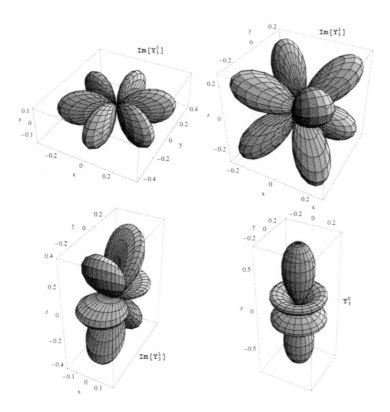

Figure 6.12: Farfield directivity patterns for the $n = 3$ spherical harmonics. The absolute value of the imaginary part of Y_n^m is plotted which leads to a $\sin(m\phi)$ dependence in ϕ.

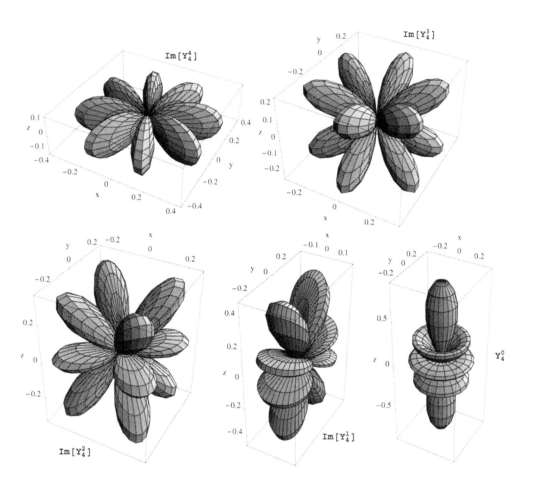

Figure 6.13: Farfield directivity patterns for the $n = 4$ spherical harmonics. The absolute value of the imaginary part of Y_n^m is plotted which leads to a $\sin(m\phi)$ dependence in ϕ.

6.7   General Solution for Exterior Problems

The exterior domain is defined as shown in Fig. 6.14. The sphere of radius a is the smallest sphere tangent to the outermost extremity of the source and does not cut through the source at any point.

Equation (6.70) applies to this problem and the radiated pressure field is given by

$$p(r,\theta,\phi,\omega) = \sum_{n=0}^{\infty} \sum_{m=-n}^{n} C_{mn}(\omega) h_n(kr) Y_n^m(\theta,\phi). \qquad (6.92)$$

The radiated pressure field is completely defined when the coefficients C_{mn} are determined. The C_{mn} are generally functions of frequency. The spherical Hankel function

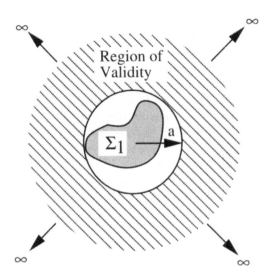

Figure 6.14: Exterior domain problem with all sources inside the spherical surface at $r = a$.

$h_n(kr)$ stands for the outgoing definition, $h_n^{(1)}(kr)$. As was the case with the plate and the cylinder, if the pressure field is known on a separable surface (a sphere) of the coordinate system then the three-dimensional field outside the source can be easily determined, as we will now show.

Assume that the pressure $p(r, \theta, \phi)$ is known on a sphere of radius $r = a$. The complete three-dimensional field is uniquely defined if the coefficients C_{mn} can be determined. We determine the coefficients by using the orthonormal property of the spherical harmonics. To this end multiply each side of Eq. (6.92) (evaluated at $r = a$) by $Y_q^p(\theta, \phi)^*$ and integrate over the surface of a unit sphere. The result is

$$C_{mn} = \frac{1}{h_n(ka)} \iint p(a, \theta, \phi) Y_n^m(\theta, \phi)^* \sin\theta \, d\theta \, d\phi. \tag{6.93}$$

Inserting this result into Eq. (6.92) yields the complete solution,

$$p(r, \theta, \phi) = \sum_{n=0}^{\infty} \frac{h_n(kr)}{h_n(ka)} \sum_{m=-n}^{n} Y_n^m(\theta, \phi) \int p(a, \theta', \phi') Y_n^m(\theta', \phi')^* d\Omega', \tag{6.94}$$

where $d\Omega' = \sin\theta' d\theta' d\phi'$. This important equation relates the pressure on a sphere of radius a to the radiated pressure on a sphere of radius r.

6.7.1 Spherical Wave Spectrum

Just as we define the plane wave spectrum for plane waves and the helical wave spectrum for cylindrical waves, we define the spherical wave spectrum $P_{mn}(r)$ at $r = r_0$ as

$$P_{mn}(r_0) \equiv \int p(r_0, \theta, \phi) Y_n^m(\theta, \phi)^* d\Omega, \tag{6.95}$$

which decomposes the pressure on the sphere of radius r_0 into its spherical wave components. For comparison the helical and plane wave spectra were given in Eq. (4.52) on page 125 and Eq. (2.52) on page 32, respectively. For the sphere the wave is defined by Eq. (6.20) and thus is composed of a traveling wave component in ϕ using $e^{im\phi}$ and a standing wave component in θ given by $P_n^m(\cos\theta)$. Because of the standing wave component we can not define wave fronts, as we did with the plane and cylindrical spectra. One could expand the Legendre functions into traveling wave components using $\cos\theta = (e^{i\theta} + e^{-i\theta})/2$ to develop expressions purely in terms of traveling waves $(e^{im\phi+in\theta})$. This may be useful in some problems but it is rarely done in the literature, and we will not pursue it any further here.

One can make an analogy to "k-space" where the wavenumbers k_x and k_y are imitated by m/r_0 and n/r_0. At times we will refer to the spherical wave spectrum as a k-space spectrum because of this analogy.

As was done for planar and cylindrical systems we can view Eq. (6.95) as a forward Fourier transform using $Y_n^m(\theta,\phi)$ as the basis function. The corresponding inverse Fourier transform (a double Fourier series) is just

$$p(r,\theta,\phi) = \sum_{n=0}^{\infty} \sum_{m=-n}^{n} P_{mn}(r) Y_n^m(\theta,\phi), \tag{6.96}$$

due to the orthonormality of the basis functions. Comparison with Eq. (6.94) reveals that

$$P_{mn}(r) = \frac{h_n(kr)}{h_n(kr_0)} P_{mn}(r_0), \tag{6.97}$$

which provides wavefield extrapolation very similar to the cylindrical wave case given in Eq. (4.58) on page 126. When $r \geq r_0$ the extrapolation is a forward one, and when $r < r_0$ it is an inverse one, equivalent in the time domain to going back in time. The latter is the realm of spherical NAH which will be discussed in the next chapter.

6.7.2   The Relationship Between Velocity and Pressure Spectra

First we define the spherical wave spectrum for the radial velocity $\dot{W}_{mn}$:

$$\dot{W}_{mn}(r) \equiv \int \dot{w}(r,\theta,\phi) Y_n^m(\theta,\phi)^* d\Omega, \tag{6.98}$$

associated with the inverse relationship,

$$\dot{w}(r,\theta,\phi) = \sum_{n=0}^{\infty} \sum_{m=-n}^{n} \dot{W}_{mn}(r) Y_n^m(\theta,\phi). \tag{6.99}$$

Euler's equation, Eq. (6.5), applied to Eq. (6.96) with $\frac{dP_{mn}(r)}{dr}$ derived from Eq. (6.97) yields

$$\dot{w}(r,\theta,\phi) = \frac{1}{i\rho_0 c} \sum_{n=0}^{\infty} \sum_{m=-n}^{n} \frac{h_n'(kr)}{h_n(kr_0)} P_{mn}(r_0) Y_n^m(\theta,\phi). \tag{6.100}$$

Comparison with Eq. (6.99) shows that the relationship between the velocity and pressure spectra is

$$\dot{W}_{mn}(r) = \frac{1}{i\rho_0 c} \frac{h'_n(kr)}{h_n(kr_0)} P_{mn}(r_0), \tag{6.101}$$

or equivalently

$$P_{mn}(r_0) = i\rho_0 c \frac{h_n(kr_0)}{h'_n(kr)} \dot{W}_{mn}(r). \tag{6.102}$$

Similar to the cylinder (Eq. (4.71), page 133), the definition of the specific acoustic impedance Z_{mn} is

$$Z_{mn} \equiv \frac{P_{mn}(r_0)}{\dot{W}_{mn}(r_0)} = i\rho_0 c \frac{h_n(kr_0)}{h'_n(kr_0)}. \tag{6.103}$$

6.7.3 Evanescent Waves

The wavelength in the ϕ direction is dictated by m since $Y_n^m \propto e^{im\phi}$. Thus m counts the number of full wavelengths in the circumferential direction around the sphere. Of course the actual wavelength is largest around the equator, and diminishes as we move towards the poles. This would seem to imply that the acoustic short circuit is greater at the poles producing stronger evanescence. On the contrary, however, the radial decay, given by the ratio of Hankel functions in Eq. (6.97), is independent of latitude. Thus the wavefunction in the polar angle must somehow compensate for this increased acoustic short circuit near the poles. That is, the wavelength in the θ direction depends on m, in such a way that the spherical harmonic Y_n^m possesses the same radial decay for all permissible values of m, and the acoustic short circuit is equalized so that it is independent of latitude.

Study of Figs 6.3 and 6.4 reveals the dependency with respect to the polar angle θ. In this case, as the figures indicate, the number of "half-wavelengths" from $\theta = 0$ to $\theta = \pi$ is given by $n - m + 1$. (Note that the leading term of the series expansion for P_n^m given in Eq. (6.30) is $\cos^{n-m}\theta$.) These "half-wavelengths" are a bit peculiar since they are not strictly sinusoidal, as the figures show. At the poles the standing waves taper smoothly to zero. The half-wavelength near the poles is greater than the half-wavelength at the equator. These two effects, or peculiarities, would appear to compensate to some degree for the increased acoustic short circuit due to the vanishing circumferential wavelength near the poles. Also note that for fixed n the number of these "half-wavelengths" increases when m decreases and vice versa. This is reminiscent of the behavior of plane and helical waves. That is, in k-space we learned that a circle defined a set of waves which have the same attenuation (if subsonic) or phase relation (if supersonic) in the normal direction to the surface. As we move around the k-space circle the wavelength in one direction diminishes while the wavelength in the perpendicular direction increases. Reference back to Fig. 6.5 should clarify these statements. In this case $n = 8$ and all the "modes" shown undergo the same radial decay when evanescent.

Returning to Eq. (6.97) we see that the ratio of Hankel functions dictates how the spherical wave spectrum components change phase and amplitude from one sphere to another. We can understand the asymptotic behavior of evanescent waves by studying the behavior when $kr_0 << n$ and $kr << n$. Using Eq. (6.64) we have

$$\frac{h_n(kr)}{h_n(kr_0)} \approx (r_0/r)^{n+1}, \tag{6.104}$$

a power law decay similar to the cylindrical evanescent wave case. This evanescent behavior is greatest when $r - r_0$ is small, since as r increases the condition $kr << n$ breaks down and we must take more terms in the small argument expansion for the Hankel function.

6.7.4 Boundary Value Problem with Specified Radial Velocity

As we have done with the plate and the cylinder, we want to find a relationship between the radial velocity on the inner sphere at $r = a$ and the pressure on any spherical surface $r > a$. We have derived the relationship in k-space between pressure and normal velocity, given in Eq. (6.102). If we insert this equation into Eq. (6.96) we obtain

$$p(r, \theta, \phi) = i\rho_0 c \sum_{n=0}^{\infty} \frac{h_n(kr)}{h'_n(ka)} \sum_{m=-n}^{n} \dot{W}_{mn}(a) Y_n^m(\theta, \phi). \tag{6.105}$$

Finally we use Eq. (6.98) for $\dot{W}_{mn}(a)$:

$$p(r, \theta, \phi) = i\rho_0 c \sum_{n=0}^{\infty} \frac{h_n(kr)}{h'_n(ka)} \sum_{m=-n}^{n} Y_n^m(\theta, \phi) \int \dot{w}(a, \theta', \phi') Y_n^m(\theta', \phi')^* d\Omega', \tag{6.106}$$

where $d\Omega' = \sin\theta' d\theta' d\phi'$.

Solving boundary value problems is a simple two-step process. With the normal surface velocity prescribed over the whole boundary, Eq. (6.98) is used to compute the velocity spectrum, $\dot{W}_{mn}(a)$. This spectrum is inserted into Eq. (6.105) to solve for the pressure on any sphere of radius $r \geq a$.

6.7.5 The Rayleigh-like Integrals

Rayleigh's first integral formula, Eq. (2.75) on page 36, relates the radiated pressure to surface velocity for plates. The expression for cylinders was given by Eq. (4.77) on page 134. For spheres we can express the Rayleigh-like integral as[8]

$$p(r, \theta, \phi) = i\rho_0 cka^2 \int G_N(r, \theta, \phi | a, \theta', \phi') \dot{w}(a, \theta', \phi') d\Omega', \tag{6.107}$$

where G_N is called a Neumann Green function which will be discussed in more detail in a later chapter. Comparison of this equation with Eq. (6.106) reveals the expression for G_N:

$$G_N(r, \theta, \phi | a, \theta', \phi') = \frac{1}{ka^2} \sum_{n=0}^{\infty} \frac{h_n(kr)}{h'_n(ka)} \sum_{m=-n}^{n} Y_n^m(\theta, \phi) Y_n^m(\theta', \phi')^*. \tag{6.108}$$

---

[8]M. C. Junger and D. Feit (1986). *Sound, Structures, and Their Interaction.* 2nd ed. MIT Press, Cambridge, MA. p. 157.

Note that Eq. (6.107) is no longer a convolution, as it was for plates.

The 2nd Rayleigh-like integral, which relates pressures on two boundaries, is

$$p(r,\theta,\phi) = -a^2 \int G_D(r,\theta,\phi|a,\theta',\phi') p(a,\theta',\phi') \, d\Omega', \qquad (6.109)$$

where G_D is the Dirichlet Green function, derived from Eq. (6.94). That is,

$$G_D(r,\theta,\phi|a,\theta',\phi') = \frac{-1}{a^2} \sum_{n=0}^{\infty} \frac{h_n(kr)}{h_n(ka)} \sum_{m=-n}^{n} Y_n^m(\theta,\phi) Y_n^m(\theta',\phi')^*. \qquad (6.110)$$

When $r = a$, the ratio of spherical Hankel functions is unity and the double sum is a delta function, Eq. (6.47), and the integral reduces to an identity.

6.7.6   Radiated Power

The radiated power is computed from the integral of the radial acoustic intensity (time averaged) over any sphere outside the source regions:

$$\Pi(\omega) = \frac{1}{2} \iint \mathrm{Re}[p(r_0,\theta,\phi) \dot{w}^*(r_0,\theta,\phi)] r_0^2 \sin\theta \, d\theta \, d\phi, \qquad (6.111)$$

where r_0 is the radius of the sphere. From Eq. (6.92) and Euler's equation, Eq. (6.5), we have

$$\dot{w}(r_0,\theta,\phi) = \frac{1}{i\rho_0 ck} \frac{\partial p}{\partial r} = \frac{1}{i\rho_0 c} \sum_{n=0}^{\infty} \sum_{m=-n}^{n} C_{mn}(\omega) h_n'(kr_0) Y_n^m(\theta,\phi). \qquad (6.112)$$

Inserting this result into Eq. (6.111), using Eq. (6.45) yields

$$\Pi = \frac{r_0^2}{2} \sum_{m,n} |C_{mn}|^2 \mathrm{Re}[\frac{h_n(kr_0) h_n'(kr_0)^*}{i\rho_0 c}].$$

This expression is simplified by writing $h_n = j_n + iy_n$ and using the Wronskian relationship given in Eq. (6.66). It reduces to

$$\Pi = \frac{1}{2\rho_0 ck^2} \sum_{n=0}^{\infty} \sum_{m=-n}^{n} |C_{mn}|^2. \qquad (6.113)$$

Thus the sum over the squared magnitude weights of each of the spherical harmonics provides the power radiated to the farfield. There is no power coupling between spherical harmonics of different orders. On the other hand, the active intensity in the nearfield is composed of coupled spherical harmonics.

6.7.7   Farfield Pressure

The farfield for Eq. (6.106) or Eq. (6.105) is easily obtained from the farfield expression for the spherical Hankel function, Eq. (6.68):

$$p(r,\theta,\phi) \approx \rho_0 c \frac{e^{ikr}}{kr} \sum_{n=0}^{\infty} \frac{(-i)^n}{h_n'(ka)} \sum_{m=-n}^{n} \dot{W}_{mn} Y_n^m(\theta,\phi). \qquad (6.114)$$

Farfield Low Frequency Result

We use the fact that (see Eq. (6.64))

$$\lim_{x \to 0} h_n(x) = -i \frac{(2n-1)!!}{x^{n+1}},$$

and

$$\lim_{x \to 0} h_0'(x) = \frac{i}{x^2}$$

in Eq. (6.114) to obtain to leading order (the $n = 0$ term dominates):

$$p(r, \theta, \phi) \approx -i\rho_0 cka^2 \frac{e^{ikr}}{r} \dot{W}_{00} Y_0^0(\theta, \phi)$$

so that

$$\lim_{k \to 0} p(r, \theta, \phi, \omega) = -i\rho_0 cka^2 \frac{e^{ikr}}{4\pi r} \int_0^{2\pi} d\phi' \int_0^\pi \dot{w}(a, \theta', \phi') \sin \theta' d\theta'. \tag{6.115}$$

Note that the double integral is proportional to the total volume flow (m^3/s) Q_s:

$$Q_s \equiv a^2 \int_0^{2\pi} d\phi' \int_0^\pi \dot{w}(a, \theta', \phi') \sin \theta' d\theta', \tag{6.116}$$

so that the radiated pressure at low frequencies is

$$\lim_{k \to 0} p(r, \theta, \phi, \omega) = -i\rho_0 ck Q_s \frac{e^{ikr}}{4\pi r}. \tag{6.117}$$

Comparison to Eq. (6.71) reveals that at low frequencies the leading term in the farfield pressure is a monopole, with source strength Q_s.

High Frequency Result

An interesting result is obtained in the high frequency limit, in which case both the spherical Hankel functions in the numerator and denominator of Eq. (6.106) can be approximated by their asymptotic expressions. Using Eq. (6.68) we find that

$$\lim_{k \to \infty} p(r, \theta, \phi) = \rho_0 ca \frac{e^{ik(r-a)}}{r} \int \dot{w}(a, \theta', \phi') \sum_{n=-\infty}^{\infty} \sum_{m=-n}^{n} Y_n^m(\theta, \phi) Y_n^m(\theta', \phi')^* d\Omega'.$$

The double sum can be simplified by use of Eq. (6.47), replacing them with delta functions. The integration of solid angle is then trivial yielding $\dot{w}(a, \theta, \phi)$. The final result is

$$\lim_{k \to \infty} p(r, \theta, \phi) = \rho_0 ca \frac{e^{ik(r-a)}}{r} \dot{w}(a, \theta, \phi). \tag{6.118}$$

This equation makes the significant statement that in the short wavelength limit the given velocity at any point on the sphere at $r = a$ maps to the farfield following the ray projecting radially outward at the same spherical angle.

6.7.8   Radiation from a Pulsating Sphere

As an example of the application of Eq. (6.106), we consider a pulsating sphere (constant radial velocity). The sphere has a radius a and is vibrating uniformly radially at a frequency ω such that

$$\dot{w}(a,\theta,\phi) = \dot{W}.$$

Recognizing the fact that $\dot{W} = \sqrt{4\pi}Y_0^0\dot{W}$ then

$$\int \dot{w}(a,\theta',\phi')Y_n^m(\theta',\phi')^* d\Omega' = \dot{W}\int \sqrt{4\pi}Y_0^0 Y_n^m(\theta',\phi')^* d\Omega' = \sqrt{4\pi}\dot{W}\delta_{m0}\delta_{n0}.$$

Then Eq. (6.106) becomes, using Eq. (6.62).

$$p(r,\theta,\phi) = i\rho_0 c\frac{h_0(kr)}{h_0'(ka)}Y_0^0\sqrt{4\pi}\dot{W} = \rho_0 c\dot{W}ka^2\frac{ka-i}{((ka)^2+1)r}e^{ik(r-a)}. \qquad (6.119)$$

When the pulsating sphere is very small compared to a wavelength, $ka \ll 1$, then the surface pressure is almost 90 degrees out of phase with the surface velocity,

$$p(r) \approx -i\rho_0 cka^2\dot{W}\cdot\frac{e^{ikr}}{r}. \qquad (6.120)$$

Clearly the pulsating sphere becomes a point source in this limit with source strength $Q_s = 4\pi a^2\dot{W}$.

At very high frequencies, $ka \gg 1$, the pressure and velocity are in phase and we have

$$p(r) \approx \rho_0 ca\dot{W}\frac{e^{ik(r-a)}}{r}. \qquad (6.121)$$

Note there is no restriction on r in these relations, except that $r \geq a$, thus they are valid in the near and farfield.

6.7.9   General Axisymmetric Source

We can use Eq. (6.106) to derive the general formula relating surface velocity to exterior pressure for an axisymmetric source which is defined through

$$\dot{w}(a,\theta,\phi) = \dot{w}(a,\theta) \qquad (6.122)$$

(is independent of ϕ). This is equivalent to specifying $m=0$ for the azimuthal functions $e^{im\phi}$. Using Eq. (6.53)

$$p(r,\theta) = i\rho_0 c\sum_{n=0}^{\infty}\frac{(2n+1)}{2}\frac{h_n(kr)}{h_n'(ka)}P_n(\cos\theta)\int_0^\pi \dot{w}(a,\theta')P_n(\cos\theta')\sin\theta' d\theta'. \qquad (6.123)$$

For a given surface velocity we solve for the radiated pressure by first expanding the surface velocity in Legendre polynomials. Since the $m=0$ polynomials form a complete set, the expansion is

$$\dot{w}(a,\theta') = \sum_{p=0}^{\infty}\dot{W}_p P_p(\cos\theta'), \qquad (6.124)$$

so that Eq. (6.123) becomes (using the orthogonality relations, Eq. (6.28))

$$p(r,\theta) = i\rho_o c \sum_{n=0}^{\infty} \dot{W}_n \frac{h_n(kr)}{h_n'(ka)} P_n(\cos\theta). \tag{6.125}$$

We determine the coefficients W_n by multiplying Eq. (6.124) by $P_p(\cos\theta)$ and integrating over θ:

$$\dot{W}_n = \frac{2n+1}{2} \int_0^{\pi} \dot{w}(a,\theta') P_n(\cos\theta') \sin\theta' \, d\theta'. \tag{6.126}$$

Farfield Pressure

The farfield is obtained as for the nonaxisymmetric case, using Eq. (6.68), so that Eq. (6.125) becomes

$$\lim_{r\to\infty} p(r,\theta) = \rho_o c \frac{e^{ikr}}{kr} \sum_{n=0}^{\infty} \dot{W}_n \frac{(-i)^n}{h_n'(ka)} P_n(\cos\theta). \tag{6.127}$$

High Frequency Case

As we did for the general case resulting in Eq. (6.118), we can use the asymptotic expansions for the Hankel functions in Eq. (6.127) so that

$$\lim_{k\to\infty} p(r,\theta) = \rho_o c \frac{a}{r} e^{ik(r-a)} \int_0^{\pi} \dot{w}(a,\theta') \sum_{n=0}^{\infty} \frac{2n+1}{2} P_n(\cos\theta) P_n(\cos\theta') \sin\theta' d\theta'.$$

We need the completeness relation, Eq. (6.46), for Legendre polynomials. In this case

$$U_n(x) = \sqrt{\frac{2n+1}{2}} P_n(x), \tag{6.128}$$

so that

$$\sum_{n=0}^{\infty} \frac{2n+1}{2} P_n(x') P_n(x) = \delta(x'-x). \tag{6.129}$$

Finally,

$$\lim_{k\to\infty} p(r,\theta) = \rho_o c \frac{a}{r} e^{ik(r-a)} \dot{w}(a,\theta), \tag{6.130}$$

which, of course, agrees with Eq. (6.118) for the general case. Again the velocity beams radially outward in the high frequency limit.

6.7.10   Circular Piston in a Spherical Baffle

As an example of an axisymmetric source we compute the radiation from a circular (spherical cap) piston on an otherwise rigid sphere. The piston is placed at the pole so that we can make the problem axisymmetric and use the axisymmetric formulation presented above. Figure 6.15 shows the geometry.

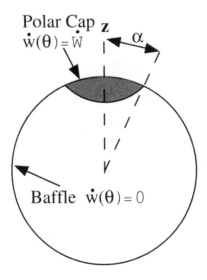

Figure 6.15: Circular piston of radius α set in a rigid sphere. The piston vibrates with a velocity $\dot{W}$. The z axis passes through the center of the piston.

The velocity on the sphere (independent of ϕ) is

$$\dot{w}(a,\theta) = \dot{W}, \qquad 0 \leq \theta \leq \alpha$$
$$= 0, \qquad \alpha < \theta \leq \pi.$$

Equation (6.125) provides the radiated pressure. We compute the coefficients $\dot{W}_n$ from Eq. (6.126):

$$\dot{W}_n = \frac{(2n+1)}{2}\dot{W}\int_0^\alpha P_n(\cos\theta)\sin\theta\,d\theta = \frac{(2n+1)}{2}\dot{W}\int_{\cos\alpha}^1 P_n(\eta)d\eta. \qquad (6.131)$$

To evaluate this integral we need one of the recurrence formulas for the Legendre polynomials,[9]

$$(2n+1)P_n(\eta) = \frac{dP_{n+1}}{d\eta} - \frac{dP_{n-1}}{d\eta}, \qquad (6.132)$$

where, when $n = 0$, only the first term on the right is used. Thus

$$\dot{W}_n = \frac{\dot{W}}{2}[P_{n-1}(\cos\alpha) - P_{n+1}(\cos\alpha)], \qquad (6.133)$$

since $P_{n-1}(1) = P_{n+1}(1) = 1$. For $n = 0$ we have $\dot{W}_0 = (1 - \cos\alpha)/2$. Finally using Eq. (6.125):

$$p(r,\theta) = \frac{i\rho_o c\dot{W}}{2}\sum_{n=0}^\infty [P_{n-1}(\cos\alpha) - P_{n+1}(\cos\alpha)]\frac{h_n(kr)}{h_n'(ka)}P_n(\cos\theta), \qquad (6.134)$$

---

[9] J. D. Jackson (1975). *Classical Electrodynamics*, 2nd ed. Wiley & Sons, p. 89.

where it is understood that for $n = 0$ the difference of Legendre polynomials is just $1 - \cos\alpha$.

6.7.11 Point Source on a Baffle

We let α become infinitesimally small so that Eq. (6.131) can be written ($P_n \approx 1$)

$$\dot{W}_n = \frac{2n+1}{2}\dot{W}\int_0^\alpha \theta\,d\theta = \frac{2n+1}{4}\alpha^2\dot{W}.$$

Given that the area of the point source is $\pi a^2\alpha^2$, the source produces a volume flow of

$$Q_s = \pi a^2\alpha^2\dot{W},$$

and we can write

$$\dot{W}_n = \frac{2n+1}{4\pi a^2}Q_s.$$

The pressure field at any range is given by Eq. (6.125):

$$p(r,\theta) = \frac{i\rho_0 c Q_s}{4\pi a^2}\sum_{n=0}^\infty (2n+1)\frac{h_n(kr)}{h_n'(ka)}P_n(\cos\theta). \tag{6.135}$$

In the farfield we use Eq. (6.127) to obtain,

$$\lim_{r\to\infty} p(r,\theta) = -\frac{i\rho_0 ck Q_s}{4\pi}\frac{e^{ikr}}{r}\sum_{n=0}^\infty \frac{(-i)^{n-1}(2n+1)}{(ka)^2 h_n'(ka)}P_n(\cos\theta). \tag{6.136}$$

We recognize the coefficient multiplying the sum as the pressure radiated from a point source in free space with no baffle:

$$p_f(r) \equiv -\frac{i\rho_0 ck Q_s}{4\pi}\frac{e^{ikr}}{r}, \tag{6.137}$$

and write Eq. (6.136) as

$$\lim_{r\to\infty} p(r,\theta) = p_f(r)\sum_{n=0}^\infty \frac{(-i)^{n-1}(2n+1)}{(ka)^2 h_n'(ka)}P_n(\cos\theta). \tag{6.138}$$

The low frequency result, keeping the first two terms, is

$$\lim_{r\to\infty} p(r,\theta) \approx p_f(r)(1 - i\frac{3ka}{2}\cos\theta), \tag{6.139}$$

which gives us some clue as to the effect of the circular baffle on the radiation from the point source. Due to the $\cos\theta$ the second term is a dipole. Comparison to Eq. (6.81) reveals that the dipole moment is $D_s = 3aQ_s/2$. At very low frequencies the second term is negligible and $p \to p_f$, that is, the baffle has no effect.

In Fig. 6.16 we plot the farfield pressure as a function of polar angle for four cases, $ka = 1, 5, 10, 20$, using Eq. (6.136).

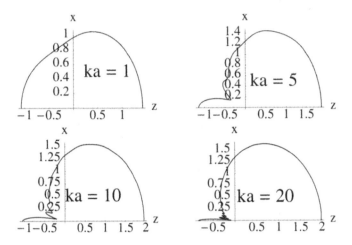

Figure 6.16: Farfield for a point source on the north pole of a rigid sphere as a function of polar angle measured from the horizontal (z) axis. The sphere shadows the point source as the frequency increases.

6.8 General Solution for Interior Problems

For the interior problem the sources are located outside a sphere of radius $r = b$ as shown in Fig. 6.17. The solution consists of radial functions which are finite at the

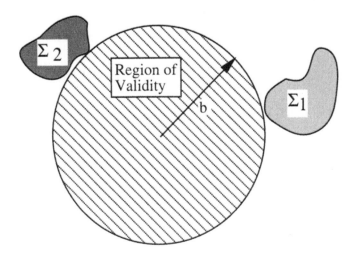

Figure 6.17: Region of validity for the interior solution. Sources are located outside the region defined by $r \geq b$.

origin, so that the internal pressure field is, in general,

$$p(r,\theta,\phi,\omega) = \sum_{n=0}^{\infty} \sum_{m=-n}^{n} A_{mn}(\omega) j_n(kr) Y_n^m(\theta,\phi). \tag{6.140}$$

As we had in Eq. (6.93),

$$A_{mn} = \frac{1}{j_n(kb)} \int \int p(b,\theta,\phi) Y_n^m(\theta,\phi)^* \sin\theta \, d\theta \, d\phi. \tag{6.141}$$

Inserting this result into Eq. (6.140) yields the complete solution,

$$p(r,\theta,\phi) = \sum_{n=0}^{\infty} \frac{j_n(kr)}{j_n(kb)} \sum_{m=-n}^{n} Y_n^m(\theta,\phi) \int p(b,\theta',\phi') Y_n^m(\theta',\phi')^* d\Omega', \tag{6.142}$$

where $d\Omega' = \sin\theta' d\theta' d\phi'$. This important equation relates the pressure on a sphere of radius b to the pressure inside.

In "k-space" the spectral components, instead of Eq. (6.97), become

$$P_{mn}(r) = \frac{j_n(kr)}{j_n(kr_0)} P_{mn}(r_0), \tag{6.143}$$

where P_{mn} is defined in Eq. (6.95). In this case, when $r \le r_0 \le b$ the problem is a forward one, and when $b \ge r > r_0$ an inverse one. We will study the latter case in detail in Chapter 7.

6.8.1 Radial Surface Velocity Specified

If the radial velocity is specified on the surface at $r = b$ then Eq. (6.5) and Eq. (6.140) lead to

$$\dot{w}(a,\theta,\phi) = \frac{1}{i\rho_0 c} \sum_{n=0}^{\infty} \sum_{m=-n}^{n} A_{mn}(\omega) j_n'(kb) Y_n^m(\theta,\phi). \tag{6.144}$$

As we did with Eq. (6.92) we invert this equation to solve for the unknown coefficients A_{mn} to obtain,

$$A_{mn} = \frac{i\rho_0 c}{j_n'(kb)} \int \dot{w}(b,\theta',\phi') Y_n^m(\theta',\phi')^* d\Omega'. \tag{6.145}$$

Finally inserting Eq. (6.145) into Eq. (6.140) yields,

$$p(r,\theta,\phi) = i\rho_0 c \sum_{n=0}^{\infty} \frac{j_n(kr)}{j_n'(kb)} \sum_{m=-n}^{n} Y_n^m(\theta,\phi) \int \dot{w}(b,\theta',\phi') Y_n^m(\theta',\phi')^* d\Omega'. \tag{6.146}$$

We can also write this in the two-step form,

$$\dot{W}_{mn}(b) \equiv \int \dot{w}(b,\theta',\phi') Y_n^m(\theta',\phi')^* d\Omega', \tag{6.147}$$

and

$$p(r, \theta, \phi) = i\rho_0 c \sum_{n=0}^{\infty} \frac{j_n(kr)}{j_n'(kb)} \sum_{m=-n}^{n} \dot{W}_{mn}(b) Y_n^m(\theta, \phi). \tag{6.148}$$

These results are very similar to the results Eq. (6.106), Eq. (6.98), and Eq. (6.105), for the exterior problem.

The spherical wave spectrum ("k-space") for the exterior problem was given in Eq. (6.102). For the interior case we have

$$P_{mn}(r) = i\rho_0 c \frac{j_n(kr)}{j_n'(kr_0)} \dot{W}_{mn}(r_0), \tag{6.149}$$

where $r_0 \leq b$ and $r \leq b$.

Using the definition for $\dot{W}_{mn}(b)$ (Eq. (6.147)) in Eq. (6.148) leads to the Rayleigh-like integral

$$p(r, \theta, \phi) = i\rho_0 c k b^2 \int G_N(r, \theta, \phi|b, \theta', \phi') \dot{w}(b, \theta', \phi') d\Omega', \tag{6.150}$$

where the Neumann Green function is defined similar to Eq. (6.107) as

$$G_N(r, \theta, \phi|b, \theta', \phi') \equiv \frac{1}{kb^2} \sum_{n=0}^{\infty} \frac{j_n(kr)}{j_n'(kb)} \sum_{m=-n}^{n} Y_n^m(\theta, \phi) Y_n^m(\theta', \phi')^*, \tag{6.151}$$

and $r \leq b$.

Although these results are very similar in form to the ones for the exterior problem, there are some fundamental differences which arise. We are now dealing with a finite domain instead of an infinite one so we expect resonances to occur in the interior space. These resonances or interior eigenfrequencies occur when the denominator of Eq. (6.151) or Eq. (6.142) is zero. To illuminate the issue we will solve the pulsating sphere problem (Section 6.7.8) again, this time for the interior field.

6.8.2   Pulsating Sphere

Consider a pulsating sphere with the boundary condition ($e^{-i\omega t}$ time dependence) on its surface,

$$\dot{w}(b, \theta, \phi) = \dot{W}.$$

Using Eq. (6.147)

$$\dot{W}_{mn}(b) = \sqrt{4\pi} \dot{W} \delta_{m0} \delta_{n0}.$$

Thus Eq. (6.148) yields, using Eq. (6.59),

$$p(r) = i\rho_0 c \frac{j_0(kr)}{j_0'(kb)} \dot{W} = -i\rho_0 c \frac{j_0(kr)}{j_1(kb)} \dot{W}. \tag{6.152}$$

The denominator of Eq. (6.152) is plotted in Fig. 6.18 below, Every zero crossing corresponds to an infinite pressure in Eq. (6.152) for all $r \leq b$, that is, an antiresonance of the interior space. In terms of the acoustic impedance, with $p(r)$ evaluated at $r = b$,

$$Z_{ac} \equiv p(b)/\dot{W}; \tag{6.153}$$

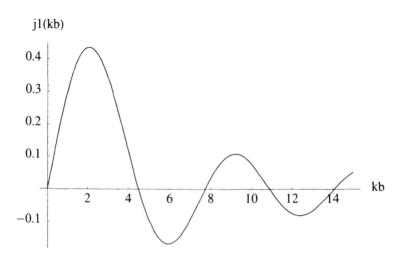

Figure 6.18: Plot of $j_1(kb)$.

the impedance becomes infinite at these zero crossings, and thus the motion of the sphere encounters an infinite force opposing its attempt to oscillate. Since the motion is blocked by the infinite force we call it an antiresonance, not a resonance. We can see from Eq. (6.60), as kb becomes large that the zeros of j_1 are solutions of $\cos(kb) = 0$ or

$$kb = \frac{\pi}{2}(2n + 1), \quad n = 1, 2, 3 \cdots.$$

This translates to antiresonance frequencies

$$\omega_n = \frac{\pi c}{2b}(2n + 1). \tag{6.154}$$

On the other hand the numerator of Eq. (6.152) has an infinite number of zeros given by the zeros of $j_0(kb)$, when $r = b$. These occur exactly when $\sin(kb) = 0$ or

$$\omega_n = \frac{\pi c n}{b}, \quad n = 1, 2, 3 \cdots,$$

and since $p = 0$ on the surface we call these frequencies resonances. That is, the surface has no force opposing its motion. The resonance and antiresonance frequencies alternate. This definition of resonance is consistent with that for a spring-mass system, if we view the fluid inside the sphere as the mass and spring, and the spherical boundary as the other end of the spring. When this end is vibrated (with velocity $\dot{w}$), the reaction force goes to zero at resonance (the mass and spring are in violent motion at the other end). Conversely, at antiresonance the force is infinite and the mass is motionless, corresponding to zero internal pressure.

Low Frequency Result

As $k \to 0$ then from Eq. (6.63)

$$\frac{j_0(kb)}{j_1(kb)} \to 3/kb,$$

and Eq. (6.152) yields

$$p(b) = -i\rho_0 c \frac{3}{kb} \dot{W} = -i\rho_0 c^2 \frac{3}{\omega b} \dot{W} = \frac{3B}{i\omega b} \dot{W}, \tag{6.155}$$

where B is the bulk modulus (stiffness of the fluid) defined by

$$c = \sqrt{B/\rho_0}. \tag{6.156}$$

Define a spring constant K through

$$F = \frac{K}{-i\omega} \dot{W},$$

where F is the outward force (in Newtons) and $\dot{W}$ the velocity, or rate of change of the compression of the spring. Since positive pressure represents compression, and thus an inward force squeezing the fluid, the force $F = -4\pi b^2 p(b)$. Thus the internal fluid represents a spring of stiffness K from Eq. (6.155) given by

$$K = 12\pi bB. \tag{6.157}$$

The "bubble" acts as a spring at low frequencies, and is completely reactive. As the frequency increases we keep the second term in the expansion of $j_0(kb)$ which gives rise to a mass term which resonates with this spring to create the first resonance at $kb = \pi$ or $\omega_1 = \pi c/b$.

6.9 Transient Radiation - Exterior Problems

Because of the simplicity of the radial functions, it is possible to solve for the time-dependent pressure radiated from a spherical source with temporal boundary conditions. Assume that the surface velocity in the time domain is given by $\dot{w}_t(a, \theta, \phi, t)$ where we use the subscript t to represent time domain quantities so as to distinguish from the frequency domain. We will assume that $\dot{w}_t(a, \theta, \phi, t) = 0$ for $t < 0$. We solve the transient problem by transforming to the frequency domain so that we can use the expressions relating pressure and velocity, derived in this chapter. Equation (1.7), page 2, defines the temporal inverse Fourier transform:

$$\dot{w}_t(a, \theta, \phi, t) = \frac{1}{2\pi} \int_{-\infty}^{\infty} \dot{w}(a, \theta, \phi, \omega) e^{-i\omega t} d\omega.$$

We can rewrite Eq. (6.105) taking the inverse Fourier transform of both sides as

$$p_t(r, \theta, \phi, t) = i\rho_0 c \sum_{n=0}^{\infty} \sum_{m=-n}^{n} Y_n^m(\theta, \phi) \frac{1}{2\pi} \int_{-\infty}^{\infty} \frac{h_n(\omega r/c)}{h_n'(\omega a/c)} \dot{W}_{mn}(\omega) e^{-i\omega t} d\omega, \tag{6.158}$$

with the "k-space" velocity given by

$$\dot{W}_{mn}(\omega) = \int d\Omega' Y_n^m{}^*(\theta', \phi') \int \dot{w}_t(a, \theta', \phi', t) e^{i\omega t} dt. \tag{6.159}$$

The frequency integral in Eq. (6.158) is not difficult to evaluate since the Hankel functions are just polynomial-like expressions. We illustrate this by a simple case.

6.9.1   Radiation from an Impulsively Moving Sphere

Given v and α constants, let the surface velocity be given by

$$\dot{w}_t(a,\theta,t) = ve^{-\alpha t}Y_1^0 \;\; = \;\; \sqrt{\frac{3}{4\pi}}ve^{-\alpha t}\cos\theta, \quad t \geq 0, \tag{6.160}$$

$$= \;\; 0, \quad t < 0.$$

The magnitudes of the velocity $\dot{w}_t$ and displacement w_t of the sphere are shown in Fig. 6.19.

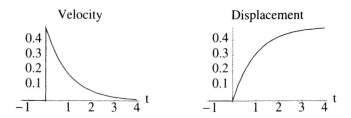

Figure 6.19: Velocity and displacement for the sphere, for the case where $\alpha = 1$ and $v=1$.

In this case Eq. (6.159) yields

$$\dot{W}_{mn}(\omega) = \delta_{n1}\delta_{m0}\frac{v}{\alpha - i\omega}. \tag{6.161}$$

The frequency integral in Eq. (6.158) to be evaluated is

$$I(\omega) \equiv \frac{1}{2\pi}\int_{-\infty}^{\infty} \frac{h_1(\omega r/c)}{h_1'(\omega a/c)}\frac{v}{(\alpha - i\omega)}e^{-i\omega t}d\omega \tag{6.162}$$

and Eq. (6.158) is written as

$$p_t(r,\theta,\phi,t) = i\rho_0 c\sum_{n=0}^{\infty}\sum_{m=-n}^{n} Y_n^m(\theta,\phi)I(\omega)\delta_{n1}\delta_{m0}.$$

Using Eq. (6.62)

$$I(\omega) = \frac{va^3}{2\pi r^2}\int_{-\infty}^{\infty}\frac{-e^{ik(r-a)}(i+kr)}{(2i + 2ka - i(ka)^2)(\alpha - i\omega)}e^{-i\omega t}d\omega.$$

We can evaluate this integral by contour integration, closing the contour in the lower half plane as shown in Fig. 6.20. The integrand will go to zero on the semicircle as long as $r - a - ct < 0$, so that the quantity

$$e^{i\omega(r-a-ct)/c}$$

is decaying on the semicircle since ω has a negative imaginary part. For times such that $r - a - ct > 0$ then the contour must be closed in the upper half plane. Since there are no poles in the upper half plane, then the result of the integration is zero, that is, the sound field has not yet reached the field position r $(r - a > ct)$. To communicate this fact to the solution we introduce the Heaviside step function multiplying the solution. Let $H(t)$ be the Heaviside step function defined by

$$
\begin{aligned}
H(t) &= 0 \quad t < 0 \\
&= 1 \quad t > 0.
\end{aligned}
\tag{6.163}
$$

Thus $H(ct - (r - a))$ with the discontinuity at $r - a - ct = 0$ is correct multiplier.

The zeros of the denominator are at $ka = \pm 1 - i$, or $\omega = (\pm 1 - i)c/a$; and $\omega = -i\alpha$. Thus

$$
I(\omega) = \frac{vac^2}{2\pi r^2} \int_{-\infty}^{\infty} \frac{e^{i\omega(r-a-ct)/c}(i + \omega r/c)}{[\omega - (1 - i)c/a][\omega - (-1 - i)c/a][\omega + i\alpha]} d\omega.
\tag{6.164}
$$

The contour in the complex plane is shown in Fig. 6.20. The residue theorem states

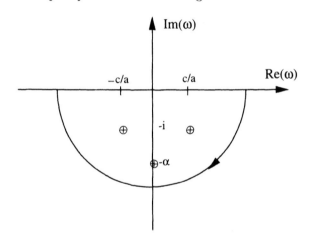

Figure 6.20: Complex contour in the ω plane for the integral in Eq. (6.164).

that

$$
I(\omega) = -2\pi i \sum_{j=1}^{3} R_j,
$$

where R_j is the residue of the jth pole. Thus

$$
p_t(r, \theta, \phi, t) = i\rho_0 c \sum_{n=0}^{\infty} \sum_{m=-n}^{n} Y_n^m(\theta, \phi) I(\omega) \delta_{n1} \delta_{m0} = 2\pi \rho_0 c Y_1^0 \sum_{j=1}^{3} R_j.
$$

For simplicity we set $\alpha = 0$ so that the motion of the sphere corresponds to an impulse in velocity. After calculation of the residues we find that (the residue from the pole at $\omega = -i0$ being zero)

$$
p_t(r, \theta, \phi, t) = \frac{\rho_0 cva}{r} e^{-\beta}(\cos\beta - \sin\beta + \frac{a}{r}\sin\beta)Y_1^0(\theta)H(ct - (r - a)),
\tag{6.165}
$$

where we have applied the Heaviside step function to the solution and

$$\beta \equiv \frac{ct - (r - a)}{a}.$$

The result, pressure versus r, is shown in Fig. 6.21 for four different times at $\theta = 0$. Note that a wake follows the wavefront (sharp edge to the right of the wake). This effect

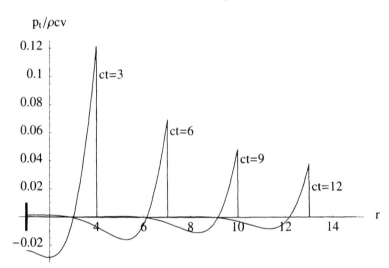

Figure 6.21: Normalized pressure from Eq. (6.161) for four different times and $\theta = 0$. For simplicity, $c = a = 1$ and thus $\beta = t - (r - 1)$. The surface of the sphere, shown by the short vertical bar on the left, is at $r = 1$. The impulse occurs on the sphere at $t = 0$.

is due to signals from the various regions of the sphere arriving at the observation point r at different times.

6.10 Scattering from Spheres

Although we have not studied scattering from plates or cylinders in this book, we will fill the void with scattering of plane waves from spheres. Important concepts are presented which are needed for practical and theoretical work in acoustics.

6.10.1 Formulation

Most scattering problems are formulated for plane wave insonification. The incident field p_i is then represented by a plane wave in the direction $\mathbf{k_i}$ of magnitude P_0 as

$$p_i(\mathbf{r}, t) = P_0 e^{i(\mathbf{k_i} \cdot \mathbf{r} - \omega t)} \tag{6.166}$$

where $\mathbf{r}$ is the position vector in a spherical coordinate system, as shown in Fig. 6.22. $\mathbf{k_i}$ or $\mathbf{r}$ may be expanded in terms of its vector components using a spherical or Cartesian coordinate system. For mathematical simplicity we choose the direction of the plane wave to be along the z axis (which will allow us to use the axisymmetric formulation for the sphere, Section 6.7.9, since the incident field is axisymmetric). This corresponds

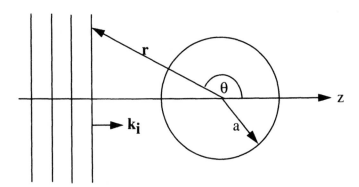

Figure 6.22: Direction of the incident plane wave field for scattering from a sphere of radius a.

to a point source located on the z axis at $z = -\infty$. Thus $\mathbf{k_i} = k\hat{\mathbf{k}}$ where $k = \omega/c$. We have in the frequency domain then

$$p_i(r, \theta, \phi, \omega) = P_0 e^{ikz} = P_0 e^{ikr \cos \theta} \qquad (6.167)$$

which provides the mathematical description of our plane wave in spherical coordinates.

The formulation of the scattering problem involves separating the total pressure field $p_t(r, \theta, \phi, \omega)$ into two parts:

$$p_t = p_s + p_i, \qquad (6.168)$$

where p_i is the incident field given above, and p_s is a new quantity called the scattered field. p_t is the total field which corresponds to the pressure which would be measured if one were to carry out a steady state experiment. In Eq. (6.168) p_i remains fixed and independent of the scatterer. In other words, in an experimental sense, the incident field is the field measured without the target (sphere) in place. Then p_s measures the change in the incident field which results when the target is placed in the insonification field. If p_s is zero then the target is completely transparent.

In order to solve the scattering problem mathematically we need to express the *incident* field, Eq. (6.167), on the surface of the sphere; that is, to find an expansion in terms of spherical harmonics which mathematically describes the incident field. We return to our general equation, Eq. (6.21), which provides a solution of the wave equation in terms of spherical harmonics and spherical Bessel functions. Actually, since we know that our source is located at infinity, then the formulation for the interior problem is appropriate as given in Section 6.8. Thus the incident pressure field must be expressible as given in Eq. (6.140):

$$p_i(r, \theta, \omega) = P_0 \sum_{n=0}^{\infty} \sum_{m=-n}^{n} A_{mn}(\omega) j_n(kr) Y_n^m(\theta, \phi), \qquad (6.169)$$

which for the axisymmetric case, given arbitrary coefficients A_n, is just

$$p_i(r,\theta,\omega) = P_0 e^{ikr\cos\theta} = P_0 \sum_{n=0}^{\infty} A_n j_n(kr) P_n(\cos\theta). \tag{6.170}$$

The relationship between A_n in Eq. (6.170) and A_{mn} in Eq. (6.169) is obtained by multiplication of both sides of Eq. (6.169) by $e^{-im'\phi}$, integrating over ϕ and using Eq. (6.53):

$$A_n = A_{0n}\sqrt{\frac{(2n+1)}{4\pi}}.$$

Equation (6.169) expresses the fact that the incident field must be finite at the origin.

To solve for A_n we multiply Eq. (6.170) by P_n, integrate over θ and use the orthogonality, Eq. (6.28), of P_n on the right hand side:

$$\int_0^\pi e^{ikr\cos\theta} P_n(\cos\theta)\sin\theta\, d\theta = \frac{2A_n}{2n+1} j_n(kr).$$

Using a change of variables, this equation becomes

$$\int_{-1}^1 e^{ikr\eta} P_n(\eta) d\eta = \frac{2A_n}{2n+1} j_n(kr). \tag{6.171}$$

If we let $r = 0$ then since $P_n(\eta)$ is orthogonal to 1 (i.e. to $P_0(\eta)$) we find immediately that $A_0 = 1$. To get the other n terms we differentiate with respect to r and evaluate at $r = 0$:

$$\frac{d^n}{dr^n}(e^{ikr\eta}) = (ik\eta)^n e^{ikr\eta}\Big|_{r=0} = (ik\eta)^n.$$

The right hand side of Eq. (6.171) becomes, from Eq. (6.63),

$$\frac{d^n}{dr^n} j_n(kr)\Big|_{r=0} = \frac{d^n}{dr^n}\left(\frac{(kr)^n}{(2n+1)!!}\right) = k^n\frac{n!}{(2n+1)!!}. \tag{6.172}$$

Equating left and right hand sides, and using the relation

$$(2n+1)!! = \frac{(2n+1)!}{2^n n!},$$

we have

$$(ik)^n \int_{-1}^1 \eta^n P_n(\eta) d\eta = \frac{2A_n}{2n+1} k^n \frac{n!}{(2n+1)!!} = \frac{2A_n}{2n+1} k^n \frac{2^n (n!)^2}{(2n+1)!}.$$

We need the integral relation,

$$\int_{-1}^1 \eta^n P_n(\eta) d\eta = \frac{2^{n+1}(n!)^2}{(2n+1)!}, \tag{6.173}$$

so our final (and simple) result is,

$$A_n = i^n(2n+1).$$

nserting this result into Eq. (6.170) we have the important equation

$$e^{ikr\cos\theta} = \sum_{n=0}^{\infty} i^n(2n+1)j_n(kr)P_n(\cos\theta).$$

(6.174)

For future reference it is useful to also show the expression for the general case, where the direction of the incident field is given by (θ_i, ϕ_i). In this case the expansion of the incident field is

$$e^{i\vec{k}_i\cdot\vec{r}} = 4\pi\sum_{n=0}^{\infty} i^n j_n(kr) \sum_{n=-m}^{m} Y_n^m(\theta,\phi)Y_n^m(\theta_i,\phi_i)^*.$$

(6.175)

6.10.2 Scattering from a Pressure Release Sphere

Armed with the incident field expanded in spherical harmonics, or Legendre polynomials, we can solve for the pressure field scattered from a sphere of radius a which is pressure release, that is, the *total* pressure on the surface of the sphere must vanish. Thus the boundary condition is

$$p_i(a,\theta) + p_s(a,\theta) = 0.$$

(6.176)

Now we must expand the scattered field in terms of spherical harmonics. Since the scattered field represents spherical waves emanating from the surface of the sphere due to scattering from the incident wave, the scattered field is given by outgoing spherical Hankel functions:

$$p_s(r,\theta) = \sum_{n=0}^{\infty} C_n(\omega)h_n^{(1)}(kr)P_n(\cos\theta).$$

(6.177)

Inserting Eq. (6.177) and Eq. (6.174) into Eq. (6.176) leads to

$$C_n(\omega) = -P_0(2n+1)i^n\frac{j_n(ka)}{h_n^{(1)}(ka)}$$

(6.178)

so that

$$p_s(r,\theta) = -P_0\sum_{n=0}^{\infty}(2n+1)i^n\frac{j_n(ka)}{h_n^{(1)}(ka)}h_n^{(1)}(kr)P_n(\cos\theta).$$

(6.179)

The total field (dropping the superscript on the Hankel function) is then

$$p_t(r,\theta) = P_0\sum_{n=0}^{\infty}(2n+1)i^n\left(j_n(kr) - \frac{j_n(ka)}{h_n(ka)}h_n(kr)\right)P_n(\cos\theta).$$

(6.180)

In the farfield we use Eq. (6.68) to obtain the scattered field from Eq. (6.179):

$$\lim_{r\to\infty} p_s(r,\theta) = iP_0\frac{e^{ikr}}{kr}\sum_{n=0}^{\infty}(2n+1)\frac{j_n(ka)}{h_n(ka)}P_n(\cos\theta).$$

(6.181)

Note it does not make sense to try to write the total field in the farfield since the incident field has a singularity there. At low frequencies using Eq. (6.63) and Eq. (6.64),

$$\lim_{ka \to 0} \frac{j_n(ka)}{h_n(ka)} = ika,$$

so that the scattered pressure becomes

$$p_s(r, \theta) \approx -P_0 a \frac{e^{ikr}}{r}. \tag{6.182}$$

6.10.3   Scattering from a Rigid Sphere

For a rigid sphere the radial velocity vanishes on the surface of a sphere at $r = a$. This is the total radial velocity (including incident and scattered). The boundary condition is then

$$\dot{w}_t(a, \theta) = \dot{w}_i(a, \theta) + \dot{w}_s(a, \theta) = 0. \tag{6.183}$$

By Euler's equation Eq. (6.183) can be written as

$$\frac{\partial}{\partial r} \left(p_i(r, \theta) + p_s(r, \theta) \right) \Big|_{r=a} = 0. \tag{6.184}$$

Again, as in Eq. (6.177), the scattered field is expanded in outgoing waves:

$$p_s(r, \theta) = \sum_{n=0}^{\infty} C_n(\omega) h_n^{(1)}(kr) P_n(\cos \theta),$$

with the incident field given by Eq. (6.174),

$$p_i(r, \theta) = P_0 \sum_{n=0}^{\infty} i^n (2n + 1) j_n(kr) P_n(\cos \theta).$$

Using these expressions in Eq. (6.184) yields,

$$C_n = -P_0(2n + 1) i^n \frac{j_n'(ka)}{h_n'(ka)}$$

so that

$$p_s(r, \theta) = -P_0 \sum_{n=0}^{\infty} (2n + 1) i^n \frac{j_n'(ka)}{h_n'(ka)} h_n(kr) P_n(\cos \theta). \tag{6.185}$$

The total field is then

$$p_t(r, \theta) = P_0 \sum_{n=0}^{\infty} (2n + 1) i^n \left(j_n(kr) - \frac{j_n'(ka)}{h_n'(ka)} h_n(kr) \right) P_n(\cos \theta). \tag{6.186}$$

The farfield scattered pressure, using Eq. (6.68), is

$$\lim_{r \to \infty} p_s(r, \theta) = i P_0 \frac{e^{ikr}}{kr} \sum_{n=0}^{\infty} (2n + 1) \frac{j_n'(ka)}{h_n'(ka)} P_n(\cos \theta). \tag{6.187}$$

We now consider the farfield pressure at low frequencies. We will keep the first two terms of the series in Eq. (6.187). Thus,

$$p_s \approx i P_0 \frac{e^{ikr}}{kr}\left(\frac{j_0'(ka)}{h_0'(ka)} + \frac{3j_1'(ka)}{h_1'(ka)}\cos\theta\right) \approx P_0 \frac{e^{ikr}}{r}\frac{k^2 a^3}{3}\left(\frac{3}{2}\cos\theta - 1\right). \tag{6.188}$$

Now we see why we took two terms of the series. Both the $n = 0$ and the $n = 1$ terms are of the same order in ka. In the low frequency limit the scattered field from a rigid sphere consists of a monopole term and dipole term.

Returning to the general solution, Eq. (6.185), we define the normalized farfield scattered pressure $\bar{p}_s$ by

$$\bar{p}_s(\theta) \equiv \lim_{r\to\infty} p_s(r,\theta)\frac{r}{P_0 a}e^{-ikr}, \tag{6.189}$$

with $\lim_{r\to\infty} p_s(r,\theta)$ given by Eq. (6.187). This leads to

$$\bar{p}_s(\theta) \equiv i \sum_{n=0}^{\infty} (2n+1)\frac{j_n'(ka)}{(ka)h_n'(ka)}P_n(\cos\theta). \tag{6.190}$$

This normalization provides a dimensionless quantity with e^{ikr}/r removed. Some results for $\bar{p}_s(\theta)$ are plotted in Fig. 6.23. In the figure the plane wave is incident from the left. The large peaks along the z axis for $ka = 10$ and 20 are due to the shadowing effect of the sphere which causes the *total* pressure to diminish behind it. At low frequencies the backscattering ($\theta = \pi$) dominates over the forward scatter. For high frequencies the scattering becomes omnidirectional reaching a value of 0.5 for all angles except forward scatter ($\theta = 0$).

The differential cross-section $d\sigma/d\Omega$ is defined as the square of the magnitude of the directivity function of the farfield pressure normalized to the incident pressure:

$$d\sigma/d\Omega \equiv |D(\theta,\phi)/P_0|^2 \tag{6.191}$$

where D was defined, as in Eq. (2.85) on page 39 (given by the farfield pressure with the term e^{ikr}/r factored out), as

$$\lim_{r\to\infty} p_s(r,\theta,\phi) = \frac{e^{ikr}}{r}D(\theta,\phi).$$

Thus for the farfield scattered pressure field, we remove the term e^{ikr}/r to arrive at the differential cross-section:

$$d\sigma/d\Omega = \lim_{r\to\infty}\left|\frac{r}{P_0 e^{ikr}}p_s(r,\theta,\phi)\right|^2. \tag{6.192}$$

Another definition of use in underwater acoustics is the target strength, TS. It is defined by

$$\mathrm{TS} = 10\log_{10}\left(\frac{d\sigma/d\Omega}{r_{\mathrm{ref}}^2}\right), \tag{6.193}$$

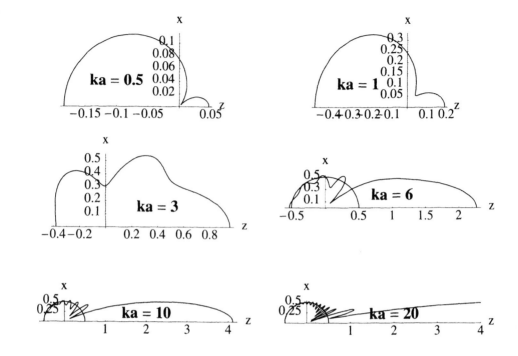

Figure 6.23: The magnitude of the normalized scattered farfield pressure for a rigid sphere, using Eq. (6.190), for various ka. The plane wave is incident from the left. Note that as $ka \to \infty$ that $\bar{p}_s(\theta) = 0.5$ except at $\theta = 0$ (forward scatter) where it is infinite.

where r_{ref} is a reference distance taken as 1 m. (Older definitions used a one yard reference.) Comparison of Eq. (6.192) with Eq. (6.189) indicates that

$$d\sigma/d\Omega = \left| a\bar{p}_s(\theta, \phi) \right|^2. \tag{6.194}$$

At low frequencies, from Eq. (6.188), we see that

$$\frac{1}{a}\sqrt{\frac{d\sigma}{d\Omega}} = \bar{p}_s = \frac{k^2 a^2}{3}\left|\frac{3}{2}\cos\theta - 1\right|. \tag{6.195}$$

Note that the quantity plotted in Fig. 6.23, $\bar{p}_s$, is the square root of the differential cross-section divided by the radius. Thus at high frequencies since $\bar{p}_s$ approaches $1/2$, so the target strength approaches

$$\text{TS} = 10\log_{10}(a/2)^2. \tag{6.196}$$

6.10.4 Scattering from an Elastic Body

There is a very powerful formulation for the solution to the scattering from bodies (not necessarily spherical) which are elastic. We present this formulation in this section for the spherical body. This technique essentially turns the scattering problem into a radiation problem by formulating a "blocked" pressure. We now describe this formulation.[10]
As usual the total pressure is made up of the incident and scattered pressure,

$$p_t = p_i + p_{se}, \tag{6.197}$$

where we have added the e subscript on the scattered term to indicate an elastic body. As always the total normal velocity at the body surface is given by Euler's equation,

$$\dot{w}_t = \frac{1}{i\rho_0 ck} \frac{\partial p_t}{\partial r} = \frac{1}{i\rho_0 ck} \left(\frac{\partial p_i}{\partial r} + \frac{\partial p_{se}}{\partial r} \right). \tag{6.198}$$

At this point we break the scattered field into two components, the first being the field produced if the sphere were rigid, $p_{s\infty}$ (the ∞ subscript indicates infinite impedance), and the second, p_r; what is left over as a result of elasticity. Thus we write the scattered pressure as

$$p_{se} = p_{s\infty} + p_r, \tag{6.199}$$

where $p_{s\infty}$ does not depend on the sphere's elasticity, and p_r now contains the elastic dependencies. With these definitions, Eq. (6.198) becomes

$$\dot{w}_t = \frac{1}{i\rho_0 ck} \left(\underbrace{\frac{\partial p_i}{\partial r} + \frac{\partial p_{s\infty}}{\partial r}}_{=0} + \frac{\partial p_r}{\partial r} \right), \tag{6.200}$$

where the first two terms, evaluated at $r = a$, are zero due to Eq. (6.184) (definition of a rigid body). These terms correspond to the first two terms (at $r = a$) of

$$p_t = \underbrace{p_i + p_{s\infty}}_{\text{rigid}} + p_r = p_b + p_r, \tag{6.201}$$

where $p_b \equiv p_i + p_{s\infty}$ is defined as the *blocked pressure*. Thus the total velocity at the surface of the sphere depends only on p_r. That is,

$$\dot{w}_t(a, \theta, \phi) = \frac{1}{i\rho_0 ck} \frac{\partial p_r(a, \theta\phi)}{\partial r}. \tag{6.202}$$

Now we ask, "to what problem do these last two equations provide a solution?" The key is in the fact that in Eq. (6.202) only p_r contributes to the total velocity at the surface of the sphere, and this velocity is independent of the gradient of the blocked pressure. In view of Eq. (6.106), once the radial surface velocity is specified (through Eq. (6.202)), then the pressure from the surface to the farfield is uniquely determined, and that pressure must then be $p_r(r, \theta, \phi)$. But at the same time, in order to satisfy Eq. (6.201) which represents the total force per unit area on the surface of the sphere,

---

[10] Junger and Feit, *Sound, Structures, and their Interactions*, Chapter 11.

we must think of p_b not as an acoustic pressure, but as an internal force generated at the boundary, like a layer of infinitesimally small shakers driving the elastic sphere from the inside with the prescribed force distribution p_b. These shakers excite the elastic sphere causing it to radiate a pressure field p_r.

To sum up the formulation for scattering from an elastic body,

(1) Solve the rigid body problem, calculating the total pressure at the surface (the blocked pressure p_b) and the scattered pressure over all space $p_{s\infty}(r, \theta, \phi)$.

(2) Solve the radiation problem with $-p_b$ the forcing function at the surface of the elastic body, to obtain the radiated pressure $p_r(r, \theta, \phi)$.

(3) The solution to the scattering problem p_{se} is now the sum $p_{s\infty} + p_r$ and the total field is just $p_t = p_i + p_{s\infty} + p_r$.

Problems

6.1 Find the general expression for the pressure inside a sphere of radius a if the pressure on the surface is given as $p(a, \theta, \phi)$.

6.2 Find the pressure between concentric spheres of radii $b > a$ if the pressure is zero on the inner sphere and $p(b, \theta, \phi)$ on the outer.

6.3 Small spherical caps (small enough to be considered point sources) with area A at the North and South poles of a sphere of radius a vibrate 180 degrees out of phase while the rest of the sphere is rigid. Calculate the radiation pattern and the total average power radiated if $ka << 1$.

6.4 A sphere of radius a vibrates with normal velocity $V \sin^2 \theta e^{-i\omega t}$. Calculate the total average power radiated. Hint: $\sin^2 \theta = \frac{2}{3}[P_0(\cos \theta) - P_2(\cos \theta)]$.

6.5 Find the normal velocity distribution on a sphere of radius a which would produce a farfield pressure $p_0(\theta, \phi)$.

6.6 A small spherical cap with angular width θ_0 on an otherwise rigid sphere consists of two halves which vibrate 180 degrees out of phase. Thus,

$$\dot{w}(a, \theta, \phi) = +V \quad 0 \le \phi < \pi, \quad \theta < \theta_0$$
$$= -V \quad \pi \le \phi < 2\pi, \quad \theta < \theta_0.$$

Obtain approximate expressions for the farfield radiation pattern and the total average power radiated assuming $ka << 1$, where a is the radius of the sphere.

6.7 A rigid sphere is oscillating with radial velocity $\dot{w}(\theta) = \dot{w}_0 \cos \theta$. Compute the farfield pressure valid for all frequencies. What is the low frequency limit?

6.8 Use separation of variables, Eq. (6.7), to derive the 4 differential equations, Eqs (6.8 6.11).

6.9 Demonstrate by integration that $Y_3^{-2}(\theta, \phi)$ and $Y_3^3(\theta, \phi)$ are orthonormal.

6.10 Demonstrate by integration that $P_3^1(\cos\theta)$ and $P_2^1(\cos\theta)$ are orthogonal.

6.11 The pressure from a longitudinal quadrupole oriented along the z axis is given by

$$p = -i\rho_0 ckQzz\frac{\partial^2}{\partial z'^2}G(\mathbf{r}|\mathbf{r}')\bigg|_{r'=0}.$$

Perform the indicated differentiation and show that the result can be written as proportional to the weighted sum of two spherical harmonics, Y_2^0 and Y_0^0, that is, $p = A_2(r)Y_2^0 + A_0(r)Y_0^0$. Determine the weighting factors, A_2 and A_0.

6.12 Calculate the power radiated from the sphere with the vibrating cap, in Section 6.7.10 using Eq. (6.134).

6.13 Compute the farfield pressure from Eq. (6.134). Compute the farfield pressure at very low frequencies. Do you need to keep the $n = 1$ term in comparison with the $n = 0$ term?

6.14 Compute $\dot{w}$, $\dot{u}$, e_{av}, I_r, I_θ and the total power radiated, Π, for the dipole in Section 6.5.2.

6.15 For an interior problem Eq. (6.111) also provides the power transmitted to the inside. Using Eq. (6.140) prove that the power transmitted to the interior of a vibrating sphere of radius a is identically zero.

6.16 Compute the differential cross-section and the target strength for the scattering of a plane wave from a pressure release sphere.

6.17 Compute the scattering from a sphere of radius a filled with fluid of density and sound speed, ρ_1, c_1, and surrounded by fluid of density ρ_0 and sound speed, c_0. The boundary conditions at the surface of the sphere are continuity of radial velocity and pressure (no surface tension) between the two fluids. Show that your answer matches that of the pressure release and rigid spheres when $c_1 \to 0$ and $c_1 \to \infty$, respectively.

6.18 A rigid sphere of radius a with center at the origin is oscillating in the z direction so that its radial surface velocity is given by

$$\dot{w}(a, \theta) = v\cos\theta.$$

(a) Find the solution for the radiated pressure which is valid everywhere outside the sphere.

(b) Write this solution in terms of spherical harmonics.

(c) We change the direction of oscillation so that now the sphere is oscillating in the y direction. The radial velocity is now given by

$$\dot{w}(a, \theta) = v\sin\theta\sin\phi.$$

Using the nonaxisymmetric version of the exterior problem, again find the pressure outside of the sphere.

(d) Note that the radius vector to the field point (where we measure the pressure) is given by the spherical coordinates, θ and ϕ, and that the projection of that radius vector onto the y axis is simply $r \sin\theta \sin\phi$. If we define the angle between the radius vector and the y axis as γ, then we notice that the motion of the sphere could be described simply in terms of this angle as

$$\dot{w}(a,\theta) = v \cos\gamma.$$

Equating the projections of the radius vector onto the y axis yields

$$\cos\gamma = \sin\theta \sin\phi.$$

Use this fact to show that the two solutions which you derived in parts (1) and (3) are identical (interpreting the angle γ as the polar angle, and the y axis the polar axis).

6.19 A small piston set in a rigid sphere (see Fig. 6.15) vibrates with a constant radial velocity, (independent of ϕ), $\dot{w}(a,\theta)$ given by

$$\begin{aligned} \dot{w}(a,\theta) &= \dot{W}, & 0 \le \theta \le \alpha \\ &= 0, & \alpha < \theta \le \pi. \end{aligned}$$

(a) Solve for the pressure field, $p(r,\theta)$, **inside** the sphere. In your solution you can assume that α is small enough so that one can replace $P_n(\cos\alpha)$ with unity.

(b) Keeping the $n = 0$ and $n = 1$ terms what is the low frequency approximation to the pressure in the interior?

(c) The motion of the piston generates a force equal to its area times the pressure generated at its inner surface. This defines a mechanical impedance given by $F/\dot{W}$. As $k \to 0$ show that this impedance is spring like. What is the equivalent stiffness of the air spring given by this impedance? Keep your answer in terms of α (not in terms of volume flow).

6.20 Write down (no algebra required!) the value of the sum,

$$\sum_{n=m}^{\infty} \frac{2n+1}{2} \frac{(n-m)!}{(n+m)!} P_n^m(x) P_n^m(y).$$

What is the significance of this result?

Chapter 7

Spherical Nearfield Acoustical Holography

7.1   Introduction

Spherical NAH encompasses two different problems. As before we have the exterior domain problem in which all the sources are contained within the measurement surface and reconstructions are obtained from the surface out to infinity. The second problem uses the interior domain in which the sources are located completely *outside* the measurement sphere. We have not dealt with the interior NAH problem in previous chapters. However, it offers a powerful technique which can be applied to study noise in the interiors of aircraft, automobiles and rooms, to mention just a few applications. We call these two problems exterior and interior NAH, respectively.

Exterior spherical NAH has seen some application in electromagnetism, although mostly for the forward problem, that is, the prediction of the farfield from nearfield measurements on a sphere. This finds application in radar ranges for antenna calibration. An extensive review of the exterior problem for electromagnetism is published in *Spherical Near-field antenna measurements*.[1] A great deal of discussion is found there with respect to actual measurement techniques on antennas. However, there is little discussion of the inverse problem found in the electromagnetism literature.

In the acoustic regime we also find implementation and applications of the forward exterior problem. One such application for transducer calibration is noteworthy.[2] Another notable paper provides both farfield and nearfield reconstructions using a two surface holographic scanner,[3] used in pioneering experimental work on the radiation from violins. For the interior problem, almost no work has appeared in the literature

---

[1] J. E. Hansen, ed. (1988). *Spherical Near-field antenna measurements*. Peter Peregrinus Ltd., London.

[2] R. D. Mardiniak (1979), "A nearfield, underwater measurement system, J. Acoust. Soc. Am. **66**, pp. 955–964.

[3] G. Weinreich and E. B. Arnold (1980), "Method for measuring acoustic radiation fields", J. Acoust. Soc. Am. **68**, pp. 404–411.

on spherical NAH. This chapter provides a first look into this new area.

As usual our main concern here is the inverse problem, that is, reconstruction of the acoustic field between the measurement sphere and the sources for both the exterior and interior domain problems.

7.2   Formulation of the Inverse Problem - Exterior Domain

Figure 6.14 on page 207, defines the geometry for the NAH problem for the exterior domain. All the sources are contained within the minimum radius a. The hologram (measured pressure on a spherical surface) is obtained on a sphere of greater radius r_h ($r_h \geq a$). The forward problem has already been presented in the last chapter where the boundary value problem was formulated to compute the radiated pressure from the normal velocity on a sphere as given in Eq. (6.106). The inverse problem attacked by NAH provides the reconstruction of the normal velocity (or pressure) on a sphere of smaller radius given the pressure measured on a larger sphere at $r = r_h$.

Repeating the definition, Eq. (6.95) on page 207, the spherical wave spectrum of the pressure is

$$P_{mn}(r_h) \equiv \int p(r_h, \theta, \phi, \omega) Y_n^m(\theta, \phi)^* d\Omega \tag{7.1}$$

and similarly for $\dot{W}_{mn}(r)$. The relationship between the pressure and normal velocity wave spectra was (see Eq. (6.101))

$$\dot{W}_{mn}(r) = \frac{h_n'(kr)}{i\rho_0 c h_n(kr_h)} P_{mn}(r_h). \tag{7.2}$$

Since

$$\dot{w}(r, \theta, \phi, \omega) = \sum_{m,n} \dot{W}_{mn}(r) Y_n^m(\theta, \phi), \tag{7.3}$$

then using Eqs (7.2) and (7.1), the final reconstruction equation is

$$\dot{w}(r, \theta, \phi, \omega) = \sum_{n=0}^{\infty} \frac{h_n'(kr)}{i\rho_0 c h_n(kr_h)} \sum_{m=-n}^{n} Y_n^m(\theta, \phi) \int p(r_h, \theta', \phi', \omega) Y_n^m(\theta', \phi')^* d\Omega'. \tag{7.4}$$

This equation provides a reconstruction of the surface radial velocity given a pressure measurement on a larger sphere of radius r_h. This equation represents one of the fundamental reconstruction equations of exterior, spherical NAH.

The solution for the reconstructed pressure can be obtained from Eq. (6.97) on page 208. Clearly this equation is still valid if r_0 and r are interchanged. Thus we rewrite Eq. (6.97) as

$$P_{mn}(r) = \frac{h_n(kr)}{h_n(kr_h)} P_{mn}(r_h), \tag{7.5}$$

with $r_0 = r_h$ and $r < r_h$. Using Eq. (6.96) and Eq. (7.1) we obtain

$$p(r, \theta, \phi) = \sum_{n=0}^{\infty} \frac{h_n(kr)}{h_n(kr_h)} \sum_{m=-n}^{n} Y_n^m(\theta, \phi) \int p(r_h, \theta', \phi') Y_n^m(\theta', \phi')^* d\Omega'. \tag{7.6}$$

This equation is the second fundamental reconstruction equation for exterior, spherical NAH. It provides the reconstruction of the pressure on a sphere from the measured pressure on a larger sphere.

Returning to Eq. (7.4) and comparing it with Eq. (6.107) we can write explicitly the spatial form of the inverse of the Neumann Green function G_N^{-1}. That is, at $r = a$,

$$\dot{w}(a, \theta, \phi, \omega) = \frac{1}{i \rho_0 c k a^2} \int G_N^{-1}(r_h, \theta, \phi | a, \theta', \phi') p(r_h, \theta', \phi') \, d\Omega', \qquad (7.7)$$

where $a < r_h$ and

$$G_N^{-1} = k a^2 \sum_{n=0}^{\infty} \frac{h_n'(ka)}{h_n(kr_h)} \sum_{m=-n}^{n} Y_n^m(\theta, \phi) Y_n^m(\theta', \phi')^*. \qquad (7.8)$$

From our experience with the planar and cylindrical reconstruction problems, we would expect that the sums in Eq. (7.8) to be divergent. In fact as $n \to \infty$, then using Eq. (6.64) on page 197,

$$\frac{h_n'(ka)}{h_n(kr_h)} \to -\frac{n+1}{ka} (\frac{r_h}{a})^{n+1},$$

and by the ratio test this series diverges when $r_h > a$ and converges absolutely when $r_h < a$. To demonstrate this the reader should evaluate the series for $\theta = \theta' = 0$.

Of course, as we have always done in the past, we must apply a k-space-like filter in order to attain convergence. The equivalent of the k-space filter in this case would be to take only a finite number of terms in the n series, cutting it off at $n = N$. The choice of N will depend on the signal-to-noise ratio in the pressure measurement. We will pursue this issue in more detail in the section on interior, spherical NAH.

7.2.1 Tangential Components of Velocity

The two tangential components of the velocity $\dot{u}$ and $\dot{v}$ are obtained from Euler's equation, Eq. (6.5) on page 184, applied to Eq. (7.6). (Equation (7.4) can also be derived in the same way.) Thus the velocity component in the θ direction is related to the measured pressure:

$$\dot{u}(r, \theta, \phi, \omega) = \frac{1}{i \rho c k} \sum_{n=0}^{\infty} \frac{h_n(kr)}{r h_n(kr_h)} \sum_{m=-n}^{n} \frac{\partial Y_n^m(\theta, \phi)}{\partial \theta} P_{mn}(r_h). \qquad (7.9)$$

Similarly, the velocity in the circumferential direction is

$$\dot{v}(r, \theta, \phi, \omega) = \frac{1}{i \rho c k} \sum_{n=0}^{\infty} \frac{h_n(kr)}{r h_n(kr_h)} \sum_{m=-n}^{n} \frac{i m Y_n^m(\theta, \phi)}{\sin \theta} P_{mn}(r_h). \qquad (7.10)$$

In these formulas $r < r_h$. Using Eq. (7.1) for $P_{mn}(r_h)$ in these two equations provides the complete reconstruction formulas for the in plane components of velocity from a measurement of the acoustic pressure.

7.2.2 Evanescent Spherical Waves

Evanescent waves were discussed in Section 6.7.3 in the last chapter. When $kr_h << n$ we found that Eq. (7.5) became

$$P_{mn}(r) \approx (\frac{r_h}{r})^{n+1} P_{mn}(r_h). \qquad (7.11)$$

Thus the spherical evanescent waves undergo a power law increase as we move in towards the source, reminiscent of the cylindrical wave case as was given in Eq. (5.16), page 153. Note that the evanescent field is independent of the azimuthal harmonic m. This effect was discussed in Section 6.7.3.

7.3 Interior NAH

A spherical measurement surface provides an ideal tool for the reconstruction of the field when the sources are entirely outside of the measurement surface. In this case Fig. 6.17 on page 217 describes the geometry. The radius of the measurement sphere is r_h, where $r_h \leq b$.

The reconstruction of the pressure anywhere in the volume $r \leq b$ is provided by Eq. (6.142) (which does not change when it is inverted):

$$p(r, \theta, \phi) = \sum_{n=0}^{\infty} \frac{j_n(kr)}{j_n(kr_h)} \sum_{m=-n}^{n} Y_n^m(\theta, \phi) \int p(r_h, \theta', \phi') Y_n^m(\theta', \phi')^* d\Omega'. \qquad (7.12)$$

This equation is valid for $r \leq r_h$ (forward problem) and also for $b \geq r > r_h$ (inverse problem). The reconstruction sphere may not cross any of the external sources.

The reconstruction of the radial velocity is obtained from applying Euler's equation to Eq. (7.12). This results in the radial velocity reconstruction:

$$\dot{w}(r, \theta, \phi) = \frac{1}{i\rho_0 c} \sum_{n=0}^{\infty} \frac{j_n'(kr)}{j_n(kr_h)} \sum_{m=-n}^{n} Y_n^m(\theta, \phi) \int p(r_h, \theta', \phi') Y_n^m(\theta', \phi')^* d\Omega'. \qquad (7.13)$$

Again, when $r > r_h$ (outward reconstruction) the problem is an inverse one. Reconstruction of the radial velocity inside the measurement sphere is a forward problem, characterized by the condition $r \leq r_h$. The other components of velocity are obtained in the same fashion as in Eq. (7.11).

7.3.1 Evanescent Spherical Waves

The "k-space" spectral components are given by

$$P_{mn}(r) = \frac{j_n(kr)}{j_n(kr_h)} P_{mn}(r_h). \qquad (7.14)$$

When $n >> kr$ then using Eq. (6.63) on page 196,

$$\frac{j_n(kr)}{j_n(kr_h)} \approx (\frac{r}{r_h})^n, \qquad (7.15)$$

a slightly different dependence from the exterior problem which was given in Section 7.2.2. Thus the evanescent waves follow a power law decay (for the forward problem) and a power law increase in the inverse problem.

7.3.2 Effect of Measurement Noise

We assume that the noise in the experiment is random, Gaussian and uncorrelated spatially. To denote a quantity with noise, we use a tilde over the symbol. Thus the measured signal with noise is given by

$$\tilde{p}(r_h, \theta', \phi') = p(r_h, \theta', \phi') + \epsilon(\theta', \phi'), \tag{7.16}$$

where ϵ is random, Gaussian noise with variance σ^2.

At the same time we model another effect. In any practical experiment the sum over n will be limited to some maximum harmonic. This maximum may arise due to the number of sensor measurement locations in the measurement array, or due to a processing cutoff to simulate a k-space filter. We define the truncated quantity using a subscript N, without noise, from Eq. (7.13) as

$$\dot{w}_N(r, \theta, \phi) = \frac{1}{i\rho_0 c} \sum_{n=0}^{N} \frac{j_n'(kr)}{j_n(kr_h)} \sum_{m=-n}^{n} Y_n^m(\theta, \phi) P_{mn}(r_h). \tag{7.17}$$

Thus, when $N \to \infty$, $\dot{w}_\infty \equiv \dot{w}$.

The truncated reconstructed velocity using the measured pressure with noise is denoted $\tilde{\dot{w}}_N$, to correspond with Eq. (7.17), so that

$$\tilde{\dot{w}}_N(r, \theta, \phi) = \frac{1}{i\rho_0 c} \sum_{n=0}^{N} \frac{j_n'(kr)}{j_n(kr_h)} \sum_{m=-n}^{n} Y_n^m(\theta, \phi) \tilde{P}_{mn}(r_h), \tag{7.18}$$

where

$$\tilde{P}_{mn}(r_h) \equiv \int (p(r_h, \theta', \phi') + \epsilon(\theta', \phi')) \, Y_n^m(\theta', \phi')^* d\Omega' = P_{mn}(r_h) + B_{mn}, \tag{7.19}$$

and

$$B_{mn} \equiv \int \epsilon(\theta', \phi') Y_n^m(\theta', \phi')^* d\Omega'. \tag{7.20}$$

When Eq. (7.19) is inserted into Eq. (7.18) $\tilde{\dot{w}}_N$ breaks down into two components, one due to noise, which we call $\mathcal{N}_N(r, \theta, \phi)$ and the second the exact result (given just N terms, however) $\dot{w}_N$:

$$\tilde{\dot{w}}_N(r, \theta, \phi) = \dot{w}_N(r, \theta, \phi) + \mathcal{N}_N(r, \theta, \phi), \tag{7.21}$$

where

$$\mathcal{N}_N(r, \theta, \phi) \equiv \frac{1}{i\rho_0 c} \sum_{n=0}^{N} \frac{j_n'(kr)}{j_n(kr_h)} \sum_{m=-n}^{n} Y_n^m(\theta, \phi) B_{mn}. \tag{7.22}$$

In order to model in a quantitative way the effects of noise in the reconstruction, we need to determine the root mean square error (RMSE) computed by comparing the correct reconstruction with the reconstruction with noise. The square of the difference between these two quantities is given by

$$|\dot{w}_\infty(r,\Omega) - \tilde{\dot{w}}_N(r,\Omega)|^2.$$

The relevant question to ask, since we are dealing with Gaussian noise, is for a given position in space Ω_i what would be the expected result of this square difference if we repeated the experiment an infinite number of times, and averaged the result. We call the result of this thought experiment the expected value, denoted by E. For a zero mean, Gaussian process the expected value of the noise is

$$E[\epsilon(\Omega_i)] = 0.$$

Furthermore, it is sensible to determine the average error over the whole reconstruction sphere by integrating the expected value of the square difference over that sphere. We define this average using angle brackets as:

$$\langle f(\theta,\phi)\rangle \equiv \frac{1}{4\pi} \int f(\theta,\phi)\, d\Omega. \tag{7.23}$$

Thus the RMSE error $\mathcal{E}(r)$ at r is defined as:

$$\mathcal{E}(r) \equiv \left(\frac{\langle E[|\dot{w}_\infty(r,\Omega) - \tilde{\dot{w}}_N(r,\Omega)|^2]\rangle}{\langle|\dot{w}_\infty(r,\Omega)|^2\rangle} \right)^{1/2}, \tag{7.24}$$

where division by the denominator provides a dimensionless result and allows $100\mathcal{E}(r)$ to be interpreted as a percent error.

Now we proceed to evaluate Eq. (7.24). Define $\dot{w}_{\infty N}$:

$$\dot{w}_{\infty N}(r,\Omega) \equiv \dot{w}_\infty(r,\Omega) - \dot{w}_N(r,\Omega) = \frac{1}{i\rho_0 c}\sum_{n=N+1}^\infty \frac{j_n'(kr)}{j_n(kr_h)}\sum_{m=-n}^n Y_n^m(\theta,\phi)P_{mn}(r_h), \tag{7.25}$$

and insert Eq. (7.21) into Eq. (7.24) to obtain

$$\mathcal{E}(r) = \left(\frac{\langle E[|\dot{w}_{\infty N}(r,\Omega) - \mathcal{N}_N(r,\theta,\phi)|^2]\rangle}{\langle|\dot{w}_\infty(r,\Omega)|^2\rangle} \right)^{1/2}. \tag{7.26}$$

Since $E[\dot{w}_{\infty N}\mathcal{N}_N] = 0$ and $E[|\dot{w}_{\infty N}|^2] = |\dot{w}_{\infty N}|^2$, we expand the squared magnitude to obtain

$$\mathcal{E}(r) = \left(\frac{\langle|\dot{w}_{\infty N}(r,\Omega)|^2\rangle + \langle E[|\mathcal{N}_N(r,\theta,\phi)|^2]\rangle}{\langle|\dot{w}_\infty(r,\Omega)|^2\rangle} \right)^{1/2}. \tag{7.27}$$

Using the definition Eq. (7.22), we have

$$E[|\mathcal{N}_N(r,\theta,\phi)|^2] = \frac{1}{(\rho_0 c)^2}\sum_{n,m}^N \frac{j_n'(kr)}{j_n(kr_h)}\sum_{q,p}^N \frac{j_q'(kr)}{j_q(kr_h)} E[B_{mn}B_{pq}^*]Y_n^m(\theta,\phi)Y_q^p(\theta,\phi)^*, \tag{7.28}$$

where we have defined

$$\sum_{n,m} \equiv \sum_{n=0}^{N} \sum_{m=-n}^{n}.$$

Using Eq. (7.20),

$$E[B_{mn}B_{pq}^*] = \iint E[\epsilon(\theta,\phi)\epsilon(\theta',\phi')^*]Y_n^m(\theta,\phi)^*Y_q^p(\theta',\phi')\,d\Omega\,d\Omega'. \tag{7.29}$$

In order to calculate the estimated value of the noise product (cross correlation) in Eq. (7.29) we consider the discrete nature of the measurement problem and write

$$E[\epsilon(\theta_i,\phi_i)\epsilon(\theta'_j,\phi'_j)^*] = \sigma^2\delta_{ij}, \tag{7.30}$$

where σ^2 is the variance of the noise and δ_{ij} is the Kroniker delta equal to one when $i = j$ and zero otherwise. σ has the units of pressure. Continuing with discrete considerations, we assume that Q represents the total number of measurement points over the whole sphere. We still assume that orthonormality holds for the discrete case, so that

$$\sum_{i=1}^{Q} Y_n^m(\Omega_i)Y_q^{p*}(\Omega_i)\Delta\Omega = \delta_{nq}\delta_{mp}.$$

The discrete form of the orthonormality integral in practice could be implemented with a quadrature algorithm such as presented in handbooks[4] or in the literature.[5] In these algorithms the quadrature formula is exact up to some prescribed order N based on the number of quadrature points and their locations.

Also we assume equal distribution of quadrature points so that $\Delta\Omega = 4\pi/Q$. With these assumptions we get a simple result for the evaluation of the cross correlation integral of Eq. (7.29):

$$\begin{aligned}
E[B_{mn}B_{pq}^*] &= \iint E[\epsilon(\theta,\phi)\epsilon(\theta',\phi')^*]Y_n^m Y_q^{p*}\,d\Omega\,d\Omega' \\
&= \sum_{i=1}^{Q}\sum_{j=1}^{Q} E[\epsilon(\Omega_i)\epsilon(\Omega'_j)^*]Y_n^m(\Omega_i)Y_q^{p*}(\Omega'_j)\Delta\Omega^2 \\
&= \sigma^2\sum_{i=1}^{Q} Y_n^m(\Omega_i)Y_q^{p*}(\Omega_i)\Delta\Omega^2 \\
&= \sigma^2\frac{4\pi}{Q}\delta_{nq}\delta_{mp}.
\end{aligned} \tag{7.31}$$

Using this result in Eq. (7.28) yields

$$E[|\mathcal{N}_N(r,\theta,\phi)|^2] = \frac{4\pi\sigma^2}{Q(\rho_0 c)^2}\sum_{n=0}^{N}\left(\frac{j'_n(kr)}{j_n(kr_h)}\right)^2\sum_{m=-n}^{n}|Y_n^m(\theta,\phi)|^2. \tag{7.32}$$

---

[4] Abramowitz and Stegun, *Handbook of Mathematical Functions*, p. 894
[5] V. I. Lebedev (1976), Quadratures on a sphere. Zh. Vychisl. Mat. Fiz. **16**, pp. 293–306.

The average of Eq. (7.32) over the reconstruction sphere is

$$\langle E[|\mathcal{N}_N(r,\theta,\phi)|^2]\rangle = \frac{\sigma^2}{Q(\rho_0 c)^2} \sum_{n=0}^{N}(2n+1)\left(\frac{j_n'(kr)}{j_n(kr_h)}\right)^2, \tag{7.33}$$

due to the orthonormality of the spherical harmonics.

Using these results in Eq. (7.27) we obtain as the final result for the RMSE of the reconstructed velocity:

$$\mathcal{E}(r) = \left(\frac{\langle|\dot{w}_{\infty N}(r)|^2\rangle}{\langle|\dot{w}_{\infty}(r)|^2\rangle} + \frac{\sigma^2}{Q(\rho_0 c)^2\langle|\dot{w}_{\times}(r)|^2\rangle}\sum_{n=0}^{N}(2n+1)\left(\frac{j_n'(kr)}{j_n(kr_h)}\right)^2\right)^{1/2}. \tag{7.34}$$

In an identical way we can derive the error for the reconstruction of the pressure based upon Eq. (7.12). Equation (7.19) still applies but $\mathcal{N}$ of Eq. (7.22) is slightly different, with the prime and $i\rho_0 ck$ missing. The final result is

$$\mathcal{E}(r) = \left(\frac{\langle|p_{\infty N}(r)|^2\rangle}{\langle|p_{\infty}(r)|^2\rangle} + \frac{\sigma^2}{Q\langle|p_{\times}(r)|^2\rangle}\sum_{n=0}^{N}(2n+1)\left(\frac{j_n(kr)}{j_n(kr_h)}\right)^2\right)^{1/2}, \tag{7.35}$$

where

$$p_{\infty N}(r,\Omega) \equiv p_{\infty}(r,\Omega) - p_N(r,\Omega) = \sum_{n=N+1}^{\infty}\frac{j_n(kr)}{j_n(kr_h)}\sum_{m=-n}^{n}Y_n^m(\theta,\phi)P_{mn}(r_h). \tag{7.36}$$

Thus the reconstruction errors for the velocity and pressure are composed of two terms. For the velocity the first term is

$$\frac{<|\dot{w}_{\times N}|^2>}{<|\dot{w}_{\times}|^2>},$$

and for the pressure,

$$\frac{<|p_{\times N}|^2>}{<|p_{\times}|^2>}.$$

This component of the error is called the "filter error" to reflect the fact that it arises from the truncation of the series at $n = N$. This truncation is necessary either due to the quadrature algorithm used to compute the coefficients P_{mn} or due to the noise. The effects of noise are given in the second term in the RMSE reconstruction error in Eq. (7.34) and Eq. (7.35) and the error is proportional to the variance of the noise and the ratio of spherical Bessel functions in the summation over n. This ratio provides an instability as n increases due to the blow up of the evanescent waves as discussed in the last section. The sum over n is divergent when $r > r_h$ since the ratio of Bessel functions goes as $(r/r_h)^N$. We restrict the number of terms to $N + 1$ to filter the blow up of the evanescent waves. The cutoff N is perfectly analogous to the k-space filter used in planar (see Section 3.5) and cylindrical geometries, given that $k_c \equiv N/r_h$ and the filter is a rectangular one (unlike the taper shown in Fig. 3.5 on page 98). Also for a given

N and σ, as r is increased, that is, as we try to increase the size of the reconstruction sphere, the error increases. To reduce the error in this case we must decrease N or increase the signal to noise ratio in the experiment.

As an aside, note that the expected value of the squared velocity at any point on the reconstruction sphere can be shown to be

$$E[|\tilde{\dot{w}}_N(r,\theta,\phi)|^2] = |\dot{w}_N(r,\theta,\phi)|^2 + \frac{4\pi\sigma^2}{Q(\rho_0 c)^2}\sum_{n=0}^{N}\left(\frac{j_n'(kr)}{j_n(kr_h)}\right)^2\sum_{m=-n}^{n}|Y_n^m(\theta,\phi)|^2. \quad (7.37)$$

7.3.3 Plane Wave Example

It is quite informative to study a particular example which uses a known source with no evanescent field; a point source at infinity, that is, an incident plane wave. Even though the source is now at $r = \infty$ (Σ_1 and Σ_2 in Fig. 6.17, page 217, are moved out to infinity), the theory for the interior problem is perfectly valid. We will concentrate on normal velocity reconstructions on a sphere with $r > r_h$ (inverse problem), where r_h is still the radius of the measurement sphere. All the evanescent waves from the source have decayed to zero on the measurement surface. So one might naively expect that we should be able to reconstruct the pressure field almost everywhere outside of the measurement sphere, since the field is purely supersonic in wavenumber content. Unfortunately, this is quite incorrect, as the plane wave example will make clear. What we will find, in essence, is that the finite aperture of the measurement sphere restricts severely the maximum radius for accurate reconstructions.

In this example we consider a plane wave incident from the negative z axis as shown in Fig. 6.22 on page 225. The measurement sphere is located at $r = r_h$ as usual. Given a perfect measurement system the resulting $P_{mn}(r_h)$, Eq. (6.95) on page 207, is given by the expansion of the plane wave into spherical harmonics centered at the origin of the measurement sphere. This expansion was given by Eq. (6.174), page 227. Rewriting this equation in terms of spherical harmonics, where from Eq.(6.53), page 193,

$$P_n(\cos\theta) = \sqrt{\frac{4\pi}{2n+1}}Y_n^0(\theta,\phi), \quad (7.38)$$

yields

$$p(r,\theta,\phi) = e^{ikr\cos\theta} = \sum_{n=0}^{\infty}i^n\sqrt{4\pi(2n+1)}Y_n^0(\theta)j_n(kr). \quad (7.39)$$

Evaluating Eq. (6.95):

$$P_{mn}(r) = i^n j_n(kr)\sqrt{4\pi(2n+1)}\delta_{m0}. \quad (7.40)$$

An infinite number of spectral coefficients P_{0n} exist in the representation of a single plane wave (represented by a single point in the planar angular spectrum). However, it can be shown that only the coefficients up to $n = kr$ are significant. The following figure will make this clear. In Fig. 7.1 the quantity $20\log_{10}|j_n(x)|$ is plotted versus n for three values of $x = kr$; $kr = 2, 10, 20$. Thus we conclude that $P_{0n}(r)$ oscillates up

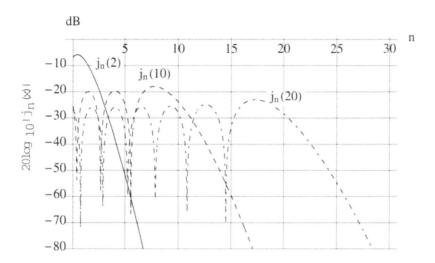

Figure 7.1: Plot of $20\log_{10}|j_n(x)|$ versus n for $x = (2, 10, 20)$. There is a rapid decay when $n > kr$ in every case. (To smooth the display non-integer values of n are included in the plot.)

to about $kr = n$, and then drops off exponentially above this. As a rule of thumb, the significant spherical wave spectrum components are limited to $n \leq kr$. For $n > kr$ the spherical wave components are evanescent (even though the plane wave is not) and are small and negligible in amplitude.

If we assume that the measurement sphere is at $r = r_h$, then P_{mn} is specified at this radius. We add noise to the experiment and reconstruct the normal velocity at a different surface $r > r_h$ and compare the reconstruction to the known result, that is,

$$\dot{w}_\infty(r, \theta, \phi) = \frac{1}{i\rho_0 ck} \frac{\partial p(r, \theta, \phi)}{\partial r} = \frac{\cos\theta}{\rho_0 c} e^{ikr\cos\theta}, \tag{7.41}$$

derived from differentiating Eq. (7.39). Equation (7.34) provides the error in the reconstruction of the normal velocity over the sphere at r. Before we apply this formula note that $|\dot{w}_\infty| = \cos\theta/\rho_0 c$ so that

$$(\rho_0 c)^2 \left\langle |\dot{w}_\infty(r)|^2 \right\rangle = \frac{1}{3}. \tag{7.42}$$

Again from Eq. (7.39) we have that

$$\dot{w}_{\infty N} = \frac{1}{\rho_0 c} \sum_{n=N+1}^{\infty} i^{n-1} \sqrt{(2n+1)} Y_n^0(\theta) j_n'(kr),$$

so that

$$(\rho_0 c)^2 \left\langle |\dot{w}_{\infty N}(r)|^2 \right\rangle = \sum_{n=N+1}^{\infty} (2n+1)(j_n'(kr))^2. \tag{7.43}$$

Given these results the RMSE, Eq. (7.34), becomes

$$\mathcal{E}(r) = \left(3 \sum_{n=N+1}^{\infty} (2n+1)(j_n'(kr))^2 + \frac{3\sigma^2}{Q} \sum_{n=0}^{N} (2n+1) \left(\frac{j_n'(kr)}{j_n(kr_h)} \right)^2 \right)^{1/2}. \qquad (7.44)$$

The first term in this equation can be written in another form. If we insert Eq. (7.42) into Eq. (7.43) we obtain

$$3 \sum_{n=0}^{\infty} (2n+1)(j_n'(kr))^2 = 1,$$

and, breaking up this sum into two parts, Eq. (7.44) becomes

$$\mathcal{E}(r) = \left(1 - 3 \sum_{n=0}^{N} (2n+1)(j_n'(kr))^2 + \frac{3\sigma^2}{Q} \sum_{n=0}^{N} (2n+1) \left(\frac{j_n'(kr)}{j_n(kr_h)} \right)^2 \right)^{1/2}. \qquad (7.45)$$

The two error terms in Eq. (7.44) behave in an opposite sense as N increases. When $N > kr$ consideration of Fig. 7.1 indicates that the first term will be very small since the spherical Bessel function is in its evanescent region. However the opposite is true of the second term, as long as $kr > kr_h$, since the ratio of Bessel functions "blows up" as we have already discussed.

7.4 Scattering Nearfield Holography

The measurement of scattered fields from objects within a spherical measurement surface (not necessarily spheres) can be carried out using scattering holography. Radiation holography assumes that the source is either within or totally outside a given closed measurement surface. A scattering experiment violates both of these assumptions, for the sources lie both inside (the scatterer) and outside (the incident field) of the measurement surface. In a general sense we can model the incident field as a source (Σ_2) located outside a given sphere of radius $r = b$ as shown in Fig. 7.2. The scattering body is labeled Σ_1.

7.4.1 The Dual Surface Approach

Define two concentric measurement surfaces within the region of validity, placed as close as possible to the interior sources/reflectors at the radii $r = a_1$ and $r = a_2$. To uniquely define the pressure field in the region of validity we must use two of the independent solutions of the radial Bessel's equation, Eq. (6.56) on page 193, chosen from Eq. (6.57). Incoming and outgoing waves would be represented by $h_n^{(2)}$ and $h_n^{(1)}$, respectively, but also by j_n and y_n since, by Eq. (6.57),

$$\begin{aligned} j_n &= (h_n^{(1)} + h_n^{(2)})/2, \\ y_n &= (h_n^{(1)} - h_n^{(2)})/2i. \end{aligned} \qquad (7.46)$$

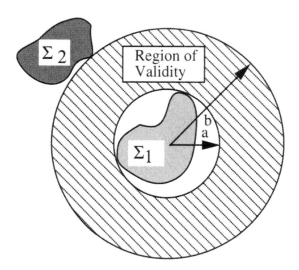

Figure 7.2: Definition of the region of validity for scattering holography, between the two spheres at $r = a$ and $r = b$. For plane wave incident scattering problems the source region Σ_2 moves to the farfield along with the sphere at $r = b$.

It is perhaps algebraically a bit simpler to use the j_n and y_n to formulate the problem. Thus we represent the pressure field in the region between the two spheres in Fig. 7.2 as

$$p(r, \theta, \phi) = \sum_{n=0}^{\infty} \sum_{m=-n}^{n} (A_{mn} j_n(kr) + B_{mn} y_n(kr)) Y_n^m(\theta, \phi). \tag{7.47}$$

The holographic experiment consists of the measurement of the pressure field on two concentric surfaces at $r = a_1$ and $r = a_2$ contained in the volume between $r = a$ and $r = b$. Writing Eq. (7.47) at the measurement surfaces leads to two equations in two unknowns A_{mn} and B_{mn}:

$$p(a_1, \theta, \phi) = \sum_{n=0}^{\infty} \sum_{m=-n}^{n} (A_{mn} j_n(ka_1) + B_{mn} y_n(ka_1)) Y_n^m(\theta, \phi), \tag{7.48}$$

$$p(a_2, \theta, \phi) = \sum_{n=0}^{\infty} \sum_{m=-n}^{n} (A_{mn} j_n(ka_2) + B_{mn} y_n(ka_2)) Y_n^m(\theta, \phi). \tag{7.49}$$

Multiplication of each of these equations by Y_n^{m*} and integration over the solid angle Ω yields two simple equations in "k-space".

$$P_{mn}(a_1) = (A_{mn} j_n(ka_1) + B_{mn} y_n(ka_1)), \tag{7.50}$$

$$P_{mn}(a_2) = (A_{mn} j_n(ka_2) + B_{mn} y_n(ka_2)), \tag{7.51}$$

where, as before,

$$P_{mn}(r) \equiv \int p(r, \theta, \phi) Y_n^m(\theta, \phi)^* d\Omega. \tag{7.52}$$

The solution to Eqs (7.50) and (7.51) is

$$A_{mn} = \frac{\begin{vmatrix} P_{mn}(a_1) & y_n(ka_1) \\ P_{mn}(a_2) & y_n(ka_2) \end{vmatrix}}{\Delta}, \tag{7.53}$$

$$B_{mn} = \frac{\begin{vmatrix} j_n(ka_1) & P_{mn}(a_1) \\ j_n(ka_2) & P_{mn}(a_2) \end{vmatrix}}{\Delta}, \tag{7.54}$$

where

$$\Delta \equiv j_n(ka_1)y_n(ka_2) - j_n(ka_2)y_n(ka_1). \tag{7.55}$$

The coefficients P_{mn} on the right hand side of Eq. (7.53) and Eq. (7.54) are determined from the measurement using Eq. (7.52). The resulting values for A_{mn} and B_{mn} are inserted back into Eq. (7.48) and Eq. (7.49) to determine the pressure in the region of validity.

We note that since Eq. (7.55) is a real function it has zeros for particular choices of a_1, a_2, and k. When Δ vanishes we have no solution and the pressure field can not be extended into the volume. However, as long as the two surfaces are taken to be close enough together, we can show that Δ will not go to zero. Let $d = a_2 - a_1$ and assume that this spacing is much less than an acoustic wavelength, $kd << 1$. We expand $j_n(ka_2)$ and $y_n(ka_2)$ in Taylor series about ka_1,

$$j_n(ka_2) \approx j_n(ka_1) + kd\,j'_n(ka_1)$$
$$y_n(ka_2) \approx y_n(ka_1) + kd\,y'_n(ka_1).$$

Under these conditions, using the Wronskian relation for the spherical Bessel functions,

$$j_n(x)y'_n(x) - j'_n(x)y_n(x) = \frac{1}{x^2}, \tag{7.56}$$

we find that

$$\Delta \approx kd[j_n(ka_1)y'_n(ka_1) - j'_n(ka_1)y_n(ka_1)] = \frac{d}{ka_1^2}. \tag{7.57}$$

It can be seen that Eq. (7.57) will never vanish under the assumptions made.

If the solution is desired in terms of Hankel functions instead of Bessel functions, then

$$p(r, \theta, \phi) = \sum_{n=0}^{\infty} \sum_{m=-n}^{n} (C_{mn}h_n^{(1)}(kr) + D_{mn}h_n^{(2)}(kr))Y_n^m(\theta, \phi). \tag{7.58}$$

This solution can be constructed from the Eq. (7.48) and Eq. (7.49) by using Eq. (7.46) to arrive at

$$C_{mn} = \frac{iA_{mn} + B_{mn}}{2i}, \tag{7.59}$$

$$D_{mn} = \frac{iA_{mn} - B_{mn}}{2i}. \tag{7.60}$$

Equation (7.58) was used by Weinreich and Arnold[6] in some pioneering experimental work on the radiation from violins. The incoming fields were generated by reflections of the radiated field from the walls of the test chamber.

7.4.2 Holography Using an Intensity Probe

The method presented in the last section poses measurement problems when the two surfaces are very close together, due to the fact that the pressure will not be very different on the two surfaces and thus very accurate measurement of the phase of the pressure field is necessary. We can reformulate the solution given an intensity measuring system which is used to provide the pressure and the radial velocity on a measurement surface (we don't actually use the intensity). We can rewrite Eq. (7.48) and Eq. (7.49) with the measurement surface at $r = a_1$ using Euler's equation as

$$p(a_1, \theta, \phi) = \sum_{n=0}^{\infty} \sum_{m=-n}^{n} (A_{mn} j_n(ka_1) + B_{mn} y_n(ka_1)) Y_n^m(\theta, \phi), \qquad (7.61)$$

and

$$\dot{w}(a_1, \theta, \phi) = \frac{1}{i\rho_0 c} \sum_{n=0}^{\infty} \sum_{m=-n}^{n} (A_{mn} j_n'(ka_1) + B_{mn} y_n'(ka_1)) Y_n^m(\theta, \phi). \qquad (7.62)$$

Instead of using Eq. (7.51), we write the spherical wave spectrum of the velocity and use Eq. (7.62):

$$\dot{W}_{mn}(a_1) \equiv \int \dot{w}(a_1, \theta, \phi) Y_n^m(\theta, \phi)^* d\Omega = \frac{1}{i\rho_0 c} (A_{mn} j_n'(ka_1) + B_{mn} y_n'(ka_1)). \qquad (7.63)$$

Now the coefficients are determined from Eq. (7.50) and Eq. (7.63):

$$A_{mn} = \frac{\begin{vmatrix} P_{mn}(a_1) & y_n(ka_1) \\ i\rho_0 c \dot{W}_{mn}(a_1) & y_n'(ka_1) \end{vmatrix}}{\Delta} \qquad (7.64)$$

$$B_{mn} = \frac{\begin{vmatrix} j_n(ka_1) & P_{mn}(a_1) \\ j_n'(ka_1) & i\rho_0 c \dot{W}_{mn}(a_1) \end{vmatrix}}{\Delta}, \qquad (7.65)$$

where

$$\Delta \equiv j_n(ka_1) y_n'(ka_1) - j_n'(ka_1) y_n(ka_1) = \frac{1}{(ka_1)^2}. \qquad (7.66)$$

Note that Δ is never zero.

Inserting these results into Eq. (7.47) provides the final reconstruction of the pressure field in the region of validity:

$$p(r, \theta, \phi) = (ka_1)^2 \sum_{n,m} \Big(P_{mn}(a_1) [y_n'(ka_1) j_n(kr) - j_n'(ka_1) y_n(kr)]$$

[6]Gabriel Weinreich and Eric B. Arnold (1980), "Method for measuring acoustic radiation fields", J. Acoust. Soc. Am. **68**, pp. 404–411.

$$-i\rho_0 c \dot{W}_{mn}(a_1)[y_n(ka_1)j_n(kr) - j_n(ka_1)y_n(kr)]\Big) Y_n^m(\theta,\phi), \qquad (7.67)$$

where P_{mn} and $\dot{W}_{mn}$ are determined from the experiment.

Problems

7.1 Show that the series for the inverse of the Neumann Green function given in Eq. (7.8) diverges using the ratio test when $a < r_h$. For simplicity only consider convergence at the point $\theta = \theta' = 0$. You should find Eq. (6.30) useful in your work.

Chapter 8

Green Functions and the Helmholtz Integral Equation

8.1 Introduction

The Helmholtz integral equation and Green functions are important to the study of acoustics. Though it is not strictly in the realm of Fourier acoustics, it is crucial for the extension of NAH to arbitrary geometries. There is much work in many fields of acoustics today which relies heavily on the Helmholtz integral equation. In particular, the field of boundary elements has as its starting point this important equation. It is very relevant to nearfield acoustical holography applied to arbitrary geometries, and instead of being based in Fourier acoustics, the singular value decomposition becomes the medium for the eigenfunctions of the problem.

We will derive the Helmholtz integral equation for both the interior problem, applicable to problems in which the acoustic field is confined within boundaries, and the exterior problem applicable to radiation and scattering problems. The derivations are presented in detail so that the reader can appreciate their subtleties.

8.2 Green's Theorem

Let V be a three-dimensional volume bounded by a surface S. A point inside this volume is located by the radius vector $\mathbf{r}$. Inside the volume V we have two unknown functions $\Psi(\mathbf{r})$ and $\Phi(\mathbf{r})$ which are finite and continuous along with their first and second partial derivatives. Given these conditions then Green's theorem (Green's second identity) applies:

$$\iiint_V (\Phi \nabla^2 \Psi - \Psi \nabla^2 \Phi)\, dV = \iint_S (\Phi \frac{\partial \Psi}{\partial n} - \Psi \frac{\partial \Phi}{\partial n})\, dS, \qquad (8.1)$$

where $\frac{\partial}{\partial n}$ is the derivative with respect to the outward normal (the rate of change of the function in the direction perpendicular to the surface). The geometry is shown in Fig. 8.1.

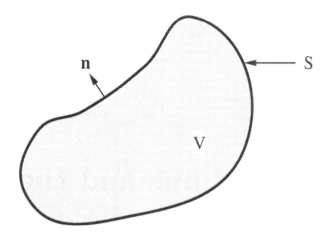

Figure 8.1: Shaded region of volume V with bounding surface S and outward normal **n**. Green's theorem applies to this volume.

Let us assume that these two functions also satisfy the homogeneous Helmholtz equation on the surface and in the volume:

$$\nabla^2 \Phi + k^2 \Phi = 0$$
$$\nabla^2 \Psi + k^2 \Psi = 0. \tag{8.2}$$

The integrand on the left hand side of Eq. (8.1) becomes

$$(\Phi \nabla^2 \Psi - \Psi \nabla^2 \Phi) = \Phi(-k^2 \Psi) - \Psi(-k^2 \Phi) = 0,$$

so that the right hand side must vanish:

$$\iint_S (\Phi \frac{\partial \Psi}{\partial n} - \Psi \frac{\partial \Phi}{\partial n}) \, dS = 0. \tag{8.3}$$

Thus if the functions Ψ and Φ have no singularities within or on the surface S, then Eq. (8.3) must be satisfied. This equation forms the basis for the derivation of the Helmholtz integral equation.

8.3   The Interior Helmholtz Integral Equation

Consider now the case in which the function Ψ has a monopole singularity within the volume V, with Φ continuous as before. The singularity is located at $\mathbf{r} = \mathbf{r}'$ and $\Psi(\mathbf{r})$ is a solution of the inhomogeneous Helmholtz equation,

$$\nabla^2 \Psi + k^2 \Psi = -\delta(\mathbf{r} - \mathbf{r}'), \tag{8.4}$$

where $\mathbf{r}$ spans all space; there are no boundary conditions on Ψ, except that it satisfy the Sommerfeld radiation condition at infinity (Eq. (8.28)). The general solution of Eq. (8.4) is

$$\Psi = \Psi_p + \Psi_h$$

where Ψ_p is the particular solution and Ψ_h the homogeneous solution. The particular solution is the free space Green function $\Psi_p = G(\mathbf{r}|\mathbf{r}')$ which was presented in Section 6.5.1 on page 198. For the sake of clarity consider the evaluation point $\mathbf{r}'$ to be fixed while the field point (observation point) $\mathbf{r}$ is allowed to vary over the volume V, as if we had a microphone scanner measuring the pressure at $\mathbf{r}$. We set the homogeneous solution Ψ_h to zero so that the solution of Eq. (8.4) is simply

$$\Psi = \Psi_p = G(\mathbf{r}|\mathbf{r}') = \frac{e^{ikR}}{4\pi R}, \tag{8.5}$$

where $R \equiv |\mathbf{r} - \mathbf{r}'|$. If the point $\mathbf{r}'$ is contained within the volume V we can no longer apply Green's theorem because one of the functions in the integrand violates the continuity requirements. However, we can modify the volume so that we can still apply the theorem by excluding the singular point from the volume. We do this by surrounding the point by a vanishingly small sphere with surface S_i connected to the outer surface by a "canal" as shown in Fig. 8.2. We rename the outer surface S_o. The functions Ψ and Φ are now continuous in the volume between the surfaces S_i and S_o and Green's theorem can be applied to this new volume. Note that the outward normals to the surface S_i point towards the evaluation point.

We will apply Green's theorem to this new volume (shaded area in the figure) and then let the width of the "canal" connecting the surfaces S_i and S_o go to zero, as well as the radius of the spherical surface, S_i. Since the normals to both sides of the canal are in opposition, then the surface integral over one of these sides cancels the another.

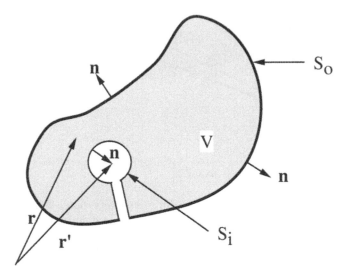

Figure 8.2: Shaded region of volume V with outside bounding surface S_0, inside bounding surface S_i, and outward normals $\mathbf{n}$ as shown. The volume excludes the evaluation point at $\mathbf{r} = \mathbf{r}'$. Green's theorem applies to this volume between S_i and S_0.

Applying Eq. (8.3) to this new volume with surface S defined by

$$S = S_o + S_i,$$

and replacing Ψ with G yields.

$$\iint_S \left(\Phi \frac{\partial G(\mathbf{r}|\mathbf{r}')}{\partial n} - G(\mathbf{r}|\mathbf{r}') \frac{\partial \Phi}{\partial n} \right) dS_o + \lim_{\epsilon \to 0} \iint \left(\Phi \frac{\partial G(\mathbf{r}|\mathbf{r}')}{\partial n} - G(\mathbf{r}|\mathbf{r}') \frac{\partial \Phi}{\partial n} \right) dS_i = 0. \quad (8.6)$$

Note that Φ is still a function of $\mathbf{r}$ and does not depend upon $\mathbf{r}'$.

What is the mathematical meaning of $\frac{\partial G(\mathbf{r}|\mathbf{r}')}{\partial n}$? Consider the following figure, which is a blow up of the spherical region in Fig. 8.2. First consider the meaning of the partial

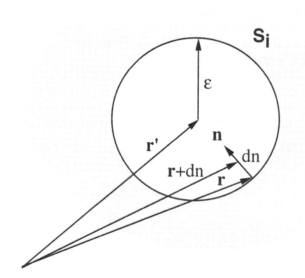

Figure 8.3: Blow up of the interior region for application of Green's theorem, and discussion of the normal derivative.

with respect to the normal. Let dn be an infinitesimally small step along the normal direction from a point on the surface S_i. The derivative definition requires

$$\frac{\partial G(\mathbf{r}|\mathbf{r}')}{\partial n} \approx \frac{G(\mathbf{r} + dn\hat{n}|\mathbf{r}') - G(\mathbf{r}|\mathbf{r}')}{dn}.$$

In other words the function G is evaluated at a point infinitesimally removed from the sphere (in the normal direction) in evaluating the normal derivative. Note that the normal derivative is taken with respect to the field point $\mathbf{r}$ not the evaluation point. Furthermore, since ϵ is the radius of the sphere measured from the sphere center, then

$$\frac{\partial}{\partial n} = -\frac{\partial}{\partial \epsilon}. \quad (8.7)$$

Proceeding with Eq. (8.6) we take the limit as $\epsilon \to 0$, with $dS_i = \epsilon^2 d\Omega$, where Ω is the solid angle in spherical coordinates ($d\Omega \equiv \sin\theta \, d\theta \, d\phi$), and on S_i

$$G(\mathbf{r}|\mathbf{r}') = \frac{e^{ikR}}{4\pi R} = \frac{e^{ik\epsilon}}{4\pi\epsilon}.$$

The second term of the second integral in Eq. (8.6) becomes (since $\frac{\partial\Phi}{\partial n}$ is continuous about $\mathbf{r} = \mathbf{r}'$ it can be taken outside the integral)

$$
\begin{aligned}
\lim_{\epsilon\to 0} \iint \left(-G(\mathbf{r}|\mathbf{r}')\frac{\partial\Phi}{\partial n}\right) dS_i &= \lim_{\epsilon\to 0} \frac{\partial\Phi(\mathbf{r})}{\partial\epsilon}\bigg|_{\mathbf{r}=\mathbf{r}'} \iint G(\mathbf{r}|\mathbf{r}')\,\epsilon^2 d\Omega \\
&= \lim_{\epsilon\to 0} \epsilon\frac{\partial\Phi(\mathbf{r})}{\partial\epsilon}\bigg|_{\mathbf{r}=\mathbf{r}'} = 0. \qquad (8.8)
\end{aligned}
$$

For the first term of the second integral in Eq. (8.6) we need

$$\frac{\partial G(\mathbf{r}|\mathbf{r}')}{\partial n} = -\frac{1}{4\pi}\frac{\partial e^{ik\epsilon}/\epsilon}{\partial\epsilon} = -\frac{e^{ik\epsilon}}{4\pi\epsilon}(ik - 1/\epsilon) \approx \frac{e^{ik\epsilon}}{4\pi\epsilon^2}.$$

Using this result in the first term of the second integral in Eq. (8.6), noting that the function Φ is continuous within the sphere,

$$\lim_{\epsilon\to 0} \iint (\Phi(\mathbf{r})\frac{\partial G(\mathbf{r}|\mathbf{r}')}{\partial n})dS_i = \lim_{\epsilon\to 0} \Phi(\mathbf{r}) \iint \frac{\partial G(\mathbf{r}|\mathbf{r}')}{\partial n}\epsilon^2 d\Omega = \Phi(\mathbf{r})|_{\mathbf{r}=\mathbf{r}'} = \Phi(\mathbf{r}'). \quad (8.9)$$

Inserting the results of Eq. (8.8) and Eq. (8.9) into Eq. (8.6) results in the Helmholtz integral equation,

$$\Phi(\mathbf{r}') = \frac{1}{4\pi} \iint_{S_o} \left(\frac{e^{ik|\mathbf{r}-\mathbf{r}'|}}{|\mathbf{r}-\mathbf{r}'|}\frac{\partial\Phi}{\partial n} - \Phi\frac{\partial}{\partial n}(\frac{e^{ik|\mathbf{r}-\mathbf{r}'|}}{|\mathbf{r}-\mathbf{r}'|})\right)dS_o, \qquad (8.10)$$

or writing the result in terms of the free space Green function,

$$\Phi(\mathbf{r}') = \iint_{S_o} \left(G(\mathbf{r}|\mathbf{r}')\frac{\partial\Phi}{\partial n} - \Phi\frac{\partial}{\partial n}G(\mathbf{r}|\mathbf{r}')\right)dS_o. \qquad (8.11)$$

Now consider the case when the evaluation point is located on the boundary of S_o, as shown in Fig. 8.4. In this case only half the sphere is contained within the original volume V so that $\iint_{S_i} d\Omega = \iint_{S_i} \sin\theta \, d\theta d\phi = 2\pi$. As $\epsilon \to 0$ re-evaluation of Eq. (8.9) leads to

$$\lim_{\epsilon\to 0} \iint (\Phi\frac{\partial G(\mathbf{r}|\mathbf{r}')}{\partial n})dS_i = \frac{1}{2}\Phi(\mathbf{r}'), \qquad (8.12)$$

and Eq. (8.10) results with a $1/2$ on the left hand side, so that the HIE is now

$$\Phi(\mathbf{r}') = \frac{1}{2\pi} \iint_{S_o} \left(\frac{e^{ik|\mathbf{r}-\mathbf{r}'|}}{|\mathbf{r}-\mathbf{r}'|}\frac{\partial\Phi}{\partial n} - \Phi\frac{\partial}{\partial n}(\frac{e^{ik|\mathbf{r}-\mathbf{r}'|}}{|\mathbf{r}-\mathbf{r}'|})\right)dS_o. \qquad (8.13)$$

Finally, we consider the case where the evaluation point is totally outside of the surface S_o. Since there is no longer any singularity within or on the surface then Eq. (8.3) applies so that

$$0 = \frac{1}{4\pi} \iint_{S_o} \left(\frac{e^{ik|\mathbf{r}-\mathbf{r}'|}}{|\mathbf{r}-\mathbf{r}'|}\frac{\partial\Phi}{\partial n} - \Phi\frac{\partial}{\partial n}(\frac{e^{ik|\mathbf{r}-\mathbf{r}'|}}{|\mathbf{r}-\mathbf{r}'|})\right)dS_o. \qquad (8.14)$$

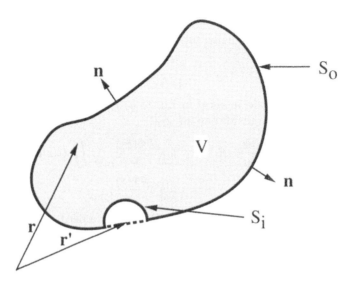

Figure 8.4: Derivation of Helmholtz integral equation when the evaluation point, $\mathbf{r}'$, is located on the boundary of the outer surface, S_o.

This last equation is valid for the evaluation point located *anywhere* outside of the surface S_o.

Thus combining Eq. (8.10). Eq. (8.13), and Eq. (8.14). the full statement of the Helmholtz integral equation is

$$\alpha\Phi(\mathbf{r}') = \iint_{S_o} \left(G(\mathbf{r}|\mathbf{r}')\frac{\partial\Phi(\mathbf{r})}{\partial n} - \Phi(\mathbf{r})\frac{\partial}{\partial n}G(\mathbf{r}|\mathbf{r}') \right)dS_o, \tag{8.15}$$

where

$$\alpha = \begin{cases} 1 & \text{if } \mathbf{r}' \text{ is inside } S_o \\ 1/2 & \text{if } \mathbf{r}' \text{ is on } S_o \\ 0 & \text{if } \mathbf{r}' \text{ is outside } S_o, \end{cases}$$

and

$$G(\mathbf{r}|\mathbf{r}') = \frac{e^{ik|\mathbf{r}-\mathbf{r}'|}}{4\pi|\mathbf{r}-\mathbf{r}'|}.$$

Returning to the solution of Eq. (8.4), if we assume the homogeneous equation solution $\Psi_h \neq 0$, Ψ_h satisfies Eq. (8.3) and thus we can replace G in Eq. (8.15) with $G + \Psi_h$ and the equality still holds; Eq. (8.15) is still valid. This fact will be important later in this chapter when we study Dirichlet and Neumann Green functions.

In solving acoustic problems, we can write Eq. (8.15) in terms of pressure and normal velocity. Thus we equate $\Phi = p$ and using Euler's equation, defining v_n as the velocity in the normal (outward) direction so that

$$\frac{\partial\Phi(\mathbf{r})}{\partial n} = i\rho_o c k v_n(\mathbf{r}),$$

Eq. (8.15) becomes

$$\alpha p(\mathbf{r}') = \iint_{S_o} \left(i\rho_o c k G(\mathbf{r}|\mathbf{r}') v_n(\mathbf{r}) - p(\mathbf{r}) \frac{\partial}{\partial n} G(\mathbf{r}|\mathbf{r}') \right) dS_o. \tag{8.16}$$

Note that the integration is over the field points $\mathbf{r}$ with the evaluation point providing the location of where the interior pressure is evaluated. Equation (8.16) is an alternative form of the interior Helmholtz integral equation for an interior domain.

The HIE, Eq. (8.15) or Eq. (8.16), can *not* be used to determine the pressure outside of the surface S_o, because the exterior pressure in Eq. (8.16) is multiplied by zero. Since the sources are also outside, determining the exterior pressure is an inverse problem, like the kind solved by NAH. The HIE does not solve the inverse problem. It can only provide the forward solution.

8.3.1 Example with Sphere

To help solidify understanding of the HIE we present an example which should accomplish this purpose if it is understood thoroughly. To begin, note that the surface in the integral of the HIE can be an invisible boundary whose location we choose according to the problem which we want to solve. It is not necessarily the boundary of an object. As an example of the mechanics of the application of the HIE, we choose an invisible spherical boundary. This boundary is placed in a known field arising from a source outside the boundary. The HIE, Eq. (8.16), provides a mathematical tool to compute the pressure at a point inside this spherical boundary from a knowledge of the pressure and its normal derivative on the boundary.

Assume that a point source is located at point P on the z axis at $z = -d$, outside of the spherical boundary centered at the origin as shown in Fig. 8.5. Assume the pressure generated from the source is, with p_0 a constant,

$$p(\mathbf{r}) = p_0 \frac{e^{ik|\mathbf{r}-\mathbf{d}|}}{ik|\mathbf{r}-\mathbf{d}|}$$

where $\mathbf{d} = -d\hat{\mathbf{k}}$. Note that we have not violated the assumption that Φ (or equivalently, p) satisfies a homogeneous Helmholtz equation inside the boundary since the point source is outside. This is perfectly consistent with Eq. (8.4). We use a spherical coordinate system with origin at the center of the sphere which has a radius a. The sphere, S_0, is the HIE surface. Let $\xi \equiv |\mathbf{r} - \mathbf{d}|$ so that

$$p(\mathbf{r}) = p_0 \frac{e^{ik\xi}}{ik\xi} = p_0 h_0(k\xi) \tag{8.17}$$

where, using Eq. (6.62) on page 196, we represent the source by a spherical Hankel function. In order to facilitate solution we need to write ξ as a function of (r, θ) where on the sphere S_0, $p(\mathbf{r}) = p(a, \theta)$. Note that the field is independent of ϕ, since the point source is on the z axis and the field is therefore axisymmetric.

We proceed by using the addition theorem to expand $h_0(k\xi)$ about the origin of the sphere, so that the mathematics is more tractable. The addition theorem provides the

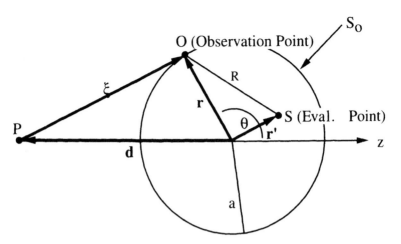

Figure 8.5: Illustration of the HIE using a spherical surface. A test field is generated by a point source located at P outside the sphere, and the field it generates on the surface is used in the HIE to predict the pressure field in the interior at the evaluation point S.

following formula,[1]

$$h_0(k|\mathbf{r} - \mathbf{d}|) = \sum_{n=0}^{\infty} (-1)^n (2n+1) j_n(kr_<) h_n(kr_>) P_n(\cos\theta), \qquad (8.18)$$

where $r_<$ is the lesser of the r and d, and $r_>$ is the greater of the two (see Fig. 8.5). We assume here that $r < d$ so that $r_< = r$ and $r_> = d$. Let R be the distance between $\mathbf{r}$ and $\mathbf{r}'$. Then Eq. (8.15) becomes ($\Phi = p$)

$$\alpha p(\mathbf{r}') = \frac{1}{4\pi} \iint_{S_o} \left(\frac{e^{ikR}}{R} \frac{\partial p(\mathbf{r})}{\partial r} - p(\mathbf{r}) \frac{\partial}{\partial r} \left[\frac{e^{ikR}}{R} \right] \right) a^2 \sin\theta d\theta d\phi, \qquad (8.19)$$

where $p(\mathbf{r})$ is to be replaced with p_0 times Eq. (8.18). Furthermore, $\frac{\partial p(\mathbf{r})}{\partial r}$ is just the radial derivative of Eq. (8.18) with respect to $r_<$ times p_0:

$$\frac{\partial p(\mathbf{r})}{\partial r} = k p_0 \sum_{n=0}^{\infty} (-1)^n (2n+1) j_n'(kr) h_n(kd) P_n(\cos\theta). \qquad (8.20)$$

Now consider the Green function and its normal derivative which appear in Eq. (8.19), which do not depend on the point source field in any way. First note that

$$\frac{\partial}{\partial r} \left[\frac{e^{ikR}}{R} \right] = \frac{e^{ikR}}{R} \left(ik - \frac{1}{R} \right) \frac{\partial R}{\partial r}, \qquad (8.21)$$

and that $\frac{\partial R}{\partial r}$ is a fairly complicated relationship. R is not axisymmetric. To understand this further, and to continue with the evaluation of Eq. (8.19), we draw upon the

---

[1]I. S. Gradshteyn and I. M. Ryzhik (1965). *Tables of integrals, series and products*, Academic Press, New York and London, p. 980.

generalization of the addition theorem which does not depend on axisymmetry, valid for all $\mathbf{r}$ and $\mathbf{r}'$:[2]

$$G \equiv \frac{e^{ik|\mathbf{r}-\mathbf{r}'|}}{4\pi|\mathbf{r}-\mathbf{r}'|} = ik \sum_{n=0}^{\infty} j_n(kr_<)h_n(kr_>) \sum_{m=-n}^{n} Y_n^m(\theta',\phi')^* Y_n^m(\theta,\phi), \qquad (8.22)$$

where $\mathbf{r} = (r,\theta,\phi)$ and $\mathbf{r}' = (r',\theta',\phi')$, $r_<$ is the lesser of (r,r') and $r_>$ the greater, and both r and r' are measured from the center of the sphere. This important formula provides an expansion of the free space Green function in spherical coordinates. Thus we see that this complicated relationship involves spherical harmonics for both the observation point and the evaluation point. Now the normal derivative is easily found, since $\frac{\partial}{\partial n} = \frac{\partial}{\partial r}$ and $R = |\mathbf{r}-\mathbf{r}'|$, Eq. (8.22) reveals that

$$\frac{\partial}{\partial r}\left[\frac{e^{ikR}}{4\pi R}\right] = ik^2 \sum_{n=0}^{\infty} \left\{ \begin{array}{c} j_n(kr')h_n'(kr) \\ j_n'(kr)h_n(kr') \end{array} \right\}\bigg|_{r=a} \sum_{m=-n}^{n} Y_n^m(\theta',\phi')^* Y_n^m(\theta,\phi), \qquad (8.23)$$

where the upper row is taken when $r' < r$ (evaluation point inside the sphere) and the lower when $r' > r$ (evaluation point outside the sphere).

Returning to Eq. (8.19), we need to insert the results from Eq. (8.18), Eq. (8.20), Eq. (8.22) and Eq. (8.23) into it. The first step is to carry out the integration over ϕ. Since p does not depend on ϕ then only the spherical harmonics in the expansions of the Green function have ϕ dependence. This leads to

$$\int Y_n^m(\theta,\phi)d\phi = \sqrt{\pi(2n+1)}P_n(\cos\theta)\delta_{m0},$$

so that the summations over m disappear. The integration over θ picks out the same value of n in the summations over n for p and G, due to the orthonormality of $P_n(\cos\theta)$. After some simplification Eq. (8.19) becomes

$$\begin{aligned} \alpha p(\mathbf{r}') &= i(ka)^2 p_0 \sum_{n=0}^{\infty} (-1)^n(2n+1)P_n(\cos\theta') \\ &\quad \times h_n(kd)\left\{ \begin{array}{c} j_n(kr')[h_n(ka)j_n'(ka) - h_n'(ka)j_n(ka)] \\ h_n(kr')[j_n(ka)j_n'(ka) - j_n'(ka)j_n(ka)] \end{array} \right\}, \end{aligned}$$

the upper row for r' inside and the lower for r' outside. In the latter case the right hand side is zero, which is consistent with the fact that α is zero. In other words, the HIE produces a null field when the evaluation point is outside the surface of integration (on the same side as the point source). Using the Wronskian relationship for Bessel functions (Eq. (6.61) on page 196), the term in square brackets becomes $-i/(ka)^2$ and we arrive at

$$\alpha p(\mathbf{r}') = p_0 \sum_{n=0}^{\infty} (-1)^n(2n+1)h_n(kd)j_n(kr')P_n(\cos\theta').$$

The right hand side is just $p_0 h_0(k|\mathbf{r}' - \mathbf{d}|)$, using Eq. (8.18), which we recognize as the incident field given by Eq. (8.17) evaluated at $\mathbf{r} = \mathbf{r}'$. Thus since $\alpha = 1$ the HIE reproduces the incident field inside.

---

[2] J. D. Jackson (1975). *Classical Electrodynamics*, 2nd ed. Wiley & Sons, p. 742.

8.4   Helmholtz Integral Equation for Radiation Problems (Exterior Domain)

In order to obtain the pressure outside of a given surface such as a vibrating sphere, we need to redefine the regions for the application of Green's theorem. The following choice of surfaces, as shown in Fig. 8.6, will allow us to derive an integral equation useful for the exterior domain. The surface of the body of interest is labeled S_o in

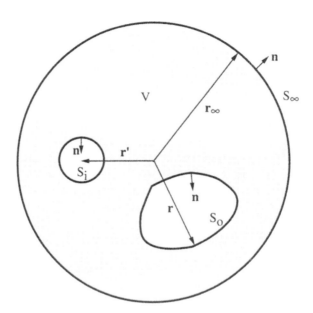

Figure 8.6: Region definitions for application of Green's theorem for the derivation of the exterior Helmholtz integral equation, useful for solving radiation problems. The shaded region defines the volume V for Green's theorem.

the figure, and it is arbitrarily shaped. The outer surface S_∞ is a sphere of radius r_∞ which tends towards infinity. Finally, the vanishingly small sphere labeled S_i surrounds an evaluation point, in exactly the same sense as it did for the interior problem in Fig. 8.2. These three surfaces border a volume V shown as the shaded area in Fig. 8.6, which defines the volume in which the functions Φ and Ψ are continuous along with their first and second derivatives. Equation (8.3) applies, where S is composed of the three bordering surfaces. We choose $\Psi = G$, the free space Green function, defined in Eq. (8.5) and satisfying Eq. (8.4), where the evaluation point $\mathbf{r}'$ is shown in Fig. 8.6.

The application of Green's theorem proceeds exactly as in Section 8.3, with the problem being identical for the two surfaces S_o and S_i except that the normal to S_o is pointing in the opposite direction. The total surface over which Green's theorem, Eq. (8.3), is applied is, in this case,

$$S = S_o + S_i + S_\infty.$$

Thus to Eq. (8.6) we must add a term which involves the integral over the surface S_∞:

$$\lim_{r=r_\infty \to \infty} \iint (\Phi \frac{\partial G(\mathbf{r}|\mathbf{r}')}{\partial n} - G(\mathbf{r}|\mathbf{r}') \frac{\partial \Phi}{\partial n}) \, dS_\infty. \qquad (8.24)$$

We will evaluate this integral after we consider the integrals over S_o and S_i.

A restatement of Eq. (8.3) symbolically becomes

$$\left(\iint_{S_o} + \iint_{S_i} + \iint_{S_\infty} \right) \left[\Phi \frac{\partial G}{\partial n} - G \frac{\partial \Phi}{\partial n} \right] dS = 0. \qquad (8.25)$$

The second integral is identical to the internal case and is the sum of Eq. (8.8) and Eq. (8.9) if the evaluation point is outside, and the sum of Eq. (8.8) and Eq. (8.12) if the evaluation point is on the boundary. Thus, since ϵ is the radius of the sphere with surface S_i,

$$\lim_{\epsilon \to 0} \iint_{S_i} \left(\Phi \frac{\partial}{\partial n} G(\mathbf{r}|\mathbf{r}') - G(\mathbf{r}|\mathbf{r}') \frac{\partial \Phi}{\partial n} \right) dS_i = \alpha \Phi(\mathbf{r}'), \qquad (8.26)$$

where

$$\alpha = \begin{cases} 1 & \text{if } \mathbf{r}' \text{ is outside } S_o \\ 1/2 & \text{if } \mathbf{r}' \text{ is on } S_o \\ 0 & \text{if } \mathbf{r}' \text{ is inside } S_o \end{cases} .$$

The third integral must be evaluated as $r = r_\infty \to \infty$:

$$\iint_{S_\infty} = \lim_{r \to \infty} \frac{1}{4\pi} \iint_{S_\infty} \left(\Phi \frac{\partial}{\partial r} (\frac{e^{ik|\mathbf{r}-\mathbf{r}'|}}{|\mathbf{r}-\mathbf{r}'|}) - \frac{e^{ik|\mathbf{r}-\mathbf{r}'|}}{|\mathbf{r}-\mathbf{r}'|} \frac{\partial \Phi}{\partial r} \right) dS_\infty.$$

In the limit we note that

$$\frac{e^{ik|\mathbf{r}-\mathbf{r}'|}}{|\mathbf{r}-\mathbf{r}'|} \approx \frac{e^{ik|\mathbf{r}-\mathbf{r}'|}}{r},$$

and

$$\frac{\partial}{\partial r} (\frac{e^{ik|\mathbf{r}-\mathbf{r}'|}}{|\mathbf{r}-\mathbf{r}'|}) \approx ik \frac{e^{ik|\mathbf{r}-\mathbf{r}'|}}{r},$$

so that

$$\iint_{S_\infty} = - \lim_{r \to \infty} e^{ik|\mathbf{r}-\mathbf{r}'|} r \left[\frac{\partial \Phi(\mathbf{r})}{\partial r} - ik\Phi(\mathbf{r}) \right]. \qquad (8.27)$$

Thus, in order for the integral over S_∞ to vanish, we must require that

$$\lim_{r \to \infty} r \left[\frac{\partial p(\mathbf{r})}{\partial r} - ikp(\mathbf{r}) \right] = 0, \qquad (8.28)$$

where we have replaced Φ with the pressure p. This latter condition is called the **Sommerfeld radiation condition** and it provides a boundary condition at infinity.

Finally, invoking the Sommerfeld radiation condition, Eq. (8.25) leads to

$$\iint_{S_o} \left[\Phi \frac{\partial G}{\partial n} - G \frac{\partial \Phi}{\partial n} \right] dS_o + \alpha \Phi(\mathbf{r}') = 0,$$

or

$$\iint_{S_o} \left[G \frac{\partial \Phi}{\partial n} - \Phi \frac{\partial G}{\partial n} \right] dS_o = \alpha \Phi(\mathbf{r}'), \tag{8.29}$$

identical to the interior result, Eq. (8.11). In terms of pressure and normal velocity,

$$\alpha p(\mathbf{r}') = \iint_{S_o} \left(i\rho_o ckG(\mathbf{r}|\mathbf{r}')v_n(\mathbf{r}) - p(\mathbf{r})\frac{\partial}{\partial n}G(\mathbf{r}|\mathbf{r}') \right) dS_o, \tag{8.30}$$

with

$$\alpha = \begin{cases} 1 & \text{if } \mathbf{r}' \text{ is outside } S_o \\ 1/2 & \text{if } \mathbf{r}' \text{ is on } S_o \\ 0 & \text{if } \mathbf{r}' \text{ is inside } S_o. \end{cases}$$

Comparison to the result for the interior HIE, shown in Eq. (8.16), we see that the results for the exterior and interior cases are identical. However, it is important to note the direction of the normals to the surface S_o as shown in Fig. 8.6 and Fig. 8.2. In both cases the normal points towards the actual sources. Also in both cases when the evaluation point is on the opposite side of surface S_o from the actual sources, then the HIE provides the radiated pressure at the evaluation point. On the other hand, when the actual sources and the evaluation point are on the same side of the boundary S_o then the HIE provides a null field at the evaluation point.

8.5   Helmholtz Integral Equation for Scattering Problems

The exterior domain problem can easily be modified to account for scattering from a body located within or coincident with the surface S_o. The incident field is provided by a point source located at $\mathbf{r} = \mathbf{r_p}$, which will simulate a plane wave as the point source moves off towards infinity. Figure 8.7 shows the volume (shaded) for the application of Green's theorem. It is the same as Fig. 8.6 except for the addition of the point source located at $\mathbf{r} = \mathbf{r_p}$. For the application of Green's theorem, we need to surround this point by a small sphere, as we did at $\mathbf{r} = \mathbf{r}'$. Thus the two following equations are a statement of our problem:

$$\nabla^2 \Phi(\mathbf{r}) + k^2 \Phi(\mathbf{r}) = -\Phi_0 \delta(\mathbf{r} - \mathbf{r_p}), \tag{8.31}$$

$$\nabla^2 G(\mathbf{r}|\mathbf{r}') + k^2 G(\mathbf{r}|\mathbf{r}') = -\delta(\mathbf{r} - \mathbf{r}'), \tag{8.32}$$

where Eq. (8.32) is the same as before and Eq. (8.31) is no longer equal to zero, rather it is now equal to a point source at $\mathbf{r} = \mathbf{r_p}$ with source strength Φ_0. We apply Green's theorem in the form of Eq. (8.3) to the surfaces shown in Fig. 8.7 with $\Psi \equiv G$ as before, which leads to an equation almost identical to Eq. (8.25) (we reversed the terms in the square brackets):

$$\left(\underbrace{\iint_{S_o}}_{-\alpha\Phi(\mathbf{r}')} + \underbrace{\iint_{S_i}}_{} + \underbrace{\iint_{S_\infty}}_{0} + \underbrace{\iint_{S_p}}_{p_i} \right) \left[G \frac{\partial \Phi}{\partial n} - \Phi \frac{\partial G}{\partial n} \right] dS = 0. \tag{8.33}$$

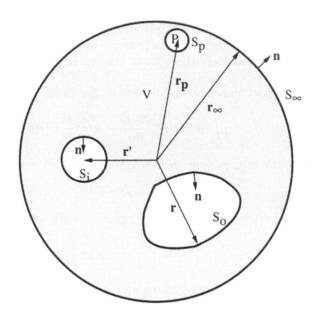

Figure 8.7: Region definitions for application of Green's theorem for a scattering formulation. Identical to Fig. 8.6 except that a point source is added at point P. The shaded region still defines the volume V for Green's theorem.

As the underbraces indicate, the second integral is $-\alpha\Phi(\mathbf{r}')$ as was determined for the exterior HIE, and the third integral is zero if the Sommerfeld radiation condition, Eq. (8.28) is satisfied.

In order to evaluate $\iint_{S_p}$ (radius of sphere S_p is ϵ), we note that as we approach the point source at P the total pressure field is not affected by the boundaries, being dominated by the nearly infinite field from the source. Thus in the nearfield of the source the pressure field behaves as though this source was in a free field and the solution to Eq. (8.31) (as $\epsilon \to 0$) is just

$$\lim_{\epsilon \to 0} \Phi(\mathbf{r}) = \Phi_0 \frac{e^{ikR}}{4\pi R},$$

where $R = |\mathbf{r} - \mathbf{r_p}|$. Given that the contribution of $\iint_{S_p}$ as $\epsilon \to 0$ will lead to the same results given in Eq. (8.8) and Eq. (8.9), except that the non-zero contribution to the integral arises from $\frac{\partial \Phi}{\partial n}$ instead of $\frac{\partial G}{\partial n}$ (note that $G(\mathbf{r}|\mathbf{r}')$ is finite and continuous at $\mathbf{r} = \mathbf{r_p}$), then

$$\lim_{\epsilon \to 0} \iint_{S_p} \left(G(\mathbf{r}|\mathbf{r}') \frac{\partial \Phi}{\partial n} - \Phi \frac{\partial}{\partial n} G(\mathbf{r}|\mathbf{r}') \right) dS_p = G(\mathbf{r_p}|\mathbf{r}') \lim_{\epsilon \to 0} \iint_{S_p} \frac{\partial \Phi}{\partial n} \epsilon^2 d\Omega_p$$

$$= \Phi_0 G(\mathbf{r_p}|\mathbf{r}'). \qquad (8.34)$$

The quantity on the right hand side of Eq. (8.34) we call the incident field p_i

$$p_i \equiv \Phi_0 G(\mathbf{r_p}|\mathbf{r}') \equiv \Phi_0 \frac{e^{ik|\mathbf{r}' - \mathbf{r_p}|}}{4\pi|\mathbf{r}' - \mathbf{r_p}|}, \qquad (8.35)$$

which we notice satisfies our definition of the incident field for a scattering problem, that is, the incident field is unaffected by the presence of the scattering body. The field given by p_i is a pure monopole everywhere in the volume, not just in the vicinity of the source center at $\mathbf{r} = \mathbf{r_p}$.

Finally, given the results shown by the underbars, Eq. (8.33) can be written as

$$\alpha \Phi(\mathbf{r}') = p_i(\mathbf{r}') + \iint_{S_o} \left(G(\mathbf{r}|\mathbf{r}') \frac{\partial \Phi(\mathbf{r})}{\partial n} - \Phi(\mathbf{r}) \frac{\partial}{\partial n} G(\mathbf{r}|\mathbf{r}') \right) dS_o. \tag{8.36}$$

Since $\Phi(\mathbf{r})$ represents the total pressure field which we call $p(\mathbf{r}')$, $p(\mathbf{r}') \equiv \Phi$, and

$$p(\mathbf{r}') = p_i(\mathbf{r}') + p_s(\mathbf{r}'),$$

where $p_s(\mathbf{r}')$ is the scattered pressure. Thus we can rewrite Eq. (8.36) as

$$\alpha p(\mathbf{r}') = p_i(\mathbf{r}') + \iint_{S_o} \left(G(\mathbf{r}|\mathbf{r}') \frac{\partial p(\mathbf{r})}{\partial n} - p(\mathbf{r}) \frac{\partial}{\partial n} G(\mathbf{r}|\mathbf{r}') \right) dS_o. \tag{8.37}$$

In Eq. (8.37) n is the inward normal (pointing away from the fluid) and

$$\alpha = \begin{cases} 1 & \text{if } \mathbf{r}' \text{ is outside } S_o \\ 1/2 & \text{if } \mathbf{r}' \text{ is on } S_o \\ 0 & \text{if } \mathbf{r}' \text{ is inside } S_o \end{cases},$$

A popular form of this equation is provided by reversing the direction of the normal, and defining v_n as the velocity on S_o with positive displacement into the fluid;

$$\alpha p(\mathbf{r}') = p_i(\mathbf{r}') + \iint_{S_o} \left(p(\mathbf{r}) \frac{\partial}{\partial n} G(\mathbf{r}|\mathbf{r}') - i\rho_o ck G(\mathbf{r}|\mathbf{r}') v_n \right) dS_o, \tag{8.38}$$

where n is the normal pointing into the fluid, opposite that shown in Fig. 8.7.

In closing we reiterate that the active and/or scattering surface is contained within S_o, on the opposite side of S_o from the source of the incident field.

8.6 Green Functions and the Inhomogeneous Wave Equation

Throughout this book we have dealt with sources in an indirect way, by studying the pressure and velocity generated by the source on a given boundary. For example, the Helmholtz integral equation replaces the actual sources with a pressure and normal velocity field on the HIE boundary surface, as we have seen in the material above. We now show how one may include sources directly using an inhomogeneous Helmholtz equation. Solution of the inhomogeneous equation provides an important key towards the solution of more complex problems.

The inhomogeneous Helmholtz equation in three dimensions with a given source distribution $Q(\mathbf{r}, \omega)$ which is assumed to be confined in space to a volume V, is given by

$$(\nabla^2 + k^2) p(\mathbf{r}, \omega) = -Q(\mathbf{r}, \omega), \tag{8.39}$$

where $\mathbf{r}$ is the radius vector measured from the origin. We can easily construct the solution of Eq. (8.39). First, consider the solution of

$$(\nabla^2 + k^2)G(\mathbf{r}|\mathbf{r}') = -\delta(\mathbf{r} - \mathbf{r}'), \tag{8.40}$$

where $\mathbf{r}'$ is a second vector measured from the same origin as $\mathbf{r}$. The solution was presented in Section 6.5.1, page 198: a monopole located at $\mathbf{r}'$,

$$G(\mathbf{r}|\mathbf{r}') = \frac{e^{ik|\mathbf{r}-\mathbf{r}'|}}{4\pi|\mathbf{r} - \mathbf{r}'|}. \tag{8.41}$$

$G(\mathbf{r}|\mathbf{r}')$ is called the **three-dimensional, free space Green function**. Second, we multiply both sides of Eq. (8.40) by $Q(\mathbf{r}', \omega)$ and integrate over the volume V', which is identical to V except that it is associated with $\mathbf{r}'$. Since ∇^2 does not depend on $\mathbf{r}'$, the result is

$$(\nabla^2 + k^2)\int_{V'} Q(\mathbf{r}', \omega)G(\mathbf{r}|\mathbf{r}') \, dV' = -\int_{V'} \delta(\mathbf{r} - \mathbf{r}')Q(\mathbf{r}', \omega) \, dV' = -Q(\mathbf{r}, \omega), \tag{8.42}$$

where the second equality is obtained using the sifting property of the delta function, Eq. 1.37. Comparison of Eq. (8.42) to Eq. (8.39) reveals that

$$p(\mathbf{r}, \omega) = \int_{V'} Q(\mathbf{r}', \omega)G(\mathbf{r}|\mathbf{r}') \, dV'. \tag{8.43}$$

Equation (8.43) is a very important result basic to the theory of Green functions. It indicates that once the Green function is known (the solution of Eq. (8.40)), then solutions to the general inhomogeneous wave equation, Eq. (8.39), are easily obtained by integration over the Green function.

8.6.1 Two-dimensional Free Space Green Function

One can derive the free space Green function for two dimensions by considering a special case of Eq. (8.39) where $Q(\mathbf{r}, \omega)$ is an infinite line source in the z direction located at $x = x''$, $y = y''$:

$$Q(\mathbf{r}, \omega) = \delta(x - x'')\delta(y - y'').$$

With this source distribution the solution of Eq. (8.39) is the two-dimensional, free space Green function. By the foregoing analysis the two-dimensional Green function G_{2D} is given by Eq. (8.43):

$$G_{2D}(x, y|x'', y'') = \int_{V'} \delta(x' - x'')\delta(y' - y'')G(\mathbf{r}|\mathbf{r}') \, dx' dy' dz'. \tag{8.44}$$

Letting $\zeta = (z - z')$, using Eq. (8.41) and integrating over $dx' dy'$, this integral becomes

$$G_{2D}(x, y|x'', y'') = \int_{-\infty}^{\infty} \frac{e^{ik\sqrt{(x-x'')^2+(y-y'')^2+\zeta^2}}}{4\pi\sqrt{(x - x'')^2 + (y - y'')^2 + \zeta^2}} d\zeta. \tag{8.45}$$

To solve this integral we need one of the integral definitions of the Hankel function:[3]

$$H_0^{(1)}(k\xi) = -\frac{i}{\pi} \int_{-\infty}^{\infty} \frac{e^{ik\sqrt{\xi^2+t^2}}}{\sqrt{\xi^2+t^2}} dt. \tag{8.46}$$

Comparison with Eq. (8.45) reveals that the **two-dimensional free space Green function** is

$$G_{2D}(x,y|x'',y'') = \frac{i}{4} H_0^{(1)}(k\sqrt{(x-x'')^2+(y-y'')^2}). \tag{8.47}$$

The two-dimensional Green function occurs in simplified three-dimensional acoustic problems, when it is assumed that the field is constant in one of the coordinate directions. Equation (8.46) provides an important link between three-dimensional and two-dimensional problems. For example, it can be used to derive Rayleigh's integral for one-dimensional vibrators as we will now show.

Assume that the surface velocity in the plane $z = z'$ is independent of y';

$$\dot{w}(x',y',z') = \dot{w}(x',z').$$

Rayleigh's integral, Eq. (2.75) on page 36, becomes

$$p(x,y,z) = \frac{-i\rho_0 ck}{2\pi} \int_{-\infty}^{\infty} dx' \dot{w}(x',z') \int_{-\infty}^{\infty} \frac{e^{ikR}}{R} dy', \tag{8.48}$$

where $R \equiv |\vec{r}-\vec{r'}| = \sqrt{(x-x')^2+(y-y')^2+(z-z')^2}$. From Eq. (8.46), with $t = y'-y$,

$$\int_{-\infty}^{\infty} \frac{e^{ikR}}{R} dy' = i\pi H_0^{(1)}(k\sqrt{(x-x')^2+(z-z')^2}) \tag{8.49}$$

which we note is independent of y. Using this result in Eq. (8.48) the one-dimensional form of Rayleigh's integral is obtained:

$$p(x,z) = \frac{\rho_0 ck}{2} \int_{-\infty}^{\infty} \dot{w}(x',z') H_0^{(1)}(k\sqrt{(x-x')^2+(z-z')^2}) dx'. \tag{8.50}$$

This equation is the starting point for investigations of how sound radiates in two dimensions from baffled one-dimensional radiators.

8.6.2   Conversion from Three Dimensions to Two Dimensions

A general result grows out of Eq. (8.49) for the conversion of three-dimensional problems to two dimensions. Dividing both sides by 4π we see that the integration of the free space Green function over one of the dimensions yields $iH_0/4$, that is,

$$\int_{-\infty}^{\infty} G(\mathbf{r}|\mathbf{r'})dy' = \frac{i}{4} H_0(k\sqrt{(x-x')^2+(z-z')^2}). \tag{8.51}$$

---

[3]I. S. Gradshteyn and I. M. Ryzhik (1965). *Tables of integrals, series and products*, Academic Press, New York and London, p. 957.

Note that the primed and unprimed variables can be interchanged in this equation without altering its form. Armed with this equation, any of the HIE formulations presented in this chapter for three-dimensional bodies which are infinite and have a constant cross-section, can be converted to two-dimensional forms, if we assume the surface pressure and velocity fields are independent of y (the infinite direction). This conversion is carried out by replacing G with $\frac{i}{4}H_0$ and removing the integral over y. (This fact was pointed out in a famous paper by Waterman.[4]) For example, attacking Eq. (8.38) in this way yields by inspection:

$$\alpha p(x', z') = p_i(x', z') + \frac{i}{4}\int_{C_o}\left(p(x, z)\frac{\partial}{\partial n}H_0(k\rho) - i\rho_o ck H_0(k\rho)v_n(x, z)\right)dC_o, \quad (8.52)$$

where $\rho \equiv \sqrt{(x - x')^2 + (z - z')^2}$, C_o is the closed line formed by the intersection of the original three-dimensional surface S_o with the $y = 0$ plane, and n is the outward normal to this surface in this plane. Of course, C_o is independent of y.

One final expression is useful in two-dimensional problems. Let $\mathbf{r}$ and $\mathbf{r}'$ be vectors in two dimensions (independent of y and y'), with magnitude r and r' and polar angles given by ϕ and ϕ', respectively. Then we have the following addition theorem:

$$H_0^{(1)}(k|\mathbf{r} - \mathbf{r}'|) = \sum_{n=-\infty}^{\infty} e^{in(\phi-\phi')}J_n(kr_<)H_n^{(1)}(kr_>), \quad (8.53)$$

where $r_<$ represents the lesser of r and r' and $r_>$ the greater.

8.7 Simple Source Formulation

One drawback of the Helmholtz integral equation is that both the surface pressure and surface velocity must be known. We will find, later in this chapter, that this difficulty can be avoided by use of a different Green function in place of the free space Green function in the integrand. Another technique which avoids this is called the simple source formulation.[5] Consider the domains shown in Fig. 8.8. Let $\mu(\mathbf{r})$ be an unknown source distribution on the boundary S_o. The volume outside the surface is V_o and the volume enclosed by the surface S_o is V_i. The normal to the surface points into the volume, V_i. We would like to investigate the possibility that the pressure inside or outside of the surface could be determined by the following equation,

$$p(\mathbf{r}') = \iint_{S_o} \mu(\mathbf{r})G(\mathbf{r}|\mathbf{r}')\,dS_o, \quad (8.54)$$

similar to Eq. (8.29) but with only one term appearing in the integrand. G is the free space Green function. The physical meaning of μ will become clear later in this section. Equation (8.54) states, if we consider the exterior problem, that the acoustic field in V_o

[4]P. C. Waterman (1969), "New formulation of acoustic scattering", J. Acoust. Soc. Am., **45**, pp. 1417–1429.

[5]Lawrence G. Copley (1968), "Fundamental Results Concerning Integral Representations in Acoustic Radiation", J. Acoust. Soc. Am. **44**, pp. 28–32.

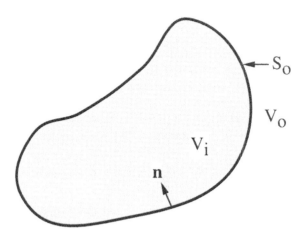

Figure 8.8: Region definitions for the simple source formulation. V_o is the exterior volume, S_o is the bounding surface and V_i is the volume interior to this surface. The normal points inward.

generated by events inside the surface S_o, can be computed uniquely by replacing these events with a distribution of simple monopole surface sources $\mu(\mathbf{r})G(\mathbf{r}|\mathbf{r}')$ and summing up their contributions over S_o.

To prove the existence of such a formula, we start with the exterior HIE, Eq. (8.29), and let $p_o(\mathbf{r}')$ be the pressure in the exterior volume V_o. Restating Eq. (8.29) we have

$$
\left.
\begin{array}{ll}
p_o(\mathbf{r}') & (\mathbf{r}' \in V_o) \\
p_o(\mathbf{r}')/2 & (\mathbf{r}' \in S_o) \\
0 & (\mathbf{r}' \in V_i)
\end{array}
\right\}
= \iint_{S_o} \left[G \frac{\partial p_o(\mathbf{r})}{\partial n} - p_o(\mathbf{r}) \frac{\partial G}{\partial n} \right] dS_o
\qquad (8.55)
$$

where the conditions on the location of $\mathbf{r}'$ are indicated in each line. The sources for Eq. (8.55) are located in the interior, and no sources are outside. Also we have assumed the external field p_o satisfies the Sommerfeld radiation condition, Eq. (8.28).

Next consider an entirely separate problem dealing with the interior HIE. Assume that the sources are located outside the surface S_o of Fig. 8.8 and produce the internal field $p_i(\mathbf{r}')$. There are no sources in V_i. Restating Eq. (8.15), but with the normal pointing in the opposite direction (which accounts for the negative signs on the left hand side),

$$
\left.
\begin{array}{ll}
0 & (\mathbf{r}' \in V_o) \\
-p_i(\mathbf{r}')/2 & (\mathbf{r}' \in S_o) \\
-p_i(\mathbf{r}') & (\mathbf{r}' \in V_i)
\end{array}
\right\}
= \iint_{S_o} \left[G \frac{\partial p_i(\mathbf{r})}{\partial n} - p_i(\mathbf{r}) \frac{\partial G}{\partial n} \right] dS_o.
\qquad (8.56)
$$

Now require that the separate problems which are solved by Eq. (8.55) and Eq. (8.56) are linked by the condition $p_o = p_i$ on the boundary surface S_o. Also we assume that on S_o, $\frac{\partial p_o}{\partial n} \neq \frac{\partial p_i}{\partial n}$. If we subtract Eq. (8.56) from Eq. (8.55) we get

$$\left.\begin{array}{ll} p_o(\mathbf{r}') & (\mathbf{r}' \in V_o) \\ p_o = p_i & (\mathbf{r}' \in S_o) \\ p_i(\mathbf{r}') & (\mathbf{r}' \in V_i) \end{array}\right\} = \iint_{S_o} \left[\frac{\partial p_o(\mathbf{r})}{\partial n} - \frac{\partial p_i(\mathbf{r})}{\partial n}\right] G \, dS_o. \tag{8.57}$$

Let the difference in normal derivatives be defined as μ,

$$\mu(\mathbf{r}) \equiv \frac{\partial p_o(\mathbf{r})}{\partial n} - \frac{\partial p_i(\mathbf{r})}{\partial n}, \tag{8.58}$$

then Eq. (8.57) becomes

$$p(\mathbf{r}') = \iint_{S_o} \mu(\mathbf{r}) G(\mathbf{r}|\mathbf{r}') dS_o, \tag{8.59}$$

where $p(\mathbf{r}')$ is the pressure anywhere in space. This is the equation we are looking for, identical to Eq. (8.54). The source distribution μ is recognized as the difference in normal derivatives of the surface pressure between the specified exterior and interior problems.

In the application of Eq. (8.59) either an external or an internal problem will be of interest, not both. Usually Eq. (8.59) is used to determine the boundary sources μ given $p(\mathbf{r}')$. If we want to apply Eq. (8.59) to an external radiation problem, then the actual sources, which must be located inside, are replaced by a distribution of fictitious surface sources μ. This distribution of monopoles, provided by Eq. (8.59), produces exactly the same field in V_o as that calculated using the HIE given by Eq. (8.55). Clearly, Eq. (8.59) provides a dramatic simplification in calculations compared to Eq. (8.55). However, when the integral in Eq. (8.59) is evaluated for $\mathbf{r}' \in V_i$ we do not obtain zero as Eq. (8.55) yields. Furthermore, there is no discontinuity in the pressure field at the surface.

In applications using Eq. (8.59) use of Euler's equation provides the velocity vector at $\mathbf{r}'$. To compute the normal velocity on the surface S_o, however, we need to be a bit careful. We return to the dual problems specified to obtain Eq. (8.57) and recognize the fact that the velocities across the boundary are discontinuous (since $\frac{\partial p_o}{\partial n} \neq \frac{\partial p_i}{\partial n}$). Equation (8.57) provides, as $\mathbf{r}'$ approaches the boundary from the outside and inside, the normal derivative of the pressure at the boundary:

$$\frac{\partial p_o}{\partial n'} \; (\mathbf{r}' \in V_o^-) \; = \iint_{S_o} \mu(\mathbf{r}) \frac{\partial G(\mathbf{r}|\mathbf{r}')}{\partial n'} dS_o \tag{8.60}$$

and

$$\frac{\partial p_i}{\partial n'} \; (\mathbf{r}' \in V_i^-) \; = \iint_{S_o} \mu(\mathbf{r}) \frac{\partial G(\mathbf{r}|\mathbf{r}')}{\partial n'} dS_o, \tag{8.61}$$

respectively. These two equations indicate that two completely different answers are obtained for the double integral depending upon which side of the boundary we approach from. The value of the double integral for $\mathbf{r}' \in S_o$ is undefined, the integral experiencing a jump discontinuity there. We resolve this problem in the usual fashion when dealing with functional discontinuities. For example, the definition of the rectangle function given in Eq. (1.41) on page 7 is defined as 1/2 at the discontinuities; the average of the function on either side. We define the value of the double integral of Eq. (8.60) and Eq. (8.61) at the discontinuity in exactly the same way:

$$\iint_{S_o} \mu(\mathbf{r}) \frac{\partial G(\mathbf{r}|\mathbf{r}')}{\partial n'} dS_o \equiv \frac{1}{2}\left(\frac{\partial p_o}{\partial n'} + \frac{\partial p_i}{\partial n'}\right) \quad (\mathbf{r}' \in S_o). \tag{8.62}$$

But since

$$\mu(\mathbf{r}') = \left(\frac{\partial p_o}{\partial n'} - \frac{\partial p_i}{\partial n'}\right),$$

then

$$\frac{1}{2}\left(\frac{\partial p_o}{\partial n'} + \frac{\partial p_i}{\partial n'}\right) = \frac{-\mu(\mathbf{r}')}{2} + \frac{\partial p_o}{\partial n'}$$

and we can rewrite Eq. (8.62) as

$$\frac{\partial p_o(\mathbf{r}')}{\partial n'} = \frac{\mu(\mathbf{r}')}{2} + \iint_{S_o} \mu(\mathbf{r})\frac{\partial G(\mathbf{r}|\mathbf{r}')}{\partial n'} dS_o \quad (\mathbf{r}' \in S_o). \tag{8.63}$$

Similarly, for solution to the interior problem, the normal derivative of the pressure on the surface is

$$\frac{\partial p_i(\mathbf{r}')}{\partial n'} = \frac{-\mu(\mathbf{r}')}{2} + \iint_{S_o} \mu(\mathbf{r})\frac{\partial G(\mathbf{r}|\mathbf{r}')}{\partial n'} dS_o \quad (\mathbf{r}' \in S_o). \tag{8.64}$$

8.7.1   Example

To illustrate the simple source formulation, Eq. (8.59), we let S_o be a spherical surface of radius a and assume that the continuous pressure field on the surface is given by a particular spherical harmonic,

$$p_i = p_o = f(r)\Big|_{r=a} Y_n^m(\theta, \phi)^*$$

where $f(r)$ is to be determined for both $r < a$ and $r \geq a$. Consider the exterior problem first. From Eq. (6.92) on page 206 we know that $f(r) = C_{mn}(\omega)h_n(kr)$ is a solution to the Helmholtz equation outside of the sphere. We make the following fortuitous choice for the constant, $C_{mn}(\omega) = j_n(ka)$, so that in the exterior region

$$p_o(r, \theta, \phi) = j_n(ka)h_n(kr)Y_n^m(\theta, \phi)^*. \tag{8.65}$$

Now we ask for the solution to the interior problem with the same surface pressure field. Since the radial dependence must now be $j_n(kr)$, we can see that the proper choice for the interior solution is

$$p_i(r, \theta, \phi) = j_n(kr)h_n(ka)Y_n^m(\theta, \phi)^*. \tag{8.66}$$

In general then $f(r) = j_n(kr_<)h_n(kr_>)$. With $\frac{\partial}{\partial n} = -\frac{\partial}{\partial r}$ the surface source density is

$$\mu(\mathbf{r}) = \left(-\frac{\partial p_o}{\partial r} + \frac{\partial p_i}{\partial r}\right)\Big|_{r=a} = \lim_{r\to a} k[j_n'(kr)h_n(ka) - h_n'(kr)j_n(ka)]Y_n^m(\theta, \phi)^*. \tag{8.67}$$

Using the Wronskian relationship Eq. (6.67) yields the surface source distribution,

$$\mu(\mathbf{r}) = \frac{-i}{ka^2} Y_n^m(\theta, \phi)^*, \tag{8.68}$$

which, when inserted into Eq. (8.59), provides

$$p(\mathbf{r}') = \iint_{S_o} \frac{-i}{ka^2} Y_n^m(\theta, \phi)^* G(\mathbf{r}|\mathbf{r}') dS_o. \tag{8.69}$$

Finally to verify that Eq. (8.69) yields Eq. (8.65) or Eq. (8.66) we need to expand G in terms of spherical harmonics, given by Eq. (8.22):

$$p(\mathbf{r}') = \iint_{S_o} \frac{-i}{ka^2} Y_n^m(\theta, \phi)^* ik \sum_{\nu=0}^{\infty} j_\nu(kr_<) h_\nu(kr_>) \sum_{\mu=-\nu}^{\nu} Y_\nu^\mu(\theta', \phi')^* Y_\nu^\mu(\theta, \phi) dS_o. \tag{8.70}$$

Due to the orthonormality of the spherical harmonics

$$\iint_{S_o} Y_n^m(\theta, \phi)^* Y_\nu^\mu(\theta, \phi) dS_o = a^2 \delta_{n\nu} \delta_{m\mu}$$

and Eq. (8.70) reduces to

$$p(\mathbf{r}') = j_n(kr_<) h_n(kr_>) Y_n^m(\theta', \phi'). \tag{8.71}$$

For $\mathbf{r}' \in V_o$, $r_< = a$ and $r_> = r'$ so that Eq. (8.65) is obtained. For $\mathbf{r}' \in V_i$, $r_< = r'$ and $r_> = a$, Eq. (8.66) is obtained. Furthermore the pressure is continuous across the boundary.

This example sheds light on the evaluation of

$$\iint_{S_o} \mu(\mathbf{r}) \frac{\partial G(\mathbf{r}|\mathbf{r}')}{\partial n'} dS_o \tag{8.72}$$

when $\mathbf{r}' \in S_o$. Approaching from the outside Eq. (8.60) leads to

$$\lim_{r' \to a^+} \iint_{S_o} \mu(\mathbf{r}) \frac{\partial G(\mathbf{r}|\mathbf{r}')}{\partial n'} dS_o = -kj_n(ka) h_n'(ka) Y_n^m(\theta', \phi')$$

and from the inside Eq. (8.61) yields

$$\lim_{r' \to a^-} \iint_{S_o} \mu(\mathbf{r}) \frac{\partial G(\mathbf{r}|\mathbf{r}')}{\partial n'} dS_o = -kj_n'(ka) h_n(ka) Y_n^m(\theta', \phi'),$$

which reveals the discontinuity. Clearly the integral, Eq. (8.72), does not have a unique value when $\mathbf{r}' \in S_o$. Thus, we must resort to our own definition at the discontinuity, as provided by Eq. (8.62).

8.8   The Dirichlet and Neumann Green Functions

The Helmholtz integral equations, Eq. (8.16), Eq. (8.30) and Eq. (8.37), require a knowledge of *both* the pressure and velocity on the surface, in order to compute the field inside the surface (or outside the surface for the exterior formulation). From our study of spheres and cylinders, we have learned that one needs either the pressure or the velocity to compute the field inside the sphere or cylinder, but not both. Unfortunately, the HIE overspecifies what is needed to solve for the interior/exterior field, requiring both the pressure and velocity, even though only one of these quantities specified on the surface is sufficient. We can escape from this overspecification of the boundary conditions by constructing special forms of the Green function which enters the HIE, as we will explain in the following analysis.

In Eq. (8.16) or Eq. (8.30) the free field Green function appears as part of the Helmholtz integral equation. It was the particular solution of Eq. (8.4) for an unbounded space, and no boundary conditions were specified, other than the Sommerfeld radiation condition. We now consider a non-zero homogeneous solution Ψ_h of Eq. (8.4). Added to the free space Green function G the sum satisfies the prescribed boundary condition. Ψ_h generally will be a function of both $\mathbf{r}$ and $\mathbf{r}'$. We define

$$g_N(\mathbf{r}|\mathbf{r}') \equiv \Psi_h,$$

and, since g_N satisfies Eq. (8.3), then

$$G_N \equiv G + g_N$$

is still a solution to Eq. (8.16):

$$\alpha p(\mathbf{r}') = \iint_{S_o} \left(i\rho_o ck G_N(\mathbf{r}|\mathbf{r}') v_n(\mathbf{r}) - p(\mathbf{r}) \frac{\partial}{\partial n} G_N(\mathbf{r}|\mathbf{r}') \right) dS_o. \tag{8.73}$$

The prescribed boundary condition for the Neumann problem is

$$\frac{\partial G_N}{\partial n} = 0$$

everywhere on the boundary, S_o. This choice eliminates one of the terms in Eq. (8.73) so that

$$\alpha p(\mathbf{r}') = i\rho_o ck \iint_{S_o} G_N(\mathbf{r}|\mathbf{r}') v_n(\mathbf{r}) dS_o. \tag{8.74}$$

This boundary condition applies only to G_N, not to $p(\mathbf{r})$. Determining the function g_N may be very difficult to do in practice, especially if the surface does not lie on the contours of one of the separable coordinate systems. G_N is called the Neumann Green function reflecting the fact that its normal derivative vanishes on the boundary.

We can choose a second Green function which we call the Dirichlet Green function, $G_D \equiv G + g_D$, which satisfies the Dirichlet boundary condition on S_o,

$$G + g_D = 0.$$

$g_D(\mathbf{r}|\mathbf{r}')$ is still a homogeneous solution of Eq. (8.4) within and on the boundary. In this case Eq. (8.16) becomes

$$\alpha p(\mathbf{r}') = -\iint_{S_o} p(\mathbf{r}) \frac{\partial}{\partial n} G_D(\mathbf{r}|\mathbf{r}') dS_o. \tag{8.75}$$

Thus with the Dirichlet Green function known, only the pressure needs to be specified on the HIE surface in order to determine the pressure in the interior. The problem of overspecification of the boundary fields is avoided. We now provide an example of determining the Neumann Green function for the interior problem for a sphere.

8.8.1 The Interior Neumann Green Function for the Sphere

In order to determine the function g_N which satisfies the homogeneous wave equation inside the sphere and together with G satisfies Neumann boundary conditions on the surface (vanishing of the normal derivative), it is necessary to expand G into spherical harmonics with origin at the center of the sphere. This powerful formula, valid for all $\mathbf{r}$ and $\mathbf{r}'$, was given on page 259 as Eq. (8.22), which we repeat here:

$$G \equiv \frac{e^{ik|\mathbf{r}-\mathbf{r}'|}}{4\pi|\mathbf{r}-\mathbf{r}'|} = ik \sum_{n=0}^{\infty} j_n(kr_<) h_n(kr_>) \sum_{m=-n}^{n} Y_n^m(\theta',\phi')^* Y_n^m(\theta,\phi), \tag{8.76}$$

where $\mathbf{r} = (r,\theta,\phi)$ and $\mathbf{r}' = (r',\theta',\phi')$, $r_<$ is the lesser of (r,r') and $r_>$ the greater, and both r and r' are measured from the center of the sphere. This equation is a generalization of Eq. (8.18).

We are looking for a function g_N which satisfies Neumann boundary conditions at $r = a$, that is,

$$\frac{\partial G_N}{\partial n}\Big|_{r=a} = \frac{\partial G_N}{\partial r}\Big|_{r=a} = \left(\frac{\partial G}{\partial r} + \frac{\partial g_N}{\partial r}\right)\Big|_{r=a} = 0.$$

Differentiating Eq. (8.76) yields

$$\frac{\partial G}{\partial r} = ik^2 \sum_{n=0}^{\infty} \left\{ \begin{array}{c} j_n(kr') h_n'(kr) \\ j_n'(kr) h_n(kr') \end{array} \right\} \sum_{m=-n}^{n} Y_n^m(\theta',\phi')^* Y_n^m(\theta,\phi),$$

where the top row is taken for $r' < r$, the case of interest here.

Consider the following possibility for g_N:

$$g_N(\mathbf{r}|\mathbf{r}') = -ik \sum_{n=0}^{\infty} j_n(kr') \frac{h_n'(ka)}{j_n'(ka)} j_n(kr) \sum_{m=-n}^{n} Y_n^m(\theta',\phi')^* Y_n^m(\theta,\phi). \tag{8.77}$$

It satisfies the requirement that $\nabla^2 g_N + k^2 g_N = 0$ *within* and on S_o since both $j_n(kr')$ and $j_n(kr)$ are finite there. When $r = a$ and $r' < a$, clearly

$$\left(\frac{\partial G}{\partial r} + \frac{\partial g_N}{\partial r}\right)\Big|_{r=a} = 0,$$

the Bessel functions canceling for each n. Thus g_N is the desired function and the Neumann Green function is

$$G_N(\mathbf{r}|\mathbf{r}') = G + g_N = ik \sum_{n=0}^{\infty} \left[j_n(kr_<)h_n(kr_>) - j_n(kr')\frac{h'_n(ka)}{j'_n(ka)}j_n(kr) \right]$$

$$\times \sum_{m=-n}^{n} Y_n^m(\theta',\phi')^* Y_n^m(\theta,\phi). \tag{8.78}$$

Note that $r' \le a$ and $r \le a$ and Eq. (8.78) provides the general result for r' and r, where they are not restricted to lie on the boundary.

Returning to Eq. (8.74), the simplified HIE, the integrand requires $\mathbf{r}$ on S_o. When $r = a$ ($r' \le a$), Eq. (8.78) is simplified, using $h_n = j_n + iy_n$ and the Wronskian relation, Eq. (6.66); the terms in square brackets become

$$j_n(kr')[h_n(ka) - j_n(ka)\frac{h'_n(ka)}{j'_n(ka)}] = \frac{-i}{(ka)^2}\frac{j_n(kr')}{j'_n(ka)} ,$$

so that Eq. (8.78) simplifies to

$$G_N(\mathbf{r}|\mathbf{r}')\Big|_{r=a} = \frac{1}{ka^2} \sum_{n=0}^{\infty} \frac{j_n(kr')}{j'_n(ka)} \sum_{m=-n}^{n} Y_n^m(\theta',\phi')^* Y_n^m(\theta,\phi). \tag{8.79}$$

This is the Neumann Green function for the integral equation, Eq. (8.74), which determines the pressure inside or on the surface of a sphere from a knowledge of the normal velocity there.

Equation (8.79) is identical to Eq. (6.151) on page 219. In the latter case the Neumann Green function was derived using the Fourier acoustics approach. Since Eq. (6.151) was valid for $\mathbf{r}'$ on S_o, we conclude that $\alpha = 1$ (not 1/2) in this case.

8.8.2 Equivalence to Scattering from a Point Source

Once the Neumann Green function has been found, we have solved an equivalent scattering problem; in the case of Eq. (8.78), the scattering of a point source within a rigid spherical cavity. To see this consider the mathematical statement of the scattering problem: the incident field, which is defined in the absence of the scatterer (see page 224), is a point source located at (r',θ',ϕ') given by

$$p_i(r,\theta,\phi) = \frac{e^{ikR}}{4\pi R}$$

with $R = |\mathbf{r} - \mathbf{r}'|$, which can be expanded using Eq. (8.22). If this source is located in a rigid spherical cavity, then we solve for a scattered field p_s where as usual the total pressure is $p_t = p_i + p_s$, and p_s is a solution of

$$\nabla^2 p_s + k^2 p_s = 0$$

in the sphere. The rigid boundary condition on the sphere of radius a is

$$\frac{\partial p_t(r,\theta,\phi)}{\partial r}\bigg|_{r=a} = 0.$$

But this equation is identical to the requirement for the Neumann Green function in the last section:

$$\frac{\partial G_N}{\partial n}\bigg|_{r=a} = \frac{\partial G_N}{\partial r}\bigg|_{r=a} = \left(\frac{\partial G}{\partial r} + \frac{\partial g_N}{\partial r}\right)\bigg|_{r=a} = 0.$$

We recognize G as the same as p_i, and thus equate g_N with p_s. In other words, the additive function which we used to construct the Neumann Green function is just the scattered pressure field for a rigid boundary. Thus,

$$p_s(r,\theta,\phi) = -ik\sum_{n=0}^{\infty} j_n(kr')\frac{h'_n(ka)}{j'_n(ka)}j_n(kr)\sum_{m=-n}^{n} Y_n^m(\theta',\phi')^* Y_n^m(\theta,\phi), \qquad (8.80)$$

as given by Eq. (8.77) in the last section.

This correspondence between construction of the Neumann Green function and the solution of a rigid scattering problem is general for any shaped boundary, not just the sphere. In fact, the construction of the Dirichlet Green function is equivalent to solving for the scattered field due to a point source within a pressure release surface.

8.8.3 Neumann and Dirichlet Green Functions for a Plane

We can use the exterior HIE to derive Rayleigh's two formulas, Eq. (2.67) on page 35 and Eq. (2.75) on page 36. First we develop the HIE for an infinite plane boundary, and then find the Neumann and Dirichlet Green functions to reduce the HIE to Rayleigh's formulas.

Modify the surfaces given in Fig. 8.6 as shown in Fig. 8.9. As we let $r_\infty \to \infty$ the

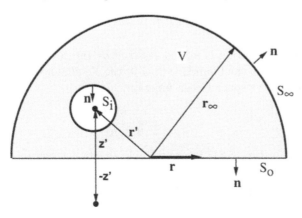

Figure 8.9: Region definitions for application of Green's theorem for the derivation of the Helmholtz integral equation for an infinite plane boundary.

base of the sphere, surface S_0 in the (x, y) plane, becomes an infinite plane and the use of Green's theorem is no different from the application in Section 8.4. Thus invoking the Sommerfeld radiation condition, we arrive at Eq. (8.30), an integral over only the infinite plane boundary. In Cartesian coordinates we have

$$G = \frac{e^{ikR}}{4\pi R}$$

and

$$R = \sqrt{(x - x')^2 + (y - y')^2 + (z - z')^2}.$$

Since $\frac{\partial}{\partial n} = -\frac{\partial}{\partial z}$ Eq. (8.30) is

$$\alpha p(\mathbf{r}') = \frac{1}{4\pi} \iint_{S_o} \left(p(x, y, 0) \frac{\partial}{\partial z} \left(\frac{e^{ikR}}{R} \right) - i\rho_o c k \frac{e^{ikR}}{R} \dot{w}(x, y, 0) \right) dS_o. \tag{8.81}$$

Again we want to find a function g_N to construct the Neumann Green function, $G_N = G + g_N$, so that

$$\frac{\partial G_N}{\partial z} = 0$$

on the surface at $z = 0$. This is carried out by taking an image source (an image of the fictitious source at $\mathbf{r}'$) imaged about the plane. This image is located at $\mathbf{r} = (x', y', -z')$ at a distance of

$$R_i = \sqrt{(x - x')^2 + (y - y')^2 + (z + z')^2}$$

from the fictitious source. This source, given by $\frac{e^{ikR_i}}{R_i}$, is finite in the upper half plane and thus is a solution to the homogeneous Helmholtz equation as required for g_N. We add this to the free space Green function:

$$G_N(x, y, z | x', y', z') = \frac{1}{4\pi} \left(\frac{e^{ikR}}{R} + \frac{e^{ikR_i}}{R_i} \right). \tag{8.82}$$

When $z = 0$, since $R = R_i$, after differentiation we find that

$$\left. \frac{\partial G_N}{\partial z} \right|_{z=0} = 0,$$

so that the boundary condition is indeed satisfied by the constructed Neumann Green function. Furthermore, on the boundary ($z = 0$) the Neumann Green function is simply twice the value of the free space Green function,

$$\left. G_N \right|_{z=0} = \frac{1}{2\pi} \frac{e^{ikR}}{R}, \tag{8.83}$$

and using G_N for G in Eq. (8.81) leads to Rayleigh's first integral,

$$\alpha p(\mathbf{r}') = \frac{-i\rho_o c k}{2\pi} \iint_{S_o} \frac{e^{ikR}}{R} \dot{w}(x, y, 0) \, dx \, dy. \tag{8.84}$$

Finally, we note that when $\mathbf{r}'$ falls on the surface of the infinite plane, the image is collocated with the evaluation point. The consequence of this is that, in the application of Green's theorem, we must account for the fact that $1/2$ of the image source now falls within the volume (its wave equation is no longer homogeneous there) and thus an extra $p(\mathbf{r}')/2$ term appears on the left hand side of Eq. (8.84). This extra term plus the original add up to just $p(\mathbf{r}')$. Thus $\alpha = 1$ when the evaluation point is on the surface and Rayleigh's integral is obtained.

When the evaluation point crosses the boundary and lies outside the Green's volume, the image lies inside the volume and we again arrive at Eq. (8.84), with $\alpha = 1$.

8.8.4 Neumann Green Function for the Exterior Problem on a Sphere

This formula is nearly identical to the interior case, Eq. (8.78), except for a slightly different combination of Bessel and Hankel functions. Note that Eq. (8.22), the expansion of the free field Green function, applies to this case as well as the interior case. The Neumann Green function for a sphere of radius a is:[6]

$$G_N(\mathbf{r}|\mathbf{r}') = G + g = ik \sum_{n=0}^{\infty} \left[j_n(kr_<)h_n(kr_>) - h_n(kr') \frac{j'_n(ka)}{h'_n(ka)} h_n(kr) \right]$$

$$\times \sum_{m=-n}^{n} Y_n^m(\theta', \phi')^* Y_n^m(\theta, \phi). \tag{8.85}$$

The first term in the square brackets represents the free space Green function G and the term subtracted from it the additive function g_N. Note that Eq. (8.85) also provides the solution to the scattering problem from a rigid body, where the first term in brackets is the incident field (from the point source at $\mathbf{r}'$) and the subtracted term is the scattered pressure field.

8.9 Construction of Interior Neumann and Dirichlet Green Functions by Eigenfunction Expansion

A classic method[7] to find Neumann and Dirichlet Green functions is based on an expansion of the eigenfunctions for the interior space, where the eigenfunctions are the three-dimensional mode shapes corresponding to the resonances of a rigid cavity for the Neumann case, and a pressure release cavity for the Dirichlet case. Figure 8.2 applies to the derivation in this section. It should be clear that, although we require the eigenfunctions to obey a particular boundary condition, we are not specifying any boundary conditions on the interior pressure.

There are no sources within or on the surface S_o, so that the pressure field inside satisfies

$$\nabla^2 p(\mathbf{r}) + k^2 p(\mathbf{r}) = 0. \tag{8.86}$$

---

[6]J. J. Bowman, T. B. A. Senior and P. L. E. Uslenghi (1987). *Electromagnetic and Acoustic Scattering by Simple Shapes*. Hemisphere Publishing Co., New York.

[7]E. Skudrzyk (1971). *Foundations of Acoustics*. Springer-Verlag, New York, p. 646.

We will provide the derivation for both Green functions in parallel using G_N to represent the Neumann, and G_D the Dirichlet Green function. Both Green functions must satisfy the inhomogeneous wave equation with the evaluation point at $\mathbf{r} = \mathbf{r}'$ (see Fig. 8.2):

$$\nabla^2 \left\{ \begin{matrix} G_N(\mathbf{r}|\mathbf{r}') \\ G_D(\mathbf{r}|\mathbf{r}') \end{matrix} \right\} + k^2 \left\{ \begin{matrix} G_N(\mathbf{r}|\mathbf{r}') \\ G_D(\mathbf{r}|\mathbf{r}') \end{matrix} \right\} = -\delta(\mathbf{r} - \mathbf{r}'), \tag{8.87}$$

satisfying the boundary condition ($\mathbf{r} \in S_o$),

$$\frac{\partial G_N(\mathbf{r}|\mathbf{r}')}{\partial n} = 0, \quad \mathbf{r} \in S_o, \tag{8.88}$$

and

$$G_D(\mathbf{r}|\mathbf{r}') = 0, \quad \mathbf{r} \in S_o, \tag{8.89}$$

for the Neumann and Dirichlet Green functions, respectively. Given these boundary conditions the corresponding term in the HIE, Eq. (8.15), is eliminated leading to the following simplified integral equations for the pressure inside ($\alpha=1$) the surface S_o:

$$p(\mathbf{r}') = \iint_{S_o} G_N(\mathbf{r}|\mathbf{r}') \frac{\partial p(\mathbf{r})}{\partial n} dS_o, \tag{8.90}$$

$$p(\mathbf{r}') = -\iint_{S_o} p(\mathbf{r}) \frac{\partial}{\partial n} G_D(\mathbf{r}|\mathbf{r}') dS_o. \tag{8.91}$$

Now assume that we know the spatial form for the three-dimensional eigenfunctions $\Psi_q(\mathbf{r})$ of the cavity which satisfy the same boundary conditions as G_N (rigid walled cavity) or G_D (pressure release walls) and have eigenfrequencies given by $k_q = \omega_q/c$ where q represents the multiple index of the eigenfunction. Of course the eigenfunctions and eigenfrequencies are different for each boundary condition, however, we will use only one symbol Ψ_q to represent both cases for the sake of economy. Assume that these eigenfunctions form a complete set and are orthogonal to one another so that any pressure distribution within S_o, even a delta function, can be represented by a sum over them. Thus we can expand G_N or G_D in terms of these eigenfunctions using unknown coefficients $A_q(\mathbf{r}')$:

$$\left\{ \begin{matrix} G_N(\mathbf{r}|\mathbf{r}') \\ G_D(\mathbf{r}|\mathbf{r}') \end{matrix} \right\} = \sum_q A_q(\mathbf{r}')\Psi_q(\mathbf{r}). \tag{8.92}$$

Since Ψ_q is an eigenfunction of the cavity it has an associated eigenvalue k_q which must satisfy

$$\nabla^2 \Psi_q + k_q^2 \Psi_q = 0. \tag{8.93}$$

Insert Eq. (8.92) into Eq. (8.87), using Eq. (8.93):

$$\sum_q A_q(-k_q^2 + k^2)\Psi_q(\mathbf{r}) = -\delta(\mathbf{r} - \mathbf{r}').$$

Multiply each side by $\Psi_m(\mathbf{r})^*$ and integrate over the volume bounded by S_o:

$$A_q(k^2 - k_q^2) \int_V \Psi_q(\mathbf{r})\Psi_q(\mathbf{r})^* dV = -\Psi_q(\mathbf{r}')^*,$$

which leads to

$$A_q = -\frac{\Psi_q(\mathbf{r}')^*}{(k^2 - k_q^2) \int_V \Psi_q(\mathbf{r})\Psi_q(\mathbf{r})^* dV}.$$

Inserting this result into Eq. (8.92) yields the final form of the Neumann and Dirichlet Green functions,

$$\left\{\begin{matrix} G_N(\mathbf{r}|\mathbf{r}') \\ G_D(\mathbf{r}|\mathbf{r}') \end{matrix}\right\} = \sum_q \frac{\Psi_q(\mathbf{r})\Psi_q(\mathbf{r}')^*}{(k_q^2 - k^2) \int_V \Psi_q(\mathbf{r})\Psi_q(\mathbf{r})^* dV}. \tag{8.94}$$

We note that the expression for G_D satisfies Eq. (8.89) since $\Psi_q(\mathbf{r})$ vanishes for all q on the surface S_o. Similarly, $\frac{\partial G_N}{\partial n}$ for $\mathbf{r} \in S_o$ vanishes for all q again by definition of the eigenfunction ($\frac{\partial \Psi_q}{\partial n} = 0$).

For the integral equation, Eq. (8.91), we need the normal derivative:

$$\frac{\partial G_D}{\partial n} = \sum_q \frac{\frac{\partial \Psi_q(\mathbf{r})}{\partial n}\Psi_q(\mathbf{r}')^*}{(k_q^2 - k^2) \int_V \Psi_q(\mathbf{r})\Psi_q(\mathbf{r})^* dV}. \tag{8.95}$$

8.9.1 Example: Cylindrical Cavity

As an example of the application of the eigenfunction expansion approach for construction of the Neumann and Dirichlet Green functions, we consider a cylindrical cavity as shown in Fig. 8.10. The surface S_o is composed of three separate surfaces; left and right endcaps, S_1 and S_3 and cylindrical surface S_2; $S_o = S_1 + S_2 + S_3$. The left endcap is

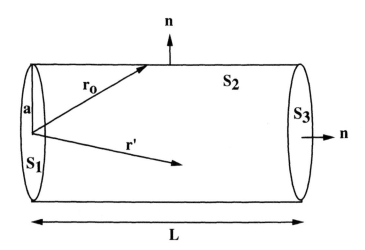

Figure 8.10: Cylindrical cavity in which the pressure is to be determined from the pressure on its surface.

located at $z = 0$ and the right endcap at $z = L$. The radius of the cavity is a. The eigenfunctions must satisfy Eq. (8.86) in cylindrical coordinates (or equivalently Eq. (4.1) on page 115) and the appropriate boundary conditions on S_o, $\Psi_q = 0$ for Dirichlet and $\frac{\partial \Psi_q}{\partial n} = 0$ for Neumann. We will restrict ourselves to the Dirichlet problem in the rest of this section. We seek a solution through separation of variables, Eq. (4.6).

We begin by seeking a solution in the axial direction to the separated differential equation, Eq. (4.8) on page 117, choosing a quantized value for k_z. The correct z dependence of Ψ_q which satisfies the Dirichlet boundary condition is $\sin(k_z z)$ (quantizing k_z to $k_z = m\pi/L$) where m is an integer, since this gives zero on both of the endcaps. Note that the sine functions form a complete set and are orthogonal:

$$\int_0^L \sin(m\pi z/L) \sin(m'\pi z/L) dz = \frac{L}{2}\delta_{mm'}. \tag{8.96}$$

The circumferential dependence satisfying Eq. (4.12) is, as usual, $e^{in\phi}$, where n is an integer. These functions are also orthogonal satisfying

$$\int_o^{2\pi} e^{in\phi} e^{-in'\phi} d\phi = 2\pi\delta_{nn'}. \tag{8.97}$$

The radial dependence must be given by a Bessel function, so that Eq. (4.13) is satisfied, with the only choice being $J_n(\kappa r)$, which is finite on the cylinder axis and has the same order as the circumferential wave functions, $e^{in\phi}$. After substitution of the complete wavefunction,

$$J_n(\kappa r) \sin(m\pi z/L) e^{in\phi}$$

into Eq. (8.86), we find that

$$\kappa = \sqrt{k^2 - (m\pi/L)^2},$$

the same relationship we had for the infinite case, Eq. (4.10), except that now the wavenumber in the axial direction is quantized. In order for the complete wavefunction to vanish on the cylindrical surface at $r = a$ we must require that

$$J_n\left(a\sqrt{k^2 - (m\pi/L)^2}\right) = 0.$$

For any given value of n this equation has an infinite number of solutions since the Bessel function is an oscillating function (see Fig. 4.2 on page 119) and has an infinite number of roots. If we define κ_{ns} as the $s'th$ root of

$$J_n(\kappa_{ns}) = 0,$$

then we conclude that the eigenfrequencies must satisfy the equation

$$a\sqrt{k^2 - (m\pi/L)^2} = \kappa_{ns}. \tag{8.98}$$

Equation (8.98) will be satisfied for discrete values of k, and these values are the eigenvalues or eigenfrequencies of the interior modes. The eigenfrequencies of the cavity are then

$$k^2 = k_q^2 \equiv \left(\frac{\kappa_{ns}}{a}\right)^2 + \left(\frac{m\pi}{L}\right)^2, \tag{8.99}$$

where q spans the indices n, s and m. Finally, the eigenfunction is

$$\Psi_q(r, \phi, z) = J_n(\kappa_{ns}r/a) \sin(m\pi z/L)e^{in\phi}. \tag{8.100}$$

For each order n $J_n(\kappa_{ns}r/a)$ forms a complete set of radial eigenfunctions with respect to s, which are orthogonal:

$$\int_0^a J_n(\kappa_{ns}r/a)J_n(\kappa_{ns'}r/a)rdr = \frac{a^2}{2}J_{n+1}^2(\kappa_{ns})\delta_{ss'}. \tag{8.101}$$

Using the orthogonality of the three functions making up the eigenfunction, Eq. (8.96), Eq. (8.97), and Eq. (8.101), the volume integral in Eq. (8.94) and Eq. (8.95) can be evaluated:

$$\int_V \Psi_q(\mathbf{r})\Psi_q(\mathbf{r})^* dV = \iiint \left(\sin(m\pi z/L)J_n(\kappa_{ns}r/a)\right)^2 dz \, rdr \, d\phi$$

$$= \frac{\pi a^2 L}{2}J_{n+1}^2(\kappa_{ns}). \tag{8.102}$$

Finally, using these results, the interior Dirichlet Green function is

$$G_D(\mathbf{r}|\mathbf{r}') = \sum_{n=-\infty}^{\infty}\sum_{s=1}^{\infty}\sum_{m=1}^{\infty} \frac{J_n(\kappa_{ns}r/a)\sin(m\pi z/L)e^{in(\phi-\phi')}J_n(\kappa_{ns}r'/a)\sin(m\pi z'/L)}{\frac{\pi a^2 L}{2}J_{n+1}^2(\kappa_{ns})(k_q^2 - k^2)} \tag{8.103}$$

and the kernel, Eq. (8.95), for the integral equation, Eq. (8.91), becomes

$$\frac{\partial G_D}{\partial n}\bigg|_{\mathbf{r}\in S_o} = \sum_{n=-\infty}^{\infty}\sum_{s=1}^{\infty}\sum_{m=1}^{\infty} \frac{\frac{\partial}{\partial n}[J_n(\kappa_{ns}r/a)\sin(m\pi z/L)]e^{in(\phi-\phi')}J_n(\kappa_{ns}r'/a)\sin(m\pi z'/L)}{\frac{\pi a^2 L}{2}J_{n+1}^2(\kappa_{ns})(k_q^2 - k^2)} \tag{8.104}$$

where

$$k_q^2 = (\kappa_{ns}/a)^2 + (m\pi/L)^2, \tag{8.105}$$

and the normal derivative applies to the first term in the square bracket when $\mathbf{r}$ is on the cylindrical section, and the second term when $\mathbf{r}$ is on the endcaps.

8.10 Evanescent Neumann and Dirichlet Green Functions

One problem with the eigenfunction expansion of the Green function, Eq. (8.103), is that it is not very physical. That is, when the wavelength of the pressure on the surface S_o is smaller than an acoustic wavelength, one would expect $\frac{\partial G_D}{\partial n}$ to reflect an evanescent behavior of the pressure close to S_o. However, the modes which make up Eq. (8.95) are never evanescent; many eigenfunctions must be summed with just the right coefficients in order to match the evanescent behavior of subsonic surface waves. We will construct an "evanescent" Green function in this section which does match the expected evanescent behavior of the field in the cavity when the surface pressure is subsonic. Furthermore we will find that one of the summations in Eq. (8.95) is eliminated and better series convergence is obtained as a result. The technique is similar to one presented in Jackson[8] for an electrostatic problem. We extend this technique to the acoustic problem.

8.10.1 Evanescent Dirichlet Green Function for a Cylindrical Cavity

Consider the general solution $G(\mathbf{r}|\mathbf{r}')$ to Eq. (8.87) valid in the interior of the cylindrical box:

$$\nabla^2 G(\mathbf{r}|\mathbf{r}') + k^2 G(\mathbf{r}|\mathbf{r}') = -\frac{\delta(\rho - \rho')}{\rho}\delta(\phi - \phi')\delta(z - z'). \qquad (8.106)$$

Once we determine the general solution we will construct the Dirichlet Green function by adding to it the homogeneous solution g_D:

$$G_D = G + g_D,$$

where g_D is chosen so that G_D is zero anywhere on the surface of the cylindrical box. First we determine the particular solution of Eq. (8.106) by expanding the delta function on the right hand side.

Using the completeness relations (see for example Eq. (6.46) on page 191), and using the axial and circumferential eigenfunctions for the Dirichlet problem, we have

$$\delta(\phi - \phi') = \frac{1}{2\pi}\sum_{n=-\infty}^{\infty} e^{in(\phi - \phi')}, \qquad (8.107)$$

$$\delta(z - z') = \frac{2}{L}\sum_{m=0}^{\infty}\sin(\frac{m\pi z}{L})\sin(\frac{m\pi z'}{L}). \qquad (8.108)$$

Substituting Eq. (8.107) and Eq. (8.108) into the right hand side of Eq. (8.106), and realizing that $e^{in\phi}$ and $\sin(m\pi z/L)$ are solutions of Eq. (8.106), leads us to consider the following form for $G(\mathbf{r}|\mathbf{r}')$:

$$G(\mathbf{r}|\mathbf{r}') = \frac{1}{\pi L}\sum_{n,m} e^{in(\phi - \phi')}\sin(\frac{m\pi z}{L})\sin(\frac{m\pi z'}{L})g_{nm}(\rho, \rho'), \qquad (8.109)$$

where, in order for Eq. (8.106) to be satisfied. $g_{nm}(\rho, \rho')$ must satisfy the resulting radial equation:

$$\left[\frac{1}{\rho}\frac{d}{d\rho}\rho\frac{d}{d\rho} + k^2 - (\frac{m\pi}{L})^2 - \frac{n^2}{\rho^2}\right]g_{nm}(\rho, \rho') = -\frac{\delta(\rho - \rho')}{\rho}. \qquad (8.110)$$

If we define

$$k_m^2 \equiv k^2 - (\frac{m\pi}{L})^2, \qquad (8.111)$$

we recognize Eq. (8.110) as Bessel's differential equation, (Eq. 4.13 on page 117), with solutions proportional to $J_n(k_m\rho)$ and $Y_n(k_m\rho)$. In order to construct the particular solution of the inhomogeneous equation, Eq. (8.110), we guess that

$$g_{nm} = AJ_n(k_m\rho_<)Y_n(k_m\rho_>), \qquad (8.112)$$

---

[8] J. D. Jackson (1975). *Classical Electrodynamics*, 2nd ed. Wiley & Sons, p. 116

where $\rho_<$ is the lesser of ρ and ρ' and $\rho_>$ is the greater of the two and A is a constant. Note that $J_n(k_m\rho_<)$ is finite when $\rho = 0$. The choice of the multiplier $Y_n(k_m\rho_>)$ is somewhat arbitrary. We could have chosen $H_n^{(1)}(k_m\rho_>)$ instead. In any case the reason behind the choice of these multipliers will become clearer as we proceed. Inserting Eq. (8.112) into Eq. (8.110), multiplying both sides by ρ and integrating from $\rho' - \epsilon$ to $\rho' + \epsilon$ (to catch the delta function inside the interval) and letting $\epsilon \to 0$, yields

$$\frac{dg_{nm}(\rho' + \epsilon)}{d\rho} - \frac{dg_{nm}(\rho' - \epsilon)}{d\rho} = -\frac{1}{\rho'} = Ak_m[J_n(k_m\rho')Y_n'(k_m\rho') - J_n'(k_m\rho')Y_n(k_m\rho')].$$
(8.113)

We recognize the latter terms as the Wronskian,

$$J_n(k_m\rho')Y_n'(k_m\rho') - J_n'(k_m\rho')Y_n(k_m\rho') = \frac{2}{\pi k_m\rho'},$$

so that Eq. (8.113) is satisfied as long as

$$A = -\pi/2.$$

Finally, the particular solution, Eq. (8.109), is given by

$$G(\mathbf{r}|\mathbf{r}') = \frac{-1}{2L} \sum_{n,m} e^{in(\phi-\phi')} \sin(\frac{m\pi z}{L}) \sin(\frac{m\pi z'}{L}) J_n(k_m\rho_<)Y_n(k_m\rho_>).$$
(8.114)

In view of the solutions already determined for the Dirichlet and Neumann Green functions for the interior problem on a sphere, (see Eq. (8.78), for example), it is not difficult to guess the form of the additive function h_{nm} (the radial part of the homogeneous solution g_D) which will provide $g_{nm} + h_{nm} = 0$ on the cylindrical surface ($\rho = a$) of the cylindrical box. The radial part is almost identical to that in Eq. (8.78), except that the derivatives are removed. We take

$$h_{nm} = \frac{\pi}{2}J_n(k_m\rho_>)Y_n(k_ma)\frac{J_n(k_m\rho_<)}{J_n(k_ma)}$$

so that

$$h_{nm} + g_{nm} = \frac{\pi}{2}(J_n(k_m\rho_>)Y_n(k_ma) - J_n(k_ma)Y_n(k_m\rho_>))\frac{J_n(k_m\rho_<)}{J_n(k_ma)},$$
(8.115)

which is clearly zero on the boundary when $r_> = a$. Note that $g_{nm} + h_{nm}$ still satisfies the jump condition for the Green function given by Eq. (8.113) since as $\epsilon \to 0$

$$\frac{dh_{nm}(\rho' + \epsilon)}{d\rho} - \frac{dh_{nm}(\rho' - \epsilon)}{d\rho} = 0.$$

Finally, adding h_{nm} to Eq. (8.109), we have the full form of the evanescent Dirichlet Green function:

$$G_D(\mathbf{r}|\mathbf{r}') = \frac{1}{2L} \sum_{n=-\infty}^{\infty} \sum_{m=1}^{\infty} e^{in(\phi-\phi')} \sin(\frac{m\pi z}{L}) \sin(\frac{m\pi z'}{L})$$

$$\times [J_n(k_m\rho_>)Y_n(k_ma) - J_n(k_ma)Y_n(k_m\rho_>)]\frac{J_n(k_m\rho_<)}{J_n(k_ma)}.$$
(8.116)

Note that G_D vanishes not only on the cylindrical section, but also on the endcaps (at $z = 0$ and $z = L$).

The pressure anywhere in the interior is given by (Eq. (8.91))

$$p(\mathbf{r}') = - \iint_{S_o} p(\mathbf{r}) \frac{\partial}{\partial n} G_D(\mathbf{r}|\mathbf{r}') dS_o, \tag{8.117}$$

where the normal derivative on the cylindrical section is obtained from Eq. (8.116) evaluated at $\rho = a$:

$$\left. \frac{\partial G_D}{\partial \rho} \right|_{\rho=a} = -\frac{1}{\pi L a} \sum_{n,m} e^{in(o-o')} \sin(\frac{m\pi z}{L}) \sin(\frac{m\pi z'}{L}) \frac{J_n(k_m \rho')}{J_n(k_m a)}, \tag{8.118}$$

and on the left endcap $(z = 0)$:

$$\frac{\partial G_D}{\partial n} = -\frac{\partial G_D}{\partial z} = -\frac{\pi}{2L^2} \sum_{n,m} e^{in(o-o')} m \sin(\frac{m\pi z'}{L}) \frac{J_n(k_m \rho')}{J_n(k_m a)}, \tag{8.119}$$

and on the right endcap $(z = L)$:

$$\frac{\partial G_D}{\partial n} = \frac{\partial G_D}{\partial z} = \frac{\pi}{2L^2} \sum_{n,m} e^{in(o-o')} \cos(m\pi) \sin(\frac{m\pi z'}{L}) \frac{J_n(k_m \rho')}{J_n(k_m a)}. \tag{8.120}$$

Compare Eq. (8.118) with the classic eigenfunction expansion given in Eq. (8.103). The summation over s has been eliminated. More importantly, however, is the fact that the arguments of the Bessel functions can now be imaginary (they were always real in Eq. (8.103)) whenever the axial wavenumber is subsonic; that is, if $(m\pi/L) > k$ then

$$k_m = \sqrt{k^2 - (m\pi/L)^2} = i\sqrt{(m\pi/L)^2 - k^2}.$$

In this case the ratio in Eq. (8.118) is

$$\frac{J_n(k_m \rho')}{J_n(k_m a)} = \frac{I_n(|k_m|\rho')}{I_n(|k_m|a)},$$

where I_n is the modified Bessel function and (see Eq. (4.33) on page 120 and Fig. 4.4)

$$I_n(z) = (-i)^n J_n(iz). \tag{8.121}$$

Furthermore, as the argument becomes large $(z \gg n)$ we have that (Eq. (4.38))

$$\frac{I_n(|k_m|\rho')}{I_n(|k_m|a)} \approx e^{-|k_m|(a-\rho')} \sqrt{a/\rho'} \tag{8.122}$$

so that the sum over m in Eq. (8.118) is rapidly convergent.

Finally, Eq. (8.118) is much more in tune with the physics of the wave propagation than Eq. (8.103). That is, when the axial wavenumber is subsonic Eq. (8.118) shows explicitly that the pressure field will decay into the interior from the cylindrical surface at $\rho = a$. The smaller the axial wavenumber the larger the decay. This is consistent with the picture which we developed in terms of helical waves for the external radiation from cylinders in Chapter 4.

To continue with this physical representation of the Dirichlet Green function, we note that when $\mathbf{r}$ is on the endcaps the evanescence is not properly modeled. For evanescent waves on the endcaps, Eq. (8.119) and Eq. (8.120) do not exhibit any decay in the axial direction as $\mathbf{r}'$ moves away from the endcap. The field merely oscillates as $\sin(m\pi z'/L)$. We can remedy this problem by rederiving the Dirichlet Green function in a way which will lead to a decay in the axial direction (when $\mathbf{r}$ is on a end cap), which will lead to $\sin(x)$ with a complex argument, instead of the Bessel function. This is the subject of the next section.

The Dirichlet Green Function near the Endcaps

We return to Eq. (8.106) and Eq. (8.107) and instead of Eq. (8.108) we use an expansion in ρ, using the completeness relation for $J_n(\kappa_{ns}\rho/a)$.

$$\frac{\delta(\rho - \rho')}{\rho} = \sum_{s=1}^{\infty} \frac{2}{a^2 J_{n+1}^2(\kappa_{ns})} J_n(\kappa_{ns}\frac{\rho}{a}) J_n(\kappa_{ns}\frac{\rho'}{a}), \qquad (8.123)$$

where

$$J_n(\kappa_{ns}) = 0, \qquad (8.124)$$

so that the Dirichlet boundary condition on the cylindrical surface is satisfied. Thus, $G(\mathbf{r}|\mathbf{r}')$ must be of the form,

$$G(\mathbf{r}|\mathbf{r}') = \frac{1}{\pi a^2} \sum_{n,s} e^{in(\phi-\phi')} \frac{J_n(\kappa_{ns}\frac{\rho}{a}) J_n(\kappa_{ns}\frac{\rho'}{a})}{J_{n+1}^2(\kappa_{ns})} g_{ns}(z,z'). \qquad (8.125)$$

Again using separation of variables (Eq. (4.6), $G = R(\rho)Z(z)\Phi(\phi)$), we obtain the axial part of the solution,

$$\frac{d^2 g_{ns}(z,z')}{dz^2} + \left(k^2 - (\frac{\kappa_{ns}}{a})^2\right) g_{ns}(z,z') = -\delta(z - z'). \qquad (8.126)$$

This equation is satisfied (jump condition satisfied) if we choose

$$g_{ns}(z,z') = \frac{\sin(k_{ns}z_<)\sin(k_{ns}(L - z_>))}{k_{ns}\sin(k_{ns}L)}, \qquad (8.127)$$

where

$$k_{ns}^2 \equiv k^2 - (\frac{\kappa_{ns}}{a})^2. \qquad (8.128)$$

Equation (8.127) can be verified by substitution back into Eq. (8.126) and integrating from $z' - \epsilon$ to $z' + \epsilon$, as we did in Eq. (8.102). We note that $g_{ns}(z,z')$ and thus G have the property that they vanish at $z = 0$ and $z = L$. Thus the constructed G is the desired Dirichlet Green function (no additive terms needed) and we have finally that $(G_D = G)$

$$G_D(\mathbf{r}|\mathbf{r}') = \frac{1}{\pi a^2} \sum_{n,s} e^{in(\phi-\phi')} \frac{J_n(\kappa_{ns}\frac{\varrho}{a})J_n(\kappa_{ns}\frac{\varrho'}{a})}{J_{n+1}^2(\kappa_{ns})} \frac{\sin(k_{ns}z_<)\sin(k_{ns}(L-z_>))}{k_{ns}\sin(k_{ns}L)}. \quad (8.129)$$

We can now replace Eq. (8.119) with the more physical form on the left endcap $(z = 0)$:

$$\frac{\partial G_D}{\partial n} = -\frac{\partial G_D}{\partial z} = -\frac{1}{\pi a^2} \sum_{n,s} e^{in(\phi-\phi')} \frac{J_n(\kappa_{ns}\frac{\varrho}{a})J_n(\kappa_{ns}\frac{\varrho'}{a})}{J_{n+1}^2(\kappa_{ns})} \frac{\sin(k_{ns}(L-z'))}{\sin(k_{ns}L)}. \quad (8.130)$$

On the right endcap $(z = L)$, Eq. (8.120) can be replaced with

$$\frac{\partial G_D}{\partial n} = \frac{\partial G_D}{\partial z} = -\frac{1}{\pi a^2} \sum_{n,s} e^{in(\phi-\phi')} \frac{J_n(\kappa_{ns}\frac{\varrho}{a})J_n(\kappa_{ns}\frac{\varrho'}{a})}{J_{n+1}^2(\kappa_{ns})} \frac{\sin(k_{ns}z')}{\sin(k_{ns}L)}. \quad (8.131)$$

The evanescent behavior of Eq. (8.130) and Eq. (8.131) is borne out by the fact that when $(\kappa_{ns}/a) > k$, k_{ns} given in Eq. (8.128) becomes purely imaginary, as does the argument of the sine function. Under this condition, the ratio of sines in Eq. (8.131) becomes

$$\frac{\sin(k_{ns}z')}{\sin(k_{ns}L)} = \frac{\sinh(|k_{ns}|z')}{\sinh(|k_{ns}|L)},$$

which becomes for large arguments

$$\frac{\sin(k_{ns}z')}{\sin(k_{ns}L)} \approx e^{-|k_{ns}|(L-z')}, \quad (8.132)$$

so that near the right end cap the pressure field decays exponentially into the volume. This is the desired physical behavior which is consistent with the exponential decay of the solution near the cylindrical surface given in Eq. (8.118). The same holds true for the left end cap. Again, as a result of this, the summation over s in Eq. (8.130) and Eq. (8.131) converges rapidly when the evanescent condition is reached in s.

Thus the complete solution for the normal derivative of the Dirichlet Green function is given by the three series, Eq. (8.130) for $\mathbf{r} \in S_1$ (the left end cap), Eq. (8.131) for $\mathbf{r} \in S_3$ (the right end cap), and Eq. (8.118) for $\mathbf{r} \in S_2$ (on the cylindrical surface). The HIE is then broken into three integrals, one over each surface with the corresponding Green functions. That is, Eq. (8.91) is written as

$$p(\mathbf{r}') = \left[\underbrace{\iint_{S_1}}_{Eq.\ (8.130)} + \underbrace{\iint_{S_2}}_{Eq.\ (8.118)} + \underbrace{\iint_{S_3}}_{Eq.\ (8.131)} \right] p(\mathbf{r}) \frac{\partial}{\partial n} G_D(\mathbf{r}|\mathbf{r}')\, dS_o, \quad (8.133)$$

where the underbraces indicate the equation number used for the Green function in the integrand.

Of course, all three formulas can be used interchangeably, since each is valid for $\mathbf{r}$ anywhere on S_o. However, many more terms will be needed to achieve the same degree of convergence if one does not use the physically meaningful solution.

One important point has not yet been made. The evanescent Green function has the interesting behavior that Eq. (8.91) is satisfied identically when $\mathbf{r}'$ is on the surface S_o, and $\alpha = 1$ as a result. Thus the integral equation does not have to be modified when the evaluation point is on the surface. This results from a rather subtle behavior of the expansions used to represent the delta functions, Eq. (8.107), Eq. (8.108), and Eq. (8.123) at the end points of the interval. The reader is encouraged to plot one of these delta function representations near the end points to verify the subtle behavior. This is also true of the evanescent Neumann Green function.

As has been noted before, G_D provides a solution to the scattering problem of a point source located at $\mathbf{r} = \mathbf{r}'$ inside a pressure release cylindrical cavity.

8.10.2   Forbidden Frequencies

When the eigenfrequency k is exactly that of one of the resonance modes, defined by Eq. (8.99), then one would expect that only one mode is excited and the amplitudes of the out-of-resonance modes are negligible. However, if we try to calculate the interior pressure from the integral equation, Eq. (8.91), we fail since $p(\mathbf{r})$ is zero on the surface S_o (since it is a mode). The evanescent Dirichlet Green function, however, is infinite at these frequencies as we will soon see, so that evaluation of Eq. (8.91) is impossible since zero times infinity is undefined. We are unable to compute the internal field in this case. These forbidden frequencies (resonances) are evident from the Green function since they appear as zeros of the denominator, that is, from Eq. (8.118)

$$J_n(k_m a) = 0.$$

Thus by Eq. (8.124) $k_m a = \kappa_{ns}$ and using Eq. (8.111) for k_m, then the forbidden frequencies are given by

$$k^2 = (m\pi/L)^2 + (\kappa_{ns}/a)^2. \tag{8.134}$$

On the other hand, from Eq. (8.130) and Eq. (8.131), the resonance occurs when

$$\sin(k_{ns}L) = 0,$$

or $k_{ns}L = m\pi$, and using Eq. (8.128) for k_{ns}, then

$$k^2 = (m\pi/L)^2 + (\kappa_{ns}/a)^2.$$

This result is identical to Eq. (8.134); the resonance frequencies are the same. The forbidden frequencies are thus given by

$$k_q = \sqrt{(m\pi/L)^2 + (\kappa_{ns}/a)^2}, \tag{8.135}$$

where q represents the triplet of numbers (n, m, s). Note that these frequencies are identical to those which occur in the eigenfunction expansion of the Dirichlet Green function, given in the denominator of Eq. (8.103) and by Eq. (8.105).

Even though the integral equation is not able to provide $p(\mathbf{r}')$ at the interior resonance, we already know the answer to within a constant and do not need the integral equation. That is, at a resonance, from Eq. (8.100),

$$p(\mathbf{r}') = p_0 J_n(\kappa_{ns}r'/a)\sin(m\pi z'/L)e^{in\phi'},$$

where p_0 is an unknown constant. Only this single constant is unknown. One can see that to overcome the forbidden frequency problem we need to specify the field in the interior at just one point (avoiding any nodal lines, however) so that the constant p_0 is determined. This specification, in fact, forms the basis of a popular technique invented by Schenck.[9]

8.10.3   Interior Evanescent Neumann Green Function for a Cylindrical Cavity

The relationship between the pressure within, and the normal velocity on, the surface of a cylindrical cavity is provided by the Helmholtz integral with the Neumann Green function, similar to the Dirichlet case, Eq. (8.91):

$$p(\mathbf{r}') = \iint_{S_o} G_N(\mathbf{r}|\mathbf{r}')\frac{\partial p(\mathbf{r})}{\partial n}dS_o, \tag{8.136}$$

where on S_o

$$\frac{\partial}{\partial n}G_N(\mathbf{r}|\mathbf{r}') = 0.$$

Equation (8.136) is important for the application of NAH to the interior noise in cylindrical-like structures such as aircraft fuselages. Derivation of the necessary Neumann Green function proceeds along the same lines as for the Dirichlet case presented above. We will not present the derivations here since the problem is solved in detail in the reference.[10] Again, three separate Neumann Green functions are derived, one for each of the endcaps and one for the cylindrical section. When the surface velocity is subsonic the evanescent Neumann Green functions exhibit a decay into the cavity.

8.11   Arbitrarily Shaped Bodies and the Neumann Green Function

8.11.1   The External Problem

The exterior formulation of the HIE was given in Eq. (8.30). We constructed the Neumann Green function for two particular examples, which corresponded to separable coordinate system geometries; the plane and the sphere, Eq. (8.82) and Eq. (8.85). In this section we consider the case in which the surface for the HIE does not correspond to a surface in a separable coordinate system. For the exterior problem we assume that all sources are contained within the surface S_o of Eq. (8.30). The construction of the Neumann Green function is quite different from before, and is carried out using a discretization of the integral.

---

[9]H. A. Schenck (1968). "Improved integral equation formulation for acoustic radiation problems", J. Acoust. Soc. Am., **44**, pp. 41–58.

[10]Earl G. Williams (1997). "On Green functions for a cylindrical cavity", J. Acoust. Soc. Am. **102**, pp. 3300–3307.

With the normals defined as pointing outward, the HIE, Eq. (8.30), becomes

$$\alpha p(\mathbf{r}') = \iint_{S_o} \left(p(\mathbf{r})\frac{\partial}{\partial n}G(\mathbf{r}|\mathbf{r}') - G(\mathbf{r}|\mathbf{r}')\frac{\partial}{\partial n}p(\mathbf{r}) \right) dS_o, \tag{8.137}$$

where $G = e^{ikR}/4\pi R$ and $R = |\mathbf{r} - \mathbf{r}'|$. Discretization of this integral is the objective of quite a deal of literature, under the name of boundary element methods.[11] We choose the simplest discretization scheme (and not the most accurate) in which the surface S_o is broken up into N small elements of area,

$$\Delta S_1, \ \Delta S_2, \cdots, \ \Delta S_k, \cdots, \ \Delta S_N,$$

with the distance R defined from the center of an element to the field point (formerly called the evaluation point) $\mathbf{r}'$ as shown in Fig. 8.11. We now approximate Eq. (8.137)

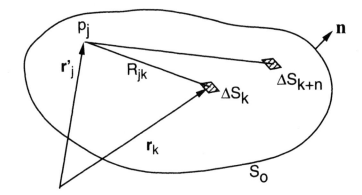

Figure 8.11: Three-dimensional surface S_o divided into N elements of area, ΔS_k for the exterior form of HIE.

by assuming that the elements of area are small enough so that the integral can be represented by a sum. Furthermore, we let the field point $\mathbf{r}'$ lie on the surface ($\alpha=1/2$) located at the center of one of these patches. Thus,

$$\frac{1}{2}p_j = \sum_{k=1}^{N} \left(p_k \frac{\partial G_{jk}}{\partial n} - G_{jk} \frac{\partial p_k}{\partial n} \right) \Delta S_k, \tag{8.138}$$

where the subscripts refer to the patch number. If we evaluate Eq. (8.138) at every patch in the discretization then we have N simultaneous equations to solve for N values of p and N values of $\frac{\partial p}{\partial n}$. We now proceed to derive a matrix form of the Neumann Green function.

We define the following $N \times N$ matrices, using a boldface type on a capital letter to indicate a matrix:

$$\mathbf{G^s} \equiv \begin{pmatrix} G_{11} & \cdots & G_{1N} \\ \vdots & G_{jk} & \vdots \\ G_{N1} & \cdots & G_{NN} \end{pmatrix}, \tag{8.139}$$

---

[11]R. D. Ciskowski and C. A. Brebbia, eds. (1991). *Boundary Element Methods in Acoustics*. Computational Mechanics Publications and Elsevier Applied Science, Southampton and London.

$$\mathbf{G}^{s\nu} \equiv \begin{pmatrix} \frac{\partial G_{11}}{\partial n} & \cdots & \frac{\partial G_{1N}}{\partial n} \\ \vdots & \frac{\partial G_{jk}}{\partial n} & \vdots \\ \frac{\partial G_{N1}}{\partial n} & \cdots & \frac{\partial G_{NN}}{\partial n} \end{pmatrix}, \tag{8.140}$$

where

$$G_{jk} \equiv \frac{e^{ikR_{jk}}}{4\pi R_{jk}},$$

$$\frac{\partial G_{jk}}{\partial n} \equiv \frac{\partial}{\partial n}\left[\frac{e^{ikR_{jk}}}{4\pi R_{jk}}\right],$$

and R_{jk} is shown in Fig. 8.11. The superscript s indicates that the field point $\mathbf{r}$ is taken on the surface. The diagonal terms are singular and must be replaced by an estimate of the integral over the patch, which is finite. The column vectors of length N, $\mathbf{p}^s$ and $\frac{\partial \mathbf{p}^s}{\partial n}$, represent the pressure and its normal derivative on the surface, each element corresponding to a patch.

We can now write a matrix equivalent of Eq. (8.138), defining the identity matrix,

$$\mathbf{I} = \begin{pmatrix} 1 & 0 & \cdots & 0 \\ 0 & 1 & \cdots & 0 \\ \vdots & \vdots & \ddots & \vdots \\ 0 & 0 & \cdots & 1 \end{pmatrix}, \tag{8.141}$$

and the discretized area matrix,

$$\mathbf{S} = \begin{pmatrix} \Delta S_1 & 0 & \cdots & 0 \\ 0 & \Delta S_2 & \cdots & 0 \\ \vdots & \vdots & \ddots & \vdots \\ 0 & 0 & \cdots & \Delta S_N \end{pmatrix}. \tag{8.142}$$

Thus, the N equations can be written as

$$\frac{1}{2}\mathbf{p}^s = (\mathbf{G}^{s\nu}\mathbf{S}\mathbf{p}^s - \mathbf{G}^s\mathbf{S}\frac{\partial \mathbf{p}^s}{\partial n}).$$

Using the identity matrix we can rewrite this equation as

$$(\mathbf{G}^{s\nu}\mathbf{S} - \frac{1}{2}\mathbf{I})\mathbf{p}^s = \mathbf{G}^s\mathbf{S}\frac{\partial \mathbf{p}^s}{\partial n}. \tag{8.143}$$

We can solve Eq. (8.143) for $\mathbf{p}^s$ by premultiplying left and right sides by the inverse of the matrix multiplying $\mathbf{p}^s$,

$$\mathbf{p}^s = (\mathbf{G}^{s\nu}\mathbf{S} - \frac{1}{2}\mathbf{I})^{-1}\mathbf{G}^s\mathbf{S}\frac{\partial \mathbf{p}^s}{\partial n}. \tag{8.144}$$

This equation provides the relationship between the pressure and its normal derivative on a surface. It is very robust except when the pressure on the surface is identically zero everywhere. This happens, as discussed in Section 8.10.2, at the forbidden frequencies of the *interior* Dirichlet problem. We will not pursue this issue here, however. It is dealt with at some length in the literature.

With the surface pressure given by Eq. (8.144), we return to the HIE and use it to construct the Neumann Green function, using the discretized version of Eq. (8.137) for the field point *off* the surface ($\alpha = 1$). Let $\mathbf{G}$ and $\mathbf{G}^\nu$ (no s superscripts) be $1 \times N$ matrices representing $G(\mathbf{r}|\mathbf{r}')$ and $\frac{\partial}{\partial n}G(\mathbf{r}|\mathbf{r}')$ of Eq. (8.137) for the field point off the surface. Thus, Eq. (8.137) discretized (for only one field point) using Eq. (8.142) is

$$p(\mathbf{r}') = (\mathbf{G}^\nu \mathbf{S} \mathbf{p}^s - \mathbf{G} \mathbf{S} \frac{\partial \mathbf{p}^s}{\partial n}). \tag{8.145}$$

The vector, $\mathbf{p}^s$ can be eliminated from this equation by using Eq. (8.144) to yield the following single matrix equation for the pressure outside of the surface,

$$
\begin{aligned}
p(\mathbf{r}') &= [\mathbf{G}^\nu \mathbf{S}(\mathbf{G}^{s\nu}\mathbf{S} - \frac{1}{2}\mathbf{I})^{-1}\mathbf{G}^s \mathbf{S}\frac{\partial \mathbf{p}^s}{\partial \mathbf{n}} - \mathbf{G}\mathbf{S}\frac{\partial \mathbf{p}^s}{\partial \mathbf{n}}], \\
&= [\mathbf{G}^\nu \mathbf{S}(\mathbf{G}^{s\nu}\mathbf{S} - \frac{1}{2}\mathbf{I})^{-1}\mathbf{G}^s \mathbf{S} - \mathbf{G}\mathbf{S}]\frac{\partial \mathbf{p}^s}{\partial n} \\
&= i\rho_o ck \underbrace{[\mathbf{G}^\nu \mathbf{S}(\mathbf{G}^{s\nu}\mathbf{S} - \frac{1}{2}\mathbf{I})^{-1}\mathbf{G}^s \mathbf{S} - \mathbf{G}\mathbf{S}]}_{\mathbf{G_N}} \mathbf{v_n}, \tag{8.146}
\end{aligned}
$$

where the normal surface velocity is

$$i\rho_0 ck\mathbf{v_n} = \frac{\partial \mathbf{p}^s}{\partial n}.$$

The Neumann Green matrix is shown by the underbrace in Eq. (8.146) which is the discretized equivalent to the Neumann Green function defined in Eq. (8.74). If we choose M locations to evaluate $p(\mathbf{r}')$ then Eq. (8.146) represents M equations, that is, $\mathbf{G}^\nu$ and $\mathbf{G}$ are now $M \times N$ matrices.

8.12 Conformal NAH for Arbitrary Geometry

We set up the NAH equation to solve for the surface velocity given a measurement of the pressure outside of a radiating surface (the exterior problem). To set up the formulation we need to write the forward equation using Eq. (8.146) with the field point taken at N different locations conformal to the reconstruction surface. We invert the forward equation through inversion of a matrix to solve for the unknown surface velocity.

Let the vector of length N, $\mathbf{p}$, be a set of measured pressures on a surface conformal to the surface S_o. The positions of $\mathbf{p}$ are taken directly above (in the sense of the normal to the surface) the desired velocity locations $\mathbf{v_n}$. The distance between the two surfaces is small so that the evanescent fields can be captured. Thus Eq. (8.146) becomes a set of N simultaneous equations, with $\mathbf{G}^\nu$ and $\mathbf{G}$ increased to $N \times N$:

$$\mathbf{p} = i\rho_o ck[\mathbf{G}^\nu \mathbf{S}(\mathbf{G}^{s\nu}\mathbf{S} - \frac{1}{2}\mathbf{I})^{-1}\mathbf{G}^s \mathbf{S} - \mathbf{G}\mathbf{S}]\mathbf{v_n} \equiv \mathbf{H}\mathbf{v_n}, \tag{8.147}$$

where $\mathbf{H}$ is an $N \times N$ matrix. Equation (8.147) can be inverted,

$$\mathbf{v_n} = \mathbf{H}^{-1}\mathbf{p}. \tag{8.148}$$

As one might expect the inversion of $\mathbf{H}$ is very ill conditioned and can not be inverted without introduction of some special methods. The ill-conditioning results from the evanescent wave information contained in the decaying pressure of the high spatial wavelength waves which blow up exponentially in the inversion process. This ill-conditioning is avoided by turning to the singular value decomposition (SVD) of the matrix $\mathbf{H}$.[12]

We quote the SVD theorem.[13]

> *Theorem:* Let $\mathbf{A} \in C^{m \times n}$. Then there exist unitary matrices $\mathbf{U} \in C^{m \times m}$ and $\mathbf{V} \in C^{n \times n}$ such that
> $$\mathbf{A} = \mathbf{U}\mathbf{\Sigma}\mathbf{V}^{\mathbf{H}} \tag{8.149}$$
> where
> $$\mathbf{\Sigma} = \begin{pmatrix} \mathbf{\Lambda} & 0 \\ 0 & 0 \end{pmatrix}$$
> and $\mathbf{\Lambda} = diag(\sigma_1, \cdots, \sigma_r)$ with
> $$\sigma_1 \geq \cdots \geq \sigma_r > 0. \tag{8.150}$$

In the theorem above, $\mathbf{V}^{\mathbf{H}}$ represents the conjugate transpose of the matrix $\mathbf{V}$ and $\in C$ indicates that the matrix may be complex. A unitary matrix is orthogonal to its conjugate transpose in the sense

$$\mathbf{U}\mathbf{U}^{\mathbf{H}} = \mathbf{U}^{\mathbf{H}}\mathbf{U} = \mathbf{I}, \tag{8.151}$$

where $\mathbf{I}$ is the diagonal identity matrix, Eq. (8.141). Thus $\mathbf{U}^{-1} = \mathbf{U}^{\mathbf{H}}$. We note that the SVD definition is not restricted to square matrices but can be applied in general to a non-square matrix.

The SVD applied to $\mathbf{H}$ in Eq. (8.148) is

$$\mathbf{H} = \mathbf{U}\mathbf{\Sigma}\mathbf{V}^{\mathbf{H}}$$

so that its inverse is

$$\mathbf{H}^{-1} = (\mathbf{U}\mathbf{\Sigma}\mathbf{V}^{\mathbf{H}})^{-1} = \mathbf{V}\mathbf{\Sigma}^{-1}\mathbf{U}^{\mathbf{H}}$$

where

$$\mathbf{\Sigma}^{-1} \equiv \begin{pmatrix} 1/\sigma_1 & 0 & \cdots & 0 \\ 0 & 1/\sigma_2 & \cdots & 0 \\ \vdots & \vdots & \ddots & \vdots \\ 0 & 0 & \cdots & 1/\sigma_N \end{pmatrix}. \tag{8.152}$$

---

[12] W. A. Veronesi and J.D. Maynard (1988), "Digital holographic reconstruction of sources with arbitrarily shaped surface," J. Acoust. Soc. Am. **85**, pp. 588–598.

[13] Virginia C. Klema and Alan J. Laub (1980). "The Singular Value Decomposition: Its Computation and Some Applications", IEEE transactions on automatic control, **AC-25**, no. 2, pp. 164–176.

Since the singular values are descending in value, the diagonal terms are increasing in value from left to right, corresponding to increasingly evanescent waves of smaller and smaller spatial wavelength. These waves are filtered out by truncating the terms in Eq. (8.152) at some point, setting the rest of the terms to zero. This is the equivalent of a k-space filter. For example, if the dynamic range of the measured pressure was 40 dB then we would filter out any singular values below this level. Thus if the nth singular value σ_n satisfied

$$20 \log_{10}(\sigma_n/\sigma_1) < -40$$

then the filter would require setting

$$1/\sigma_k = 0, \quad k = n, n+1, \cdots, N$$

in Eq. (8.152).

If we define Σ_c^{-1} as the filtered version of Σ^{-1}:

$$\Sigma_c^{-1} = diag(1/\sigma_1, 1/\sigma_2, \cdots, 1/\sigma_n, 0, 0, \cdots),$$

then the reconstructed velocity is given by

$$\hat{\mathbf{v}}_\mathbf{n} = \mathbf{V}\Sigma_c^{-1}\mathbf{U}^H\mathbf{p}, \tag{8.153}$$

where $\hat{\mathbf{v}}_\mathbf{n}$ represents the filtered velocity reconstruction vector over the entire surface of the body.

Problems

8.1 By inserting Eq. (8.22) into Eq. (8.19) and integrating for the case where z' is finite, and on the z axis, show that the result is the same as Eq. (8.18) evaluated at the evaluation point z'.

8.2 Using Eq. (8.22) determine g_D deriving the Dirichlet Green function for the field interior to a sphere. Using the Helmholtz integral equation, Eq. (8.75) show that your result is the same as Eq. (6.142). What scattering problem is this solution equivalent to?

8.3 Consider a vibrating box in an infinite medium as shown below. A measurement sphere S_o is placed as shown and the pressure and its normal derivative are given completely on the surface of the sphere. The normal direction is shown. For this problem you are to state the result of the following integral evaluation,

$$\iint_{S_o} \left(G(\mathbf{r}|\mathbf{r}')\frac{\partial p(\mathbf{r})}{\partial n} - p(\mathbf{r})\frac{\partial}{\partial n}G(\mathbf{r}|\mathbf{r}') \right) dS_o,$$

for two different locations of the vector $\mathbf{r}'$, the points labeled P_1 and P_2 as shown. (Please specify which equation number in the class notes you are using to find the answer.)

(a) CASE I: Sphere surrounding the box.

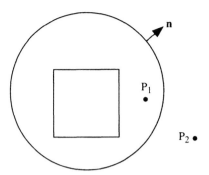

Figure 8.12: CASE I for problem 8.3(a).

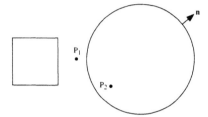

Figure 8.13: CASE II for problem 8.3(b).

(b) CASE II: Sphere not surrounding the box.

(c) CASE III: A point source is added outside of the vibrating box and labeled Q_0 as shown. What is the result of the integral evaluation for the case shown below.

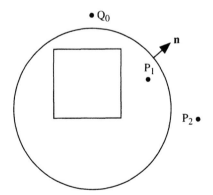

Figure 8.14: CASE III for problem 8.3(c).

(d) CASE IV: Same as CASE III except the measurement sphere is moved to the new location as shown below.

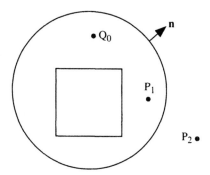

Figure 8.15: CASE IV for problem 8.3(d).

8.4 Derive the Dirichlet Green function, G_D, for a sphere of radius a for the exterior problem which satisfies $G_D(\mathbf{r}|\mathbf{r}') = 0$ on the surface of the sphere. The Dirichlet Green function satisfies

$$p(\mathbf{r}') = - \iint_S p(\mathbf{r}) \frac{\partial G_D(\mathbf{r}|\mathbf{r}')}{\partial n} \, dS_o$$

where n is the inward normal to S_o and $|\mathbf{r}'| > |\mathbf{r}|$. Show that your result is identical to Eq. (6.94). What scattering problem is this result equivalent to? (Show location of the source and receiver with respect to the sphere for the scattering problem.)

8.5 A pulsating sphere of radius a and radial velocity $\dot{w}(a, \theta, \phi) = V_0$ is placed at the origin of a measurement sphere of radius b. The pressure field, valid for $a \le r < \infty$, generated by the pulsating sphere is (Eq. (6.119))

$$p(r, \theta, \phi) = \rho_0 c V_0 k a^2 \frac{ka - i}{[(ka)^2 + 1]} \frac{e^{ik(r-a)}}{r}.$$

(a) Insert this known field and its normal derivative into the HIE and evaluate the integrals to determine the pressure at a point **outside** of the measurement sphere. Show that this result is identical to the known field at that point.

HINT: You might want to use Eq. (8.22) to expand the free space Green function.

(b) Evaluate the integrals for the case where the point is **inside** the measurement sphere, $a \le r' < b$.

8.6 Following a similar process as the one that led to Eq. (8.100), determine the eigenfunctions for the Neumann Green function for a cylindrical cavity.

Index

Printed and bound by CPI Group (UK) Ltd, Croydon, CR0 4YY

08/05/2025

01864790-0006